SERIES
ENTOMOLOGICA

EDITOR

E. SCHIMITSCHEK, GÖTTINGEN

VOLUMEN 8

SPRINGER-SCIENCE–BUSINESS MEDIA, B.V.

ISBN 978-94-015-7245-3   ISBN 978-94-015-7243-9 (eBook)
DOI 10.1007/978-94-015-7243-9

THE PHYLOGENETIC CLASSIFICATION

OF

DIPTERA CYCLORRHAPHA

with special reference to the structure
of the male postabdomen

by

G. C. D. GRIFFITHS

SPRINGER-SCIENCE+BUSINESS MEDIA, B.V.

CONTENTS

1. INTRODUCTION

This work is a specialized dissertation covering a limited field of enquiry. I deal mainly with two interrelated topics, the structure of the male postabdomen and genitalia of cyclorrhaphous flies and how these insects should be classified in a phylogenetic system. Much new information and interpretation is presented here, as well as commentary on the observations and interpretation of previous authors. The field covered by this work has long been recognized as difficult. It is my hope that I will succeed in this work in dispelling some of the difficulties. My proposal of revised terminology for certain parts of the external genitalia thus should not be regarded as innovation for its own sake, but as an attempt to remedy a situation which has been widely recognized as unsatisfactory.

In proposing a revised classification of the Schizophora, I have followed the type of analytical procedure used by such authors as CROWSON and HENNIG. In my opinion their rigorous approach to phylogenetic classification provides the best available procedure for constructing a higher classification on logically consistent principles. I discuss this question in detail below (section 2). The importance of a clear understanding of the theory of systematics for the success of any attempt at higher classification is worth stressing. The classifications which can be proposed by applying different principles become increasingly divergent as 'higher' (more extensive) groups are considered, particularly because of differences in the classification of highly modified forms. Considerations of theory are not irrelevant to the practice of biological systematics as is sometimes maintained. A good contrary example is provided by the history of the classification of the families of Schizophora, for some proposals of different authors have been discordant in the extreme. The classification of the Schizophora provides a good test for the effectiveness of phylogenetic systematics in dealing with complex patterns of character correlations which older workers were unable to interpret satisfactorily. Work on the phylogenetic classification of this group was begun by HENNIG (1958) and my present work continues this endeavour. The classification here presented is still provisional. Many additional studies could be undertaken to test the validity of my proposals, and I hope that publication of this work will stimulate some of my colleagues to undertake such research.

In some parts of this work, I have occasion to comment critically on the views of other authors. I do this in order to make clear why I have reached different conclusions. Such detailed criticism is in my opinion an important part of every scientific work which has a review content, since without such criticism it is difficult for any reader without a detailed knowledge of the subject to judge whether different conclusions have been reached arbitrarily or with good reason.

I undertook the preparation of this work at the University of Alberta (Edmonton) during the years 1967-1970. I have attempted to consider all literature relevant to the subject up to the end of 1969, although I cannot guarantee that nothing has been overlooked in the years since 1965, which have not yet been covered by the 'Zoological Record'. The material considered in this study may seem small in relation to the field covered, but it should be appreciated that my request lists were already long enough to strain the patience of museum curators. It is particularly regrettable that so little material of South American Diptera could be made available to me, because many of the genera are represented in collections only by a few type specimens. Further studies on the dipterous fauna of this region may be awaited with interest, and I hope that publication of this work will stimulate increased efforts in this field.

The names used in this work for North American species generally follow those in the most recent catalogue (STONE *et al.* 1965), except that for Platypezidae I have followed the revised generic system proposed by KESSEL & MAGGIONCALDA (1968a). The names of Palaearctic species generally follow those used in the relevant part of 'Die Fliegen der paläarktischen Region'. For material from other regions I have used the names under which the specimens were sent to me. While efforts have been made to ensure that there are no errors at generic or higher levels in the identification of specimens figured and described, it is possible that a few of the identifications may be incorrect at the species level. It was not practicable for me to check all the taxonomic literature on Schizophora, which is a group containing about five times as many species as the mammals. In most cases species identifications were accepted on trust.

The figures of insects included in this paper number about 150. To do justice to the information presented would require nearer 500. Additional figures may be sought by reference to the papers cited in the discussion of particular families.

Material and methods

I worked mainly with dried specimens. These were first relaxed in a humidified container, and the abdomen removed with needles under the binocular microscope. Then the abdomen was lightly macerated in potassium hydroxide (about five minutes in a 10% solution at 100° C), and washed for a few minutes successively in (1) tap water, (2) a weak solution of glacial acetic acid (about 20%), and (3) distilled water. After this treatment the preparations were placed on slides without coverslips in a mixture (roughly 50:50) of glycerol and polyvinyl lactophenol. Since the elasticity of chitin is restored and maintained in this medium, the preparations could be manipulated or further dissected immediately, whenever the need arose. This constant availability of the preparations for rechecking was essential to the success of the project. The medium did not dry up during the period of two years during which the preparations were kept in it. At the end of the project this temporary mounting medium was washed off with 50% ethanol in water, and the preparations taken through 98% ethanol into neutral Canada balsam for permanent storage (except for some of the preparations of specimens lent by the U.S. National Museum, which were returned in microvials of glycerol). In total one or more preparations of about 220 species were made during this study. The species of Schizophora concerned are indicated in my treatment of each family (section 6.2).

Serial sections were made of a few specimens fixed in the field in Pampel's fluid. The sections were stained in Gömöri's Chrome Alum haematoxylin and eosin according to standard procedures, and mounted in neutral Canada balsam. I was able to check from these sections the course of the ejaculatory duct and some details of the musculature. The sections remain in my possession. They represent 21 genera, as indicated in the table I.

2. PRINCIPLES AND PROCEDURES
OF CLASSIFICATION

2.1. The representation of phylogenetic relationship in the Linnaean hierarchy

The theory of systematics has been the subject of so much controversy in recent years, that all authors of major systematic works would be well advised to state their theoretical standpoint clearly at the outset and thus prevent possible misunderstanding. Therefore I state that I am concerned in this work with phylogenetic (cladistic) classification, and regard the works of HENNIG (1950, 1966a) as important clarifications of the logical foundations of this type of classification. Some of the important features of the theory of phylogenetic systematics, as affecting classification above the species level, will now be outlined in broad terms.

HENNIG asserts that all groups in the phylogenetic system should be monophyletic, including descendants of a common ancestral species. This view has of course been held by many leading systematists ever since the acceptance of Darwin's theory of evolution. HENNIG's exploration of the implications of this principle leads him to assert that, if ambiguity is to be avoided, groups should only be described as monophyletic if they include *all* the descendants of a common ancestral species (or all descendants in a particular time period, if the systems of different time periods are regarded as separate). This precise definition of the concept of monophyly is necessary if the hierarchical system of classification is to provide unambiguous information on phylogenetic relationships

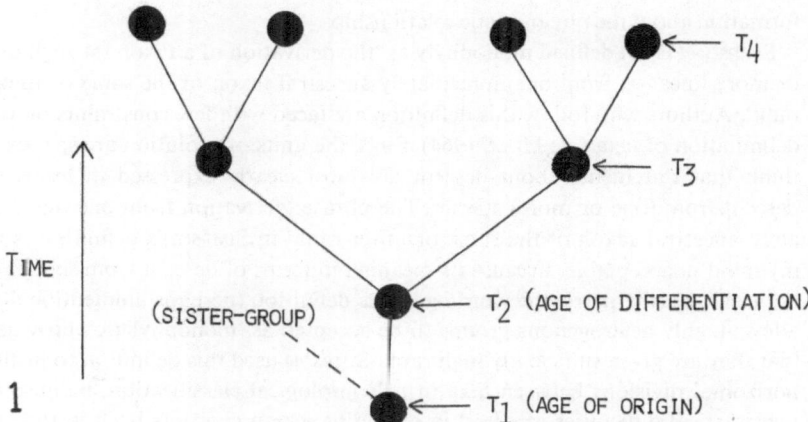

Fig. 1. Phylogeny of a monophyletic group of four species at time t₄ represented by a dendrogram.

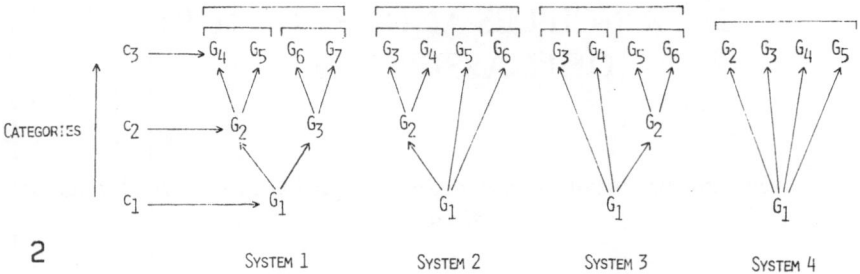

CATEGORIES

C_3 ——→ G_4 G_5 G_6 G_7 G_3 G_4 G_5 G_6 G_3 G_4 G_5 G_6 G_2 G_3 G_4 G_5

C_2 ——→ G_2 G_3 G_2 G_2

C_1 ——→ G_1 G_1 G_1 G_1

2

SYSTEM 1   SYSTEM 2   SYSTEM 3   SYSTEM 4

Fig. 2. The four possible Linnaean hierarchies which are admissible as representations of the phylogenetic relationships between the four contemporaneous species whose phylogeny is represented by the dendrogram at fig. 1, when all taxa are monophyletic according to HENNIG's definition. Two forms of presentation are used, (1) that of Gregg (1954), in which the taxa are numbered (G_1, G_2 ...) and their inclusion relations shown by arrows; and (2) the nesting bracket presentation which is more familiar to systematists. The species (members of category C_3) remain in the same order throughout.

between organisms, that is their recency of common ancestry (whether in absolute or in relative terms). To clarify the grounds for this assertion, I offer the simple hypothetical example shown in figures 1 and 2.

Of the four possible systems shown in figure 2, system 1 obviously conveys more information than the others (in fact all the information about the phylogenetic relationships between the contemporaneous species at time t_4 conveyed by the dendrogram), and is therefore to be preferred. The other systems are less complete, conveying some, but not all, of the information conveyed by system 1. None of these four systems leads to false inferences concerning phylogenetic relationships, and no other systems are possible without violating the criterion of monophyly according to HENNIG's definition. However, if a laxer definition of monophyly is accepted, other systems can be admitted (an additional six systems in the example given, thus making a total of ten). The additional six systems are all incongruent with the dendrogram, and thus do not convey information about the phylogenetic relationships.

SIMPSON (1961) defined monophyly as 'the derivation of a taxon through one or more lineages, from one immediately ancestral taxon of the same or lower rank'. Authors who follow this definition are faced with few constraints on the delimitation of taxa (see HULL 1964). Since the units of evolution are species, I think that statements about descent are most clearly expressed in terms of descent from (one or more) species. The phrase 'derivation from one immediately ancestral taxon of the same or lower rank' in SIMPSON's definition is to my mind unacceptable, because its meaning in terms of descent from species is not clear. Furthermore, the wording of the definition (perhaps unintentionally) allows highly heterogenous groups to be accepted as 'monophyletic', provided that they are given sufficiently high rank. SIMPSON used this definition to justify horizontal divisions between taxa in paleontological classification, because of the practical difficulties involved in extending group concepts back in time to the common ancestral species (vertical classification in paleontology). These difficulties are indeed serious, but I think that the way of resolving them is to

6

restrict the application of the criterion of monophyly, not to redefine the term. The proposals to divide the geological time scale for purposes of classification discussed later in this section provide a possible approach to the problem.

The example above (figs. 1 and 2) illustrates that a Linnaean hierarchy is an adequate representation of the phylogenetic relationships between recent organisms, or organisms in any particular time period. This is one of the grounds for the view of HENNIG and other phylogenetic systematists that the phylogenetic system should be expressed by revision of the traditional Linnaean system, rather than by proposal of a separate classification.

WOODGER (1952) and GREGG (1954) have studied the definition and properties of hierarchical systems from a logical standpoint. In such systems all the elements are related by unidirectional inclusion relations, emanating ultimately from the 'beginner' of the system. The Linnaean hierarchy shows an additional feature which is not present in other hierarchical systems: this is that the taxa (the elements of the hierarchy) are members of categories (such as phylum, class, order, family, etc.), which are associated with the system and indicate a sequence of ranking. A logical distinction should be made between the inclusion relations between taxa, and the membership relations between taxa and categories (BUCK & HULL 1966).* Category names are defined in terms of properties which taxa have (BUCK & HULL 1966). Clearly some of these properties are formal, involving the use of category names to indicate the sequence of inclusion relations between taxa. Thus when an individual is classified as belonging to the species *Homo sapiens*, the genus *Homo*, the family Hominidae, the order Primates and the class Mammalia, the sequence of inclusion relationships (subordination) between these five taxa is indicated by the reference of the taxa to categories. A definition of a category name may therefore specify the inclusion relations between the taxa referred to this category and taxa referred to at least two other categories. It is possible to maintain that category names have no meaning beyond this formal aspect. But most biologists assume that the taxa referred to a particular category are, or at least ought to be, equivalent to one another not only in a formal sense but also in respect of some of their empirical relations and properties. Considerable controversy has arisen over the nature of these empirical relations and properties which qualify taxa for membership of a particular category. MAYR (1969: 233) summarizes some of the criteria which have been suggested under five headings, as follows: '1. Distinctness (size of gap); 2. Evolutionary role (uniqueness of adaptive zone); 3. Degree of difference; 4. Size of taxon; 5. Equivalence of ranking in related taxa.'

However, any attempt at a consistent and precise application of these criteria would be faced with serious difficulties, which I briefly summarize as follows:
1. To speak of gaps in particular character sequences is meaningful, and taxonomists of all schools use such gaps in the normal procedure of defining character states. But to speak of gaps between whole organisms (in other words, the totality of the differences in their attributes) is quite another matter, and any attempt at precise estimation of gaps in this extended sense

---

* It is common practice to refer to taxa as 'members' of higher taxa as well as of particular categories. BUCK and HULL use the term membership in an unusually restricted sense in the interests of clarity.

must lead into a logical and mathematical morass similar to that involved in trying to estimate 'overall similarity' (see section 2.2).

2. An 'adaptive zone' is an ecological concept. Since some groups in the phylogenetic system are diverse as regards the ecological roles of their members, it seems scarcely possible to make consistent use of ecological criteria of ranking.

3. The 'degree of difference' is a phenetic concept, similar to Mayr's first criterion (size of gap). The extension of such a concept from particular character sequences to whole organisms is likewise problematical (again see section 2.2).

4. MAYR recommends that 'the size of the gap justifying the separation of a higher taxon should be inversely correlated to the size of the taxa.' The application of a criterion of this kind again depends on estimating gaps between whole organisms.

5. The application of this criterion of equivalence depends on assuming that the related taxa are correctly ranked. This may be an acceptable temporary expedient in studies of a limited range of organisms, but does not provide any general criterion for ranking.

For the reasons given above, I doubt whether the criteria listed by MAYR can be satisfactorily applied to the ranking of taxa in a phylogenetic system in HENNIG's sense. Even if the problems of measuring gaps or degrees of difference between organisms could be overcome, a further serious difficulty would arise. This is that the use of such criteria would not necessarily lead to the reference of the monophyletic taxa in a phylogenetic system to categories in a way compatible with the use of the category names to indicate the sequence of subordination. Inconsistencies would need to be admitted in the application either of the criteria for referring taxa to categories, or of the monophyly criterion by which the taxa are delimited (such as by the admission of paraphyletic groups into the system). The reason for this incompatibility may be shown by comparison of the system of taxa and the phylogenetic relationships between organisms which it represents. If the dendrogram at figure 1 is compared with the complete representation of the phylogenetic relationships in a Linnaean hierarchy (system 1 of fig. 2), the sequence of inclusion relations (subordination) which the categories indicate is clearly congruent with the sequence of branching of the phylogeny in time. Thus the categories necessarily correspond with the time dimension in a relative sense in a system in which all taxa are monophyletic according to HENNIG's definition, because the sequence of inclusion relationships (subordination) in such a system is congruent with the branching of the phylogeny in time. It is impossible within the constraints of a phylogenetic system of this kind to define categories in terms of phenetic criteria in a rigorous consistent way, because phenetic divergences and degree of phylogenetic relationship (recency of common ancestry) are not necessarily congruent.

Once it is recognized that the categories in a system of monophyletic groups necessarily indicate the relative age of the stem-species ancestral to members of each taxon, it is but a small step to suggest that categories should be related in broad terms to the absolute age of the stem-species. HENNIG (1966a: 186) has made such proposals, and these have been taken up and elaborated by CROW-SON (1970). To be sure, these proposals have been made a little in advance of their time, since the phylogeny of most groups of organisms (including those treated in this work) has not been sufficiently clarified to allow confident use of a time criterion for ranking. Nevertheless these proposals provide the only theoretically sound basis for achieving an objective equivalence between the taxa assigned to particular categories in a phylogenetic system, and thus pose a task for future phylogenetic systematics.

The question of which measurement of age is the best reference point for the assigment of taxa to categories thus arises. HENNIG has concluded that we should use the age of origin of the stem-species, that is the time at which the stem-species split from its sister-species. A logical consequence of HENNIG's conclusion is that sister-groups must have the same rank in the system, because their age of origin is the same. I have indicated the difference between age of origin and age of differentiation on the dendrogram (fig. 1).

The classification of fossil organisms poses certain problems because a Linnaean hierarchy is only an adequate representation of the phylogenetic relationships between Recent organisms, or organisms in any particular time horizon. BRUNDIN (1966) rightly states that 'we have to imagine a vast series of horizontal systems upwards through time, together reflecting successive change and the rise and fall of innumerable integrant sister-group systems'. Of course such a continuum of systems through time is unmanageable to the human mind, and we have to resort to the device of arbitrarily breaking down the continuum into discrete sections, as HENNIG (1950: 259) concluded. The Linnaean systems of different time periods can thus be regarded as different. It is a misconception to think that the demand that all taxa should be monophyletic necessarily leads to an insistence on a 'vertical' system in which all taxa are held to extend back in time to include the stem-species and all its descendants. Such vertical groups should be thought of as extensions of taxa into an additional dimension, or series of taxa in the systems of successive time-periods.

The demand that all taxa be monophyletic according to HENNIG's definition can thus be reconciled with 'horizontal' classifications of fossil organisms, if it is understood that the taxa which should be monophyletic belong to the systems of particular time periods. The extent to which the same name should be applied to a series of taxa through time thus becomes a secondary question which can be decided upon other criteria, without violating the criterion that taxa be mono-phyletic, and SIMPSON's (1961) proposal to justify horizontal classification by redefining monophyly becomes unnecessary.

CROWSON (1970: 251) has made the first specific proposals for dividing the geological time scale for the purposes of phylogenetic classification, as follows.

'The most practicable method, at the present time, of applying phylogenetic principles to the classification of fossils would probably be to divide the geological past into a number of reasonably equal periods of time and to establish a separate classification for the organisms of each of these periods. These periods would need to be defined in relation to the recognised eras and periods of the geologists, and at least initially would best be made rather long, perhaps of the order of twenty-five or thirty million years each. Thus, in addition to our classification of modern organisms, we could have one for those of the Neogene (Upper Tertiary), of the Palaeogene (Lower Tertiary), of the Upper Cretaceous and of the Lower Cretaceous, of the Jurassic, of the Upper and Lower Triassic, one for the Permian, one each for Pennsylvanian (Upper Carboniferous) and Mississippian (Lower Carboniferous), for Upper and Lower Devonian, Upper and Lower Silurian, Upper and Lower Ordovician, and Upper and Lower Cambrian. All these classifications would not, of course, be completely independent of each other. If we adopted a hierarchy containing as many categories as we have pre-Recent systems, and defined each category in relation to a specified age of origin, persisting lineages could regularly be promoted one grade in the hierarchy in each successive system in which they appeared'. (CROWSON 1970: 251).

CROWSON's proposals clearly need to be supplemented by an examination of possible alternative ways in which the nomenclature applied to fossil organisms might be amended in order to implement them. There is an obvious gap here in the theory of phylogenetic systematics which is in need of detailed exploration.

An alternative approach to the classification of fossils is to relate them to the system of Recent organisms, as has been done in HENNIG's (1969c) book on the phylogeny of insects. A consequence of this procedure is that the rank of the lowest groups to which fossils can be referred is proportional to the age of the latter. Thus, if the boundary between the Lower and Upper Cretaceous is used to define the family category, earlier fossils cannot be referred to any family in the system of Recent organisms, but only to higher groups. In the present work I am not concerned directly with the classification of fossils, and discuss them solely in relation to the classification of Recent organisms.

Some authors have interpreted HENNIG's principles as entailing (for instance) the classification of every new Precambrium species in a new phylum, which they consider to be a *reductio ad absurdum* of the principles. This is a misunderstanding. New Precambrian species can be arranged by the use of lower categories in a hierarchical system of Precambrian organisms. If discussed in relation to the system of Recent organisms, Precambrian species not referable to any Recent phylum can only be referred to a group of higher rank (e.g. the Metazoa). The description of new 'phyla' is not entailed.

Since the species in a phylogenetic system are in principle species at a particular point of time, the question of how the species concept should be extended in the time dimension is only of secondary importance for the theoretical foundations of the system. Species are discrete at any given point of time, but form part of a continuum when extended in the time dimension. The breakdown of this continuum into discrete sections for the purpose of classification is necessarily an arbitrary procedure. HENNIG argues that the best method is to delimit species in time by two successive processes of speciation, because it is these processes of speciation which determine the phylogenetic relationships between living species. This proposal is logically consistent. But serious doubts have been raised about the practicability of delimiting fossil species in this way, because of difficulty in establishing with sufficient precision when speciation has occurred. Even when continuous fossil sequences are available (as in some marine deposits), the geographical extent of the relevant record is usually limited, and therefore the times at which geographical speciation may have occurred can rarely be postulated with much confidence. The known fossil record of insects contains enormous temporal, as well as geographical, gaps and the fossil species as we know them are delimited by these gaps, not by theoretical principles which we may wish to apply. The possibility of devising a theoretically consistent approach to the delimitation of fossil species thus has little immediate priority for entomologists, and is mainly of concern to workers on marine fossils.

2.2. The comparative merits of phylogenetic systematics and other approaches to the classification of organisms

So far I have attempted to justify in terms of theoretical consistency the view that a phylogenetic (cladistic) classification of organisms can be presented in the form of the traditional Linnaean hierarchy. Of course the conclusion that the phylogenetic system can be so presented does not necessarily show that it ought to be. This also depends on the further question, whether the phylogenetic system is to be preferred over any other types of system which can be presented in this form. For purposes of the present argument the alternative approaches to biological classification which are currently advocated for representation by the Linnaean hierarchy are broadly grouped into two main types: phenetic classification ('typological classification' in the sense of HENNIG 1966a; including 'numerical taxonomy' in the sense of SOKAL & SNEATH 1963), and the kind of approach which MAYR (1969) calls 'evolutionary taxonomy', based on combining phylogenetic and phenetic principles of classification (called 'syncretistic system' by HENNIG 1966a). I refer to the latter as combined classification in this work, since the term 'evolutionary taxonomy' has been applied by some authors to phylogenetic and combined systematics without distinction.

In judging the merits of classification systems I follow the view that 'of the many properties that could be used to construct a classification, those causally connected to many others are the most fruitful because they group the elements of the classification into classes whose names function not only in the most inductive generalizations but in the most theoretically significant generalizations' (HULL 1969, summarizing the work of HEMPEL 1965). Some recent authors have claimed that information retrieval is one of the main purposes of classification systems (for instance, MAYR 1969). But in my opinion this emphasis is misplaced. Information retrieval systems should be as simple as possible in the interests of efficiency, and it hardly reasonable to suggest that biologists have developed an elaborate hierarchy with fifteen to twenty categories in common use primarily to act as an information retrieval system. Linnaeus' original five categories were already more than is needed for this purpose. Certain categories in the Linnaean hierarchy, particularly the species, genus and family, clearly have an important function in existing information retrieval systems; but this does not apply to the Linnaean system as a whole. The use of some of the categories in information retrieval systems leads to a conflict of interest among users of the classification. Those who are not directly interested in comparative (systematic) research often view revisions of the system with disfavour, however good the reasons for them, because stable nomenclature facilitates rapid information retrieval. If this conflict of interests becomes more severe in the future, it may become desirable to devise an independent information retrieval system which will not be affected by revision of taxa above the species level (for instance, by allocating numbers to each species).

PHENETIC CLASSIFICATION

Many pheneticists, like phylogenetic systematists, are striving to construct a theoretically consistent system, and there is no reason why both phenetic and phylogenetic systems, if both are possible, should not coexist and be mutually illuminating. I think it unfortunate in retrospect that bad relations were engendered between the two schools by SOKAL & SNEATH's (1963) attack on the validity of phylogenetic systematics, on the grounds that evolutionary reasoning is inherently circular. The arguments which they put forward have since been shown to be unwarranted (HULL 1967). It is doubtful whether such extreme views are still held by the authors, since one of them has subsequently published a method of reconstructing phylogeny (CAMIN & SOKAL 1965). SOKAL & SNEATH's term 'numerical taxonomy' was not sufficiently explicit. Numerical approaches are possible to many different kinds of classification (including phylogenetic classification, various ecological classifications, etc.), and the description 'numerical' is thus potentially ambiguous. Later authors have justifiably substituted the term 'phenetic taxonomy' for SOKAL and SNEATH's proposed classification of organisms on the basis of similarity. The attempt to classify organisms on the basis of similarity is by no means new, having historical antecedents in the school of idealistic morphology ('typology' in the strict sense), which had roots in pre-evolutionary thought of the 19th century (see SIMPSON 1961). Idealistic morphologists thought of species and higher groups as static entities which corresponded to unvarying 'archetypes'. Some held that the latter were equivalent to Platonic forms. But the metaphysical beliefs of Plato have long since fallen from favour in this field, and I do not consider them in this work.

The numerical approach to phenetic classification outlined by SOKAL and SNEATH (1963) presents many theoretical difficulties. The authors aimed at quantifying overall similarity, but unfortunately did not discuss the nature of this concept in detail. The only definition given (page 50) reads: 'Overall similarity (or affinity) between any two entities is a function of the similarity of the many characters in which they are being compared'. Taken literally this definition implies that overall similarity is an arbitrary function, which varies according to the characters used in any comparison. But probably this was not intended, since elsewhere the authors speak of the characters used in particular analyses as representing 'samples'. Presumably 'overall similarity' is a function of the similarity between two entities in all possible characters, and the values obtained with any particular sample of characters should be regarded as approximations to this. At any rate the claim that phenetic classification can provide a general reference system for biology depends ultimately on whether the concept 'overall similarity' corresponds to some real dimension which can be quantified either in relative or absolute terms. If 'overall similarity' is a concept which does not correspond to a single natural dimension, phenetic taxonomists will always be faced with the possibility of presenting a variety of different systems, depending on which characters are chosen, how they are coded and how they are weighted. Some of these systems may be useful for some purposes, but it will be impossible to extend any one system over an extensive array of organisms because of the limitations imposed by the applicability of the chosen character set. SOKAL and SNEATH seem to regard overall similarity as a function of genetic similarity, but this does not alter the problem since genetic similarity is a concept of the same kind (meaning the 'overall similarity' between two genomes). If overall similarity (or genetic similarity) is a real single dimension, phenetic classifications based on different classes of characters should become increasingly congruent as the number of characters used is increased. I do not think that this has ever been convincingly demonstrated, and some studies definitely suggest that it is not the case (see the discussion by EHRLICH & EHRLICH 1967). Relevant data have been presented in recent discussions of the 'nonspecificity hypothesis' ('Here we assume that there are no distinct large *classes* of genes affecting exclusively one class of characters such as morphological, physiological, or ethological, or affecting special regions of the organism such as head, skeleton, leaves'; SOKAL & SNEATH 1963: 85). But it seems to me that more than the relations between the genome and the phenotype is at stake. If classifications based on different classes of characters do not become more congruent as the number of characters is increased, then the fundamental assumption that the overall similarity between two organisms is a parameter which can be approximated by taking samples of characters is suspect. The status of the values for similarity (or dissimilarity) produced by phenetic analyses thus requires clarification. It is obviously important whether these values can be regarded as estimates of an underlying parameter ('overall similarity') or are merely values generated by the arbitrary coding of the particular character sets used in particular analyses. If the latter is the case (as I believe), the frequently voiced claims of 'objectivity' for such values and the classifications derived from them are unfounded.

JOHNSON (1970) has presented an important critical review of the mathematical assumptions of numerical pheneticists. According to JOHNSON the claims of objectivity and precision made by this school are ill-founded, and there is no hope of extending precise quantitative mathematics to describe the biological situations encountered in taxonomy. GHISELIN (1969) dismisses the concept of overall similarity as 'nonsense', because there is no finite number of characters or attributes of an organism; he maintains that statements about similarity only have meaning if all the terms of the similarity are supplied.

Apart from doubts about the objectivity of the similarity values produced by numerical phenetic studies, serious doubts can be raised about the merits of converting such values into a hierarchical system. Similarity (or dissimilarity) values constitute a matrix which can only be converted into a hierarchical system by compression, with consequent loss of information. The steps involved in converting similarity matrices into other forms of presentation have recently been discussed by CARMICHAEL & SNEATH (1969), whose views should be authoritative in view of their experience in this field. These authors state:

'Dendrograms have the advantage that they do not produce a linear ordering of the OTU's (Operational Taxonomic Units). However they achieve this by not specifying the relation between individual OTU's, except for the initial pairs in any stem... OTU's in widely separated stems may be more similar to each other than to some of the OTU's in their own stems... A dendrogram may be thought of as a 'best-fit' model compressed into one dimension by preserving the proximities between the closest pairs in each stem and averaging the remainder in a successively more general manner. It may be a successively more distorted manner if the relations among the OTU's do not permit a linear representation, or are not strongly hierarchical. Where a linear ordering is applicable and the OTU's form strongly hierarchical subsets, a dendrogram may be the simplest satisfactory display.'

Thus it seems that a dendrogram is a satisfactory representation of phenetic relations only when these relations are strongly hierarchical, for instance if they tend to correspond with the underlying phylogenetic relationships (which are necessarily hierarchical). In cases where the phenetic relations appear more reticulate and diverge strongly from phylogenetic relationships, a dendrogram is not an adequate representation. From CARMICHAEL and SNEATH's comments it appears that the only useful information about the phenetic relations between species which is always retained in a dendrogram is that the pairs of species linked at the final dichotomies of each stem are most closely related to each other. No unambiguous information is retained about the relations of any species linked to a stem before the final dichotomy. Because of this loss of information, few inferences can be made about the distribution of similarity values in the original matrix. Of course the same limitations apply to the representation of phenetic relations in a Linnaean hierarchy, whose properties are similar to those of a dendrogram.

If a hierarchical system so inadequately represents phenetic relations, it is reasonable to question the assumption of many phenetic taxonomists that such a system of classification should be the end-product of their work. This assumption seems to be linked with the belief that biologists are so used to hierarchical systems that they will reject any other type of classification. In my opinion this viewpoint obstructs real progress in the field of phenetic classification, since any objectivity or usefulness that may be credited to similarity values has been largely lost by the time these have been converted into a hierarchical system. The Linnaean hierarchy conveys information about phenetic relations inefficiently, but can be an adequate representation of phylogenetic relationships (see previous discussion). For this reason I think that progress in all fields of systematics would be better served if numerical pheneticists did not try to represent their classifications in the traditional Linnaean hierarchy, but concentrated on developing whatever forms of classification are best suited for representing phenetic relations. Mutual illumination between the different kinds of classific-

14

ation might then tend to replace the polemics which have been so prominent in recent literature.

From the considerations presented above I conclude that many of the claims made by SOKAL & SNEATH (1963) were overoptimistic, and that a widely useful hierarchical system cannot be generated from a static (phenetic) coding of character states which does not take account of their evolution.

Phenetic classifications can still be based on non-numerical methods, and indeed such attempts are faced with fewer methodological difficulties than those which face numerical phenetics. Often such classifications result from studies of a particular complex of characters, such as ROHDENDORF's (1964) classification of the Diptera, which was based on his analysis of the functional types of dipterous wings. In so far as such classifications present the results of morphological research, they may be useful and influential. The argument against preferring any such 'morphological' classification to a purely phylogenetic system is based mainly on what CROWSON (1970) calls the non-congruence principle. According to this principle the distribution of character states is rarely coincident, unless they are functionally correlated. This is why authors who have made investigations of a particular functional-morphological complex have so often been led to propose classifications which conflict with those based on earlier investigations of other characters. The well-known incongruences between imaginal and larval classification in some groups of insects illustrate this point. And in the order Diptera an excellent example of the classificatory problems posed by non-congruence is the long-standing dispute on whether the primary division of the order should be between Nematocera and Brachycera, or between Orthorrhapha and Cyclorrhapha. If the distribution of character states within a group is considered from a static (phenetic) standpoint, non-coincidence of the points where major structural differences occur cannot be resolved in the construction of a classification except by arbitrary choice. Different workers may well wish to make divisions at different points, in accordance with their varying interests. In fact there may be a large number of reasonably useful morphological-phenetic classifications which could be presented for groups in which the pattern of character state correlations is complex. For this reason I think that static presentations of the distribution of functional-morphological types should be regarded as special-purpose systems, and that the synthesis of all comparative data available on organisms for presentation in the Linnaean hierarchy should be undertaken through application of the principles of phylogenetic systematics. In phylogenetic systematics incongruences in the distribution of character states are not treated arbitrarily, because the sequence in which divisions should be made is determined by reference to the time dimension.

COMBINED CLASSIFICATION

For purposes of this discussion I take the work of MAYR (1969) as representative

15

of the views of advocates of combined ('evolutionary') classification. The controversy between this school and phylogenetic systematists should not be allowed to obscure substantial agreement on many questions (such as the definition of species). The main area of disagreement is that MAYR rejects HENNIG's proposals for achieving an unequivocal correspondence between phylogeny and classification above the species level. MAYR criticizes these proposals mainly on the grounds that no account is taken of different rates of evolutionary change in different lineages.

However the implication of such criticism that phylogenetic classifications ought to be corrected in some way to reflect varying rates of evolution raises serious logical difficulties. Rates of evolution can be estimated on a reasonably objective basis only for particular character sequences (particularly metric characters), but it is doubtful whether the application of such a concept can be extended to whole organisms (in other words, to the totality of their infinite attributes). Any attempt to estimate rates of evolution in the latter sense in a non-arbitrary manner must lead into the same logical morass which confronts attempts at estimating 'overall' similarity, gaps between whole organisms, etc. (see JOHNSON 1970). Even if this difficulty could be overcome, there would remain the further difficulty that the Linnaean hierarchy is only an adequate representation of one dimension, and that consequently ambiguity must result from attempting to use it to represent more than this (see HENNIG 1966a: 77).

The differences between HENNIG's and MAYR's views may be further illustrated by comparing how they define the categories. They are in full agreement in demanding a definition of species which reflects their role as the units of evolution. ('Species are groups of interbreeding natural populations that are reproductively isolated from other such groups'; MAYR 1969). But their views on the definition of categories above the species level are strongly divergent. MAYR (1969: 92) defines the genus as 'a taxonomic category.. which is separated from other taxa of the same rank (other genera) by a decided gap'. He then defines the family similarly. These definitions are logically unsatisfactory because they include the term to be defined *(definiendum)*. HENNIG's proposal to define categories above the species level in terms of the age of origin of the stem species of the member taxa is not open to this objection, and to the best of my knowledge stands as the only proposal for an empirical definition of such categories which is not open to logical objections. Of course there are practical difficulties in many cases in applying HENNIG's proposals at this time due to inadequate knowledge, but this does not affect their validity in the long term.

In my opinion HENNIG's thesis that the Linnaean hierarchy should be used to represent only phylogenetic relationships (in the sense of recency of common ancestry) is not weakened by criticisms that it is too restrictive, so long as no satisfactory alternative is demonstrated. I do not find in MAYR's work any sufficient discussion of the logical difficulties inherent in his concept of a classification which reflects both cladistic relationships and evolutionary divergence at the same time. The principles of 'evolutionary classification' as currently formulated do not provide sufficient guidance on how classifications above the species level should be revised as new information is obtained. Various different criteria are suggested for the ranking of taxa, with the question of which should be given weight in particular cases left to the arbitration of the classifier. As a result no satisfactory empirical definition of categories above the species level is possible.

Acceptance of HENNIG's view that the Linnaean hierarchy should represent only phylogenetic relationships (recency of common ancestry) does not necessarily imply a denial of the importance or interest of rates of evolution. Estimates of rates of evolution can be presented as a separate classification, as for instance in the diagrammatic presentations discussed by WAGNER (1969). However, there can be no one correct estimate of the rate of evolution of whole organisms because of the multidimensional nature of the data. Whether there is any other dimension except time to which rates of evolution can be referred is not clear. I do not exclude the possibility that they can also be referred to some determining factor at the genetic level, in which case an alternative kind of evolutionary classification to phylogenetic classification in HENNIG's sense could be devised. But so far this possibility has not been demonstrated.

PHYLOGENETIC CLASSIFICATION

The claim that the phylogenetic system is a useful reference system for biology rests on the fact that 'there is one dimension to which all other dimensions are referred, and that is the time dimension' (BRUNDIN 1966). Since there is only one phylogeny of any group of organisms, phylogenetic systematists are not beset with theoretical difficulties in reconciling the evidence of different classes of characters. All available comparative information can and should be considered. Because the phylogenetic system is congruent with a real common dimension, its conclusions are not logically dependent on the use of a particular suite of characters in analyses, and the system can be extended to include all the organisms of a particular time period. Herein lies one of the reasons for claiming superiority for the phylogenetic system over phenetic systems. The latter are inevitably restricted in their extent because of their dependence on particular character suites.

The usefulness of a consistent phylogenetic system in which phylogenetic relations (recency of common ancestry) are expressed without ambiguity is clear enough for those branches of biology which make historical comparisons between organisms. Historical biogeography is an obvious example. It is perhaps not so obvious that the use of a logically consistent phylogenetic system also has significant advantages for branches of biology which compare only Recent organisms. Comparative morphologists employ a concept of homology which is usually defined in terms of common origin in time (evolutionary homology). Statements about homology in this sense are best expressed through a classification of monophyletic groups, since only such groups have a unique common origin in time. In fact comparative morphology cannot be divorced from phylogenetics, if an evolutionary definition of homology is accepted.

17

Phylogenetic classifications are therefore very relevant to studies of comparative morphology, as well as morphological-phenetic classifications of particular character complexes. A more general case for the relevance of phylogenetic classification to comparisons between organisms, apart from the special question of the use of an evolutionary criterion of homology, can be made in terms of the predictive value of such classifications. This can be a question of practical importance in relation to attributes which can only be demonstrated after laborious studies. For instance, if an unusual physiological process has been demonstrated to occur in a small number of rather remotely related species, how widely is this process likely to occur? Predictive answers to questions of this kind are best made by reference to the monophyletic taxa of the phylogenetic system. Such predictions are less reliably made from morphological-phenetic classifications, unless there is a close functional correlation between the character suite used to construct the classification and the attribute whose distribution is predicted. Therefore, I think that over a wide field of biology the phylogenetic classification is the classification which best meets HEMPEL's criterion of functioning in the most inductive generalizations. Further arguments for the special position of the phylogenetic system in biological research have been given by HENNIG (1950, 1966a).

It is appropriate here to refer to the distinction between general-purpose ('natural') and special-purpose ('artificial') classifications made by GILMOUR and his followers (the 'philosophical school' of taxonomy). I accept the general view that this distinction is useful, although there are logical difficulties involved in attempting to apply it in a precise way (see JOHNSON 1970). GILMOUR & WALTERS (1964) define these terms as follows:

'General-purpose classifications consist of classes containing objects with a large number of attributes in common, thus making them useful for a wide range of purposes; special-purpose classifications consist of classes containing objects with only a few attributes in common, and hence serve a more limited range of purposes.'

If we attempt to read some operational significance into these definitions, they seem to suggest that we can form a judgement on whether a classification is general-purpose or special-purpose by enumerating attributes in common. However, the term 'attributes' is so indefinite that enumeration of attributes is only possible if the range and kind of attributes to be considered is specified: in which case the validity of any conclusions for demonstrating *general* purposiveness would be in doubt. Recourse to an all-attributes concept would lead us into a 'hopeless morass' (JOHNSON 1970). For this reason GILMOUR's distinction can only be applied in a vague qualitative way. We can try to consider what *kinds* of common attributes are possessed by the classes of objects in particular classifications, or in what fields of enquiry these classes are useful for conceptualization and communication. My attempt in the preceding paragraphs to demonstrate that the phylogenetic system is useful in many different branches of biology can thus be considered an attempt to demonstrate that the phylogenetic system is a 'general-purpose' biological classification. But it is not necessarily the *only* general-purpose classification of organisms. Probably some form of ecological classification of organisms can also be regarded as a general-purpose system. In the latter case the 'powerful factor' (in the sense of GIL-MCUR's and WALTER's seventh principle) which makes a general-purpose system possible is not phylogeny but the flow of energy through the biosphere.

The definition of general-purpose classifications given by GILMOUR and WALTERS seems to me only partial. They do not demand that the classes included in any particular classification shall be homogenous, nor do they attempt to clarify the significance of the formal relations between classes in classifications. A sufficient definition of general-purpose classifications should surely also refer to these aspects. The point may be illustrated with the words of LEWIS CARROLL's Walrus.

> 'The time has come,' the Walrus said,
> 'To talk of many things:
> Of shoes—and ships—and sealing wax—
> Of cabbages—and kings...'
>
> (LEWIS CARROLL, *Through the Looking-Glass*)

The five classes of objects in the last two lines doubtless all function in general-purpose classifications, but should not be included in the *same* classification. The definition of general-purpose classifications given by GILMOUR and WAL-TERS says nothing to discredit the Walrus' ability as a taxonomist. I conclude that the definition requires elaboration. GILMOUR and WALTERS were unable to suggest any clearly defined procedure for classifying through the Linnaean hierarchy. Their only comment on the use of categories is that comparatively few changes should be made in the ranking of taxa in the interests of stability of nomenclature. This is hardly consistent with their general thesis, since the categories are classes and ought therefore to indicate common attributes of the objects referred to them in a general-purpose classification.

The conclusion which I draw from the considerations presented above is that the phylogenetic system is a general-purpose classification widely useful in many branches of biology, and that it is the most suitable classification for representation through the Linnaean hierarchy. But this does not imply that it is the only useful kind of classification, nor the only possible general-purpose classification. Other kinds of classification are useful in most, if not all, fields of biology. If misunderstanding and conceptual confusion are to be avoided, other kinds of systems should be presented independently of the phylogenetic system. There are no grounds for supposing it possible to produce a single optimal classification which is useful for all possible purposes.

Revision of the Linnaean hierarchy to represent phylogenetic relations is not inconsistent with the recent history of systematics. In practice, most recent revisionary work has tended to change classifications so that the proportion of monophyletic groups in the system is increased, even when the authors have not adopted any consistent theoretical standpoint. For instance, if the history of changes in the ordinal classification of insects is considered, the trend to break down polyphyletic or paraphyletic groups (such as the Linnaean 'Neuroptera') is readily apparent. Nearly all the orders currently accepted in the classification

19

of Recent insects are believed to represent monophyletic groups, with possible exceptions only in areas where the phylogenetic relationships require clarification (e.g. Mecoptera).

2.3. Procedures of phylogenetic analysis

HENNIG (1950, 1966a) discusses in detail the analytical procedures involved in constructing phylogenetic classifications. He distinguishes three methods, the comparative holomorphological method, the paleontological method and the chorological method. In this work I employ the comparative holomorphological method to produce a phylogenetic classification of Recent species. My remarks are therefore concerned mainly with the use of this method.

THEORETICAL CONSIDERATIONS

The starting point of HENNIG's treatment of the comparative holomorphological method is the observation that phylogenetic relationship is not proportional to similarity, because there are demonstrated differences in the rates of character change in different phyletic lineages. Community of descent can be inferred from common possession of the apomorphous conditions of any character sequence (synapomorphy), but not from common possession of plesiomorphous conditions (symplesiomorphy).

HENNIG's terms 'plesiomorphous' and 'apomorphous' are equivalent to the more familiar terms 'ancestral' and 'derived'. I use the former terms in this work, since they have no conflicting meanings arising from usage outside biological systematics, and since they can be readily compounded to form such words as 'synapomorphous', 'autapomorphous' etc. These concepts can only be rendered clumsily using the ancestral/derivative terminology: for instance, the autapomorphous conditions of a group would have to be referred to as 'those derivative conditions of a group, which are not shown by species outside the group'. HENNIG's terminology was first proposed in the German language (as 'plesiomorpher', 'apomorpher' etc.) and has been variously rendered into English. I here follow the English forms given in HENNIG's (1966a) book. The terms 'plesiomorphous' and 'apomorphous' refer properly only to morphological characters, but similar terms are available for characters of other kinds, such as 'plesiooecous' and 'apooecous' (for ecological characters) and 'plesiochorous' and 'apochorous' for distributional characters. TUOMIKOSKI (1967) has appropriately suggested that the more general terms 'plesiotypic' and 'apotypic' may be used to embrace all kinds of characters.

The terms 'generalized' and 'specialized' will be used in this work in a functional sense, not as synonyms of 'plesiomorphous' and 'apomorphous'.

The observation that taxa should only be characterized by apomorphous (derivative) conditions in their groundplan is, of course, by no means new, and to many people seems self-evident. But HENNIG was justified in laying great emphasis on this principle, since many authors have not followed it. After exploring the implications of this principle, HENNIG proposes the 'argument-

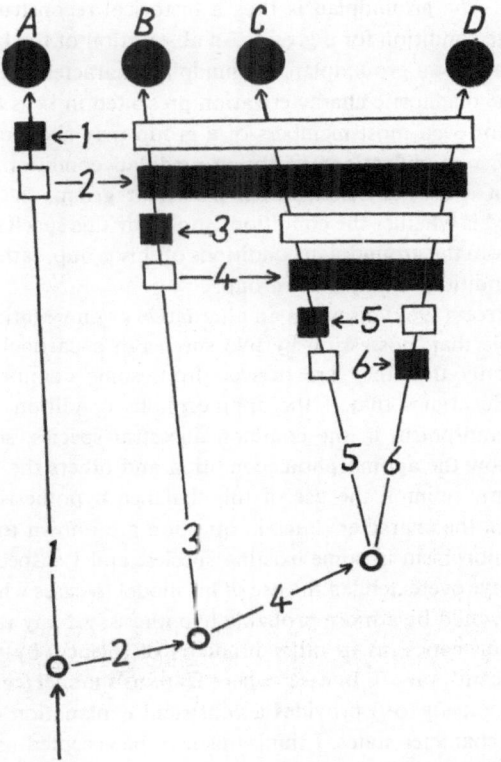

Fig. 3. The argumentation scheme of phylogenetic systematics (after Hennig 1966a). All groups regarded as monophyletic are distinguished by the possession of derived (apomorphous) stages of expression (black) of at least one pair of characters (synapomorphy of species of monophyletic groups).

ation scheme of phylogenetic systematics' (fig. 3) as a model for the performance and presentation of phylogenetic analyses.

In the argumentation scheme the grounds for a phylogenetic hypothesis are illustrated by superimposing an interpretation of character states on a dendrogram. It is conventional to use black rectangles for apomorphous states, and white rectangles for plesiomorphous states. If the number of characters considered is too large for convenient presentation in this form, the postulated branching sequence may be indicated by brackets above a table of character states (as in my fig. 17).

The terms apomorphous and plesiomorphous are relative. The apomorphous conditions which characterize a group become plesiomorphous ('outside primitives' in THROCKMORTON's terminology) from the standpoint of analysis of its subordinate groups.

In phylogenetic systematics the characterization of groups is presented in terms of their apomorphous groundplan conditions. The groundplan of a group is defined as the condition of the last common ancestral species at the age of differentiation. The groundplan is thus a historical reconstruction, not some kind of average condition for a group. An abstraction of the latter kind would be an archetype, not a groundplan. Groundplan characterization is not always the same as the diagnostic characterization presented in keys to identification, because some or even most members of a group may show conditions which represent further modifications of the groundplan condition. In considering whether or not a species belongs to a particular group, the question which should be asked is whether the conditions shown by this species are more probably derived from the groundplan conditions of this group, rather than from the groundplan conditions of any other group.

THROCKMORTON (1962) discusses an alternative argumentation model, based on the principle that 'possession by two species of a particular characteristic will indicate only that they are derived from some common heterozygous population'. He argues that if the apomorphous condition was present as part of a polymorphism in the common ancestral species, some descendant species may show the apomorphous condition and others the plesiomorphous condition. In my opinion the use of this 'balance hypothesis' model is only warranted when the character states in question are known to form part of a balanced polymorphism in some existing species, and I suspect that THROCK-MORTON may have overextended the use of his model to cases where homoiology or parallelism would be a more probable hypothesis. At any rate one must be guided in phylogenetics, as in other branches of science, by the principle of parsimony (Occam's razor). In cases where HENNIG's model (called the 'classic' model by THROCKMORTON) provides a consistent explanation of the observed distribution of character states, I think this may be accepted as the most parsimonious explanation. In studies of the relationships between groups of high rank, we are mainly concerned with the distribution of rather strongly differentiated character states to which the balance hypothesis can hardly apply. The argumentation in this work is consequently presented within the framework of HENNIG's argumentation scheme.

HENNIG's scheme has been criticized by several authors (including DARLING-TON 1970) on the grounds that it involves assuming that speciation is usually dichotomous. These criticisms raise the complex question of the nature of speciation, which I do not consider in this review. However differences of opinion on this question need not give rise to methodological disputes, if *a priori* assumptions about the extent to which evolution is dichotomous are excluded from phylogenetic analyses. I do not think that the application of HENNIG's methods depends on making *a priori* assumptions of this kind, and his argumentation scheme can be readily extended to cases of non-dichotomous speciation. If the pattern of character states expected to result from a dichotomy is found (alternance of apomorphous and plesiomorphous states), then a dichotomy should be postulated as the most parsimonious interpretation. But if such a pattern is not apparent, a dichotomous classification should not be presented.

DARLINGTON (1970) has expressed several further criticisms of the procedures outlined by HENNIG (1966a) on the grounds that they involve unacceptable simplifications of the processes of speciation and evolution. Unfortunately, some of DARLINGTON's criticisms seem directed at distorted interpretations of HENNIG's thought. For instance, I do not follow his argument that HENNIG assumes rates of evolution and divergence in related lines to be uniform, since this is contrary to HENNIG's stated opinions. Nor can I accept the argument that relative primitiveness of species or groups (not just of character states) is important to the theory of phylogenetic systematics (cladism). While the description of species or groups as relatively plesiomorphous (primitive) or apomorphous (derivative) seems meaningful in extreme cases, such statements should be avoided or at least qualified as a general rule; for non-congruent changes in different character sequences often make the extension of the concepts of plesiomorphy and apomorphy from particular character sequences to whole organisms problematical. DARLINGTON's criticisms suggest a need for comment on the nature of the simplifications inherent in HENNIG's argumentation scheme and other dendrogrammatic presentations of phylogeny and character change. That these presentations are very simplified is not disputed, for HENNIG (1966a: 88) introduces the diagram of speciation (reproduced by DARLINGTON as his figure 1) with the words 'in this simplest possibility'. Simple conceptual models are plainly useful in many fields of science, and there is no reason why systematists should be denied the use of such models. The customary presentation of phylogeny as a dendrogram is already a highly simplified conceptual model, and HENNIG's argumentation scheme differs primarily by the addition of a representation of character states. This addition improves the model, because an unadorned dendrogram is open to the misunderstanding that the differences between species can be reduced to a single dimension. If we consider the argumentation scheme (fig. 3), the simplifications inherent in it include: (1) representation of species lineages by straight lines; (2) representation of speciation by a branching point; (3) representation of character states as uniform within species or groups of species; (4) representation of the complexly interrelated differences between organisms by separate series of character states. Clearly there are purposes for which the argumentation scheme or other form of simple dendrogram is not an adequate analytical model because of these inherent simplifications (for instance, detailed analysis of a speciation process). But this restriction applies to all conceptual models, without exception (including those presented as mathematical formulae). The usefulness of models should be judged primarily in relation to the purposes for which they are proposed. It is not warranted to discuss a model in relation to some purpose for which it was not primarily intended, demonstrate its inadequacy for this purpose and then pronounce it useless for any purpose. If true, this kind of argument would invalidate much of what we now accept as theoretical biology. DARLINGTON's conclusion that many of HENNIG's models ('concepts') are of no practical use seems to rest, at least in part, on unwarranted reasoning of this kind. The diagrams from HENNIG's (1966a) book which DARLINGTON reproduces and criticizes were introduced by HENNIG in a discussion on 'the rules for evaluating morphological characters as indicators of degree of phylogenetic relationship' in a section on the delimitation and ranking of groups above the species level. The argumentation scheme and other simple dendrograms used in that section are in my opinion adequate for character evaluation and analysis of the relationships between species and groups of species, so long as the variation within species (or groups of species) which is concealed by the definition of character states is slight in relation to the differences between the character states. Although I disagree with much of what DARLINGTON says, I see value in his paper as providing a warning against our losing sight of the simplifications inherent in the analytical methods proposed by HENNIG. The only proof against the misuse of conceptual models in contexts in which they may be misleading is an awareness of their limitations.

EVALUATION OF CHARACTERS

The construction of phylogenetic classifications involves judgements about the sequence of character changes, and the distinction of probable synapomorphies from other types of resemblance (homoplasy). Such judgements require a detailed knowledge of the organisms concerned, as well as an understanding of the principles of phylogenetic analysis. The criteria which can be employed in forming such judgements have been well discussed in general terms by HENNIG (1950: 172, and 1966a: 93), and I do not give an extensive review here.

Homoplasy is defined as 'resemblance not due to inheritance from a common ancestry' (SIMPSON 1961), and is equivalent to the wide sense in which the term 'convergence' has been used by some authors. In this work I use three terms to denote types of homoplastic resemblance: convergence, parallelism and homoiology. In practice not all cases of homoplasy can be classified unequivocally with this terminology, but a rough distinction can be made. The term convergence in the restricted sense here followed is applied to homoplastic character states which are the result of different transformation sequences. The acquisition of similar character states independently by different species, but without different transformation sequences, is called parallelism. When conditions are thought to have the same genetic base and to reflect close kinship among their bearers, although the phenotypic change occurred independently, they may be called 'homoiologous'. Homoiology thus occupies an intermediate position between parallelism and true homology.

One of the main problems confronting attempts at phylogenetic analyses is the need to distinguish conditions acquired through parallelism or homoiology from truly synapomorphous conditions. Convergence in the restricted sense

24

rarely results in conditions so similar that they are likely to be mistaken for synapomophous conditions, except in some reduction sequences (such as in the wing venation of Diptera). The detection of convergence in such cases clearly depends on reconstruction of the transformation sequences. In general the most useful criteria which can be employed in evaluating character states are the following: (1) the criterion of complexity of character change ('Kriterium der Komplizierheit der Merkmale', HENNIG 1950), and (2) the criterion of compatibility with the distribution of other apomorphous character states. Doubts have been raised about the validity of Darwin's distinction betwee 'adaptive' and 'non-adaptive' characters for purposes of evolutionary evaluation, and I therefore do not employ such a criterion. Both the criteria which I employ lead to probability judgements. The justification for employing the 'criterion of complexity' is that the amount or complexity of genetic change required to bring about particular changes in phenotypic character states is not constant. The change from a plesiomorphous to an apomorphous condition may involve anything from a change at a single genetic locus to an extensive reorganization of the genome. This is why the 'non-weighting' procedure recommended by some numerical taxonomists is invalid for phylogenetic analysis. Clearly the odds against the same character change having occurred through homoiology or parallelism must increase in relation to the complexity of the required genetic changes. In cases where considerable genetic complexity is believed to be involved, such as when new and complex structures have been acquired, the possibility that the whole of the genetic changes involved may be ascribed to homoiology or parallelism becomes so negligible that it may be discounted. The same thought probably underlies REMANE's (1952) conclusions on the applicability of the law of irreversibility ('Dollo's law'). He states that the law should not be applied to proportional relationships, quantitative differences, etc., but can be applied without restriction to complex organs, because no case of the reappearance of complex organs after loss has ever been demonstrated.

The validity of the criterion of genetic complexity would be doubtful only if evolution were a predominantly convergent or equifinal, rather than divergent, process. In that case phylogeny would be relegated to a subordinate process in evolution, and a classification of monophyletic groups would be of very limited value. I follow the orthodox view that evolution has occurred within species and populations, and that when groups of species show major structural modifications in common, they are descended from a common ancestral species. This view has sound empirical foundation in the obvious divergence in many characters often shown by vicariant species or by separated populations of the same species.

The criterion of genetic complexity can at present only be applied in a crude way, because the genetic basis of most character changes is not known directly. Nevertheless many inferences about the relative genetic complexity of character changes can be made from indirect evidence. For instance, it is reasonable to suppose that character changes which can be shown by viable mutants within a species may be of relatively low complexity from a macroevolutionary point of view. For this reason I see little difficulty in inferring that parallel changes have occurred in the numbers of spermathecae and of fronto-orbital bristles, when the distribution of apomorphous states of these characters conflicts with the distribution of other character states which are believed to have resulted from more complex changes. It also appears valid to assume that structures may be readily lost through convergence or parallelism; for any complex organ or system whose efficiency ceases to be maintained by selective pressure is likely to degenerate due to the pleiotropic effects of genetic changes (as in the many unrelated cavernicolous animals which have become blind).

The criterion of genetic complexity should not be applied in isolation, but only in conjunction with the criterion of compatibility. This criterion was treated by HENNIG (1966a: 120) as the 'method of checking, correcting and rechecking'. I have applied this criterion by setting up hypotheses about the limits of monophyletic groups based on the distribution of particular apomorphous character states, and then by checking these hypotheses for compatibility with the distribution of as many other apomorphous character states as possible. There is no restriction on the kinds of characters which can and should be compared, since all the characters of organisms have evolved in time. Therefore, the true transformation sequence of all characters must be compatible. Apparent incompatibilities revealed by the analysis may often be resolved by consideration of the numbers of characters which support alternative hypotheses and their probable complexity. Many hypotheses may be rejected, or considered too doubtful to be expressed in the formal classification. In evaluating hypotheses one should always bear in mind HENNIG's 'auxiliary principle', that the presence of apomorphous conditions in different species is always reason for suspecting kinship, and that convergence or parallelism should not be assumed *a priori*. This 'auxiliary principle' is an application of the principle of parsimony (Occam's razor), based on the premise that evolution is predominantly divergent. It is warranted to postulate from character changes of low complexity that certain species constitute a monophyletic group, if this hypothesis is compatible with the evidence of other characters. Common apomorphous conditions should not be ascribed to homoiology or parallelism unless the hypothesis that they are synapomorphous is incompatible with other evidence.

CAIN & HARRISON (1960) have presented an important discussion on the detection of homoplasy ('convergence' in their sense). They argue that homoplasy between closely related groups is likely to be common, particularly in respect of 'necessary functional correlates'. Clearly there is truth in this argument at the level of parallel changes in proportions, quantitative differences, etc.; some groups of species do in fact display a reticulate pattern of resemblances, which suggests that many parallel or homoiologous changes have occurred. While accepting the opinion of CAIN and HARRISON that some cases of homoplasy are likely to remain undetected, I do not think this in conflict with the principle that the most probable approximation to a classification of monophyletic groups is that with minimal postulated homoplasy.

Recently several different methods of numerical analysis have been suggested for phylogenetic studies. EDWARDS & CAVALLI-SFORZA (1964) proposed a Prim Network forming procedure intended primarily for analysis of racial evolution. I doubt whether this method is useful for studies of macroevolution, since it incorporates some assumptions which seem unrealistic in this context (such as random drift of characters). CAMIN & SOKAL (1965) presented a method for reconstructing probable phylogenetic relationships by comparing the evolutionary patterns suggested by particular character sequences against one another. Such checking for compatibility of hypotheses suggested by particular character sequences is in my opinion a necessary part of the method of phylogenetic systematics. But I do not think that CAMIN and SOKAL have adequately treated this question. The pattern tables for the character sequences given in their paper include groupings formed on the basis of symplesiomorphy, and the compatibility matrices consequently exaggerate the apparent conflicts between the evidence of the different characters. I also do not follow the logic behind their 'monothetic method', which seems to involve grouping species on the basis of minimal number of retained plesiomorphous conditions (zeros in the matrix), irrespective of whether they show the same or different apomorphous conditions. THROCKMORTON (1968) has suggested a procedure based on complete linkage analysis, but FARRIS, KLUGE & ECKARDT (1970) have questioned the value of this method on the grounds that some useful information is lost. The latter paper probably provides the best introduction to recent work in this field.

The use of mathematical analysis may well prove of value in some circumstances, but I wish to interject a word of caution. The methods of numerical phyletics, like those of numerical phenetics, presuppose the metaphysics of logical atomism. According to logical atomists the world of experience can be analysed into basic atomic facts or sentences which are not logically subdivisible. The atomic facts relevant to the classification of organisms are now called 'unit characters', following SOKAL & SNEATH (1963). The unit characters are assumed to be all homogenous, and in order to reduce the most highly heterogenous data to a common base for mathematical handling, all that is necessary is to express the data in terms of unit characters. The frequent claim that 'non-weighting' of characters (giving a set of characters equal weight) is 'objective' clearly reveals the nature of these hidden assumptions. Unfortunately, there are no empirical grounds for the belief in the existence of atomic facts or unit characters in the sense of homogenous independent units like physical atoms. In the 1920's and 1930's some philosophers, inspired by the early works of LUDWIG WITTGENSTEIN, tried to construct model languages from atomic facts or sentences. But it is now accepted that no such unique language can be constructed. In view of the demise of logical atomist ideas in the field of philosophy, the unquestioning assumption of such metaphysics in a very simple form by some mathematically oriented biologists seems to me naive. If the characters of organisms are not analysable into homogenous atomic ('not

27

logically subdivisible') units, then the belief that there is anything 'objective' in the procedure of treating heterogenous characters as if they had a common metric base is without foundation, and it should be recognized that arbitrary decisions are taken in the choice, definition and weighting of characters. A further consequence is that the problem of evaluating characters cannot be willed away by assuming that all 'unit characters' are equal. Probably the nearest approach to a unit basis for character change lies at the genetic level, but even here there are problems of interpretation (in particular, the equifinal nature of many biological processes at the molecular level).

If my denial of logical atomism is correct, then numerical phyletics cannot supplant the traditional approach to phylogenetic analysis elaborated in HENNIG's works. At most it can provide supplementary methods, to be used with due caution in view of the arbitrary steps involved. I do not employ numerical methods in this work.

CERTAINTY AND PHYLOGENY

The end results of phylogenetic analyses represent best estimates, approximations to the underlying phylogenetic relationships which can never be known with absolute certainty. Phylogenetic classifications thus should not be regarded as static, but must be periodically revised as our knowledge of phylogenetic relationships increases. The argument that phylogenetic systematics is non-operational because phylogeny cannot be known with certainty could be applied to the whole of natural science (see HULL 1967). Our observations of the real world and the inferences which we make from these observations are never invested with absolute certainty, and it can only be assumed that authors who demand absolute certainty misunderstand the dependence of all physical and biological sciences on inductive hypotheses.

3. STRUCTURE OF THE MALE POSTABDOMEN

3.1. Preliminary treatment and explanation of terminology

SEGMENTATION

The basic segmentation of the abdomen in Cyclorrhapha has been well discussed by CRAMPTON (1942), VAN EMDEN & HENNIG (1956), STEYSKAL (1957a) and HENNIG (1958). If the andrium represents the 9th segment (as now seems almost universally accepted), the numbering of the preceding segments can be established from considerations of comparative morphology. The Cyclorrhapha contain some member groups in which all eight preceding segments are clearly defined and bear discrete sclerites. The correct numbering for groups in which sclerites have been fused or lost can be established by analysis of the sequence of morphological changes which have occurred and by ontogenetic evidence, when this is available. The interpretation of the various conditions found in particular groups is discussed below in my treatment of the groups concerned. One remaining source of dispute affecting the interpretation of the postabdominal structure of all Cyclorrhapha is whether or to what extent the 8th segment is rotated. This question is discussed in detail below in section 3.2, where I present what seems to me conclusive evidence that this segment is inverted (rotated through 180°). I therefore follow CRAMPTON in calling the dorsal sclerite of this segment the 8th sternum, and the reduced ventral sclerite the 8th tergum. Reduction of the 8th tergum to a narrow band is probably a groundplan condition of the Cyclorrhapha (see section 4). My analysis of the relationships between the families of Schizophora has led me to explore certain sequences of modifications of the postabdominal sclerites. The starting point for these sequences seems to be the condition shown by the Platypezidae (figs. 10 and 11), in which the 6th and 7th segments are more or less unmodified, with their respective terga and sterna developed as discrete sclerites in the normal dorsal and ventral positions, and the inverted 8th segment retains a large dorsal sternum and a band-like tergum vestige in ventral position.

The abdomen of male Cyclorrhapha is conventionally divided into the **preabdomen** (segments 1 to 5) and the **postabdomen** (segments 6 and following), following METCALF (1921). This distinction is one of descriptive convenience, and does not reflect a fundamental morphological change between segments 5 and 6. In some groups the 6th segment is unmodified and similar to the preceding segments; in others, such as some Syrphidae, modification of the terminal segments begins with the 5th segment. Even in forms with a highly modified postabdomen, some structures (for instance the *sensilla trichodea* (fig. 89), the intersegmental musculature and the spiracles) are often readily recognizable as homonomous ('serially homologous') as far as the 7th segment. Some authors use these terms flexibly, varying the point of division according to how many segments are unmodified. However there are difficulties in applying this criterion to some families in which the 6th tergum is unmodified but the 6th sternum modified to some extent (for instance Micropezoinae and Tanypezidae). I prefer to adhere to consistent morphological definitions (following HENNIG 1936a), to avoid possible misunderstanding. The term **protandrium** has been proposed for that part of the postabdomen which precedes the genital segment (STEYSKAL 1957a).

The term **genital pouch** may be applied, if convenient, to membranous areas of the postabdominal venter which protect the aedeagus in its rest position. However the term has no general morphological significance or implications of homology, as the extent and manner of formation of such a pouch vary between different groups.

GENITALIA (GENERAL QUESTIONS)

The homologies of some parts of the external male genitalia of holometabolous insects are still the subject of dispute, mainly because of difficulties in interpreting the ontogenetic evidence. The crux of the dispute rests on whether the large lateral clasping lobes of the genital segment (usually biarticled except in Coleoptera) are homologous with the appendage articles (precoxae and styli) found on the 9th abdominal segment of Thysanura. Authors who have accepted this homology have called these lobes 'gonopods', consisting of a basal 'basistylus' or 'coxite' and a terminal 'dististylus'. VAN EMDEN & HENNIG (1956) accepted this view, and stated that the work of ABUL NASR (1950) on the ontogenetic development of certain Nematocera had demonstrated the existence of true gonopods (limbs of the genital segment) in Diptera. However ABUL NASR's observations can be interptreted differently, and SNODGRASS (1957: 47) explicitly rejected VAN EMDEN and HENNIG's interpretation. In the same paper SNODGRASS rejected some of his own earlier opinions on the homologies of the male genitalia of insects, and in particular maintained that there are no gonopods in any holometabolous insects. According to SNODGRASS' interpretation the paired rudiments ('primary phallic lobes') which give rise to the supposed 'gonopods' of holometabolous insects, as well as to the aedeagus and other parts of the reproductive apparatus, are homologous with the rudiments which give rise to the aedeagus alone in Thysanura. SNODGRASS proposed to apply the term 'parameres' to the supposed gonopods, in accordance with the original use of the term parameres by VERHOEFF (1893).

The validity of SNODGRASS' interpretation of the homology of the genital rudiments of Holometabola is doubtful in view of the radical ontogenetic changes which have occurred in the evolution of this group, and SHAROV (1966) has criticized his work on these grounds. SHAROV states that 'the external grasping appendages... are homologous with the precoxal plates and the styli sitting on them in Thysanura', which is simply a restatement of the gonopod theory. He does not present new evidence to support this view. MATSUDA (1958) reviewed the ontogenetic evidence and raised serious objections to homologizing any part of the male genitalia of holometabolous insects with the appendage articles of Thysanura. Unless these objections can be dispelled, the gonopod theory should be regarded as not proven. Besides SNODGRASS' view and the gonopod theory, a third possibility should be considered: that the parameres may be homologous with the gonapophyses of the 9th segment in Thysanura. A firm judgement between these alternatives probably cannot be made at present, because the ontogenetic changes which occurred in the evolution of the Holometabola are not well enough understood. Controversy has continued since the 1890's, and no consensus has yet emerged. In the present work I follow SNODGRASS' (1957) proposal of applying the term **parameres** in its original sense (to the supposed 'gonopods'), and calling their component articles the **basimeres** and **telomeres** (= distimeres of CRAMPTON). This is a special terminology proposed for the genitalia of Holometabola, and has the advantage that it does not imply homologies with the structure shown by less modified insects, such as Thysanura.

E. L. SMITH (1969) has proposed a revised interpretation of insect genitalia and much new terminology. SMITH'S interpretation and terminology are not accepted in this work, since some parts of his theory are untenable. In particular, I do not understand why he assumes that interlocking gonapophyses of the 8th and 9th abdominal segments (as in females of Thysanura and some pterygote orders) were an ancestral condition in male insects, since there is no known male insect (living or fossil) which shows such a condition. His view that the gonapophyses of the 9th segment have partly fused in many male Pterygota to form an intromittent tube (aedeagus) seems tenable, but he does not demonstrate why this interpretation is to be preferred to the interpretations of the origin of the pterygote intromittent organ put forward by other authors. This question depends on the interpretation of ontogenetic evidence, which SMITH does not review in detail. I must also note that some of the statements which he makes about Diptera are misleading. For instance, the statement that female Diptera possess an 'antovipositor', defined as an 'egg-laying device where non-appendicular components dominate... homologous to ♂ aedeagus' is incorrect, because studies of intersexes in Diptera have not suggested that any structures of the female are homologous with the male aedeagus (see LAUGÉ 1968). Another incorrect statement is that 'potential homology has been seen' between the gonopods of both sexes in Drosophilidae; for irrespective of whether one supposes that gonopods are present in the male, there are certainly no structures which can be homologized with gonopods in female Drosophilidae. Because of such obvious deficiencies, I do not think that much weight should be given to SMITH'S views.

In this account of the structure of the male terminalia I include only synonyms which seem to me significant or which have been widely used. The diversity of terminology used by different authors is so great that attempts to compile complete synonymies are best left to works on individual families. This diversity has arisen not only from the dispute on the origin of the genitalia, but also because workers on cyclorrhaphous families have found difficulty in homologizing some of the structures found in these groups with those of other Diptera. As a result various provisional terminologies have been proposed in taxonomic works.

The terms **hypopygium** or **terminalia** are here used to refer to the whole of the genital (9th) segment (or **andrium**) and associated proctiger. I do not extend the application of these terms to preceding segments (as did LINDNER 1949). Objections have been raised to the term hypopygium on the grounds that its application to structures situated above the anus (which are hence epipygial rather than hypopygial) is linguistically incorrect; but this objection loses its force, if the view advanced below that all the hypopygial structures of Cyclorrhapha are ventral in origin is accepted. The genital segment (andrium) and its associated structures are also often referred to as the **external genitalia**.

THE EXTERNAL SCLERITES OF THE GENITAL SEGMENT

The genital (9th) segment or andrium is enclosed laterally and dorsally by a large sclerite which has been assumed to be the epandrium (9th tergum) by virtually all recent authors. HENNIG (1936a) maintained that this sclerite was formed by fusion of the 9th and 10th terga with the basimeres (which he called the basal articles of the gonopods). But VAN EMDEN & HENNIG (1956) followed the prevailing opinion that this sclerite is the epandrium (9th tergum). There is indeed no evidence in comparative morphology or ontogeny for a fusion of the basimeres with any tergum. But there is strong evidence that the homology of this sclerite is not with the epandrium but with the basimeres. My view is that the true epandrium (9th tergum) is completely absent in the Cyclorrhapha, and that the so-called 'epandrium' of this group is formed by upward growth of the basimeres (basal articles of the parameres) and their fusion along the centre of the dorsum. Since the term epandrium should be applied only to the 9th tergum, I therefore propose for this sclerite the new term **periandrium**. Since my interpretation is new, further justification is needed.

The structure of the terminalia of Rhagionidae (figs. 4 and 5), is taken (following KARL 1959) as close to the plesiomorphous condition for the Brachycera (of which the Cyclorrhapha are a subordinate group). The terminalia of this family can be homologized without difficulty with the terminalia of many nematocerous families and named accordingly. The genital segment bears ventrally a triangular 9th sternum (hypandrium) and biarticled parameres, which serve as claspers during copulation. In dorsal view a well-developed epandrium (9th tergum) can be seen, and the proctiger bears a distinct tergum (tergum 10), as well as cerci. The terminalia of the Eremoneura (including the Orthogenya and Cyclorrhapha) differ from those of the Rhagionidae in several respects. The

32

relevant information is found in BÄHRMANN's (1960) study on the male copulatory organs in Empididae, although the conclusions which I have drawn from the data differ in some respects from those in that author's own discussion. According to BÄHRMANN's interpretation the parameres (which he calls 'Gonopoden') only occur in those genera of Empidinae and Hemerodromiinae

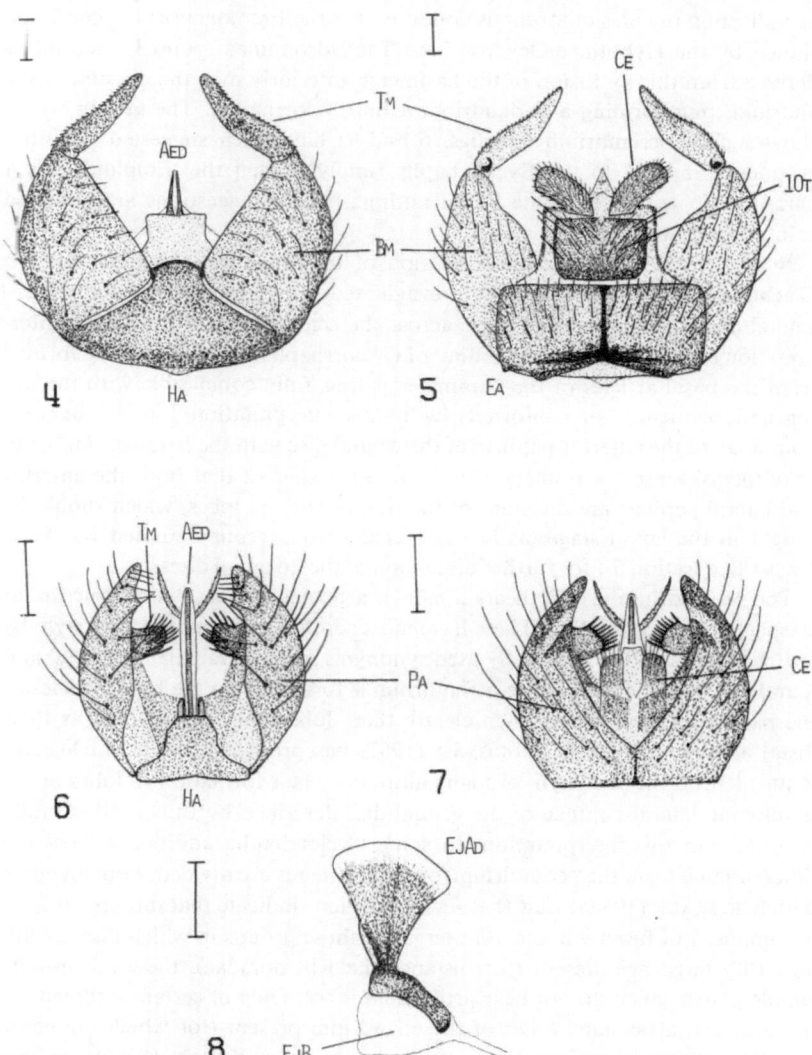

Figs. 4 - 8. 4. *Rhagio* sp. (Rhagionidae), hypopygium (♂) in ventral view. 5. *Rhagio* sp. (the same individual as shown in fig. 4), hypopygium (♂) in dorsal view. 6. *Atelestus pulicarius* (FALLÉN) (Empididae, Atelestinae), hypopygium (♂) in ventral view. 7. *Atelestus pulicarius* (FALLÉN), hypopygium (♂) in dorsal view. 8. *Agromyza phragmitidis* HENDEL (Agromyzidae), ejaculatory bulb and apodeme (♂).

33

which are characterized by a small and usually cleft epandrium. He interprets as plesiomorphous the condition shown by other groups (such as the Hybotinae, Ocydromiinae and Tachydromiinae), in which the 'epandrium' covers the greatest part of the genital segment and extends laterally to overreach the ventral surface. In my opinion the character sequence is more probably the reverse. I interpret the large laterally placed basimeres of *Empis* and its relatives as indicating the plesiomorphous condition for the Eremoneura; the condition shown by the Hybotinae/Ocydromiinae/Tachydromiinae group I interpret as derived from this by fusion of the basimeres anteriorly over the dorsum of the andrium, thus forming a periandrium as in Cyclorrhapha. The groups which show such a periandrium (see figs. 6 and 7) have been suggested on other grounds as related to the Cyclorrhapha (implying that the 'Empididae' is a paraphyletic group) by some recent authors. These suggestions are reviewed below in section 4.

BLACK (1966) describes the formation of the 'epandrium' in *Eucalliphora* (Tachinidae *s.l.*) by growth of the evaginated lateral papillae of the ventral genital disc, which eventually fuse across the centre of the dorsum. My interpretation of the so-called 'epandrium' of Cyclorrhapha as a periandrium formed from the basal articles of the parameres is thus fully compatible with the ontogenetic evidence. In conformity with this interpretation I reject BLACK's homology of the anterior papillae of the genital disc with the parameral lobes in SNODGRASS' sense (= primary phallic lobes). I suggest that both the anterior and lateral papillae are divisions of the true parameral lobes, which should be sought in the larval stage (as in the nematocerous groups studied by ABUL NASR). See section 6.1 for further discussion of the imaginal discs.

The periandrium usually bears a pair of articulated lobes which function as claspers during copulation. These have most commonly been called 'surstyli' by recent authors. Other frequently used synonyms include 'valvulae laterales' and 'paralobi'. If my view that the periandrium is formed from the basal articles of the parameres is accepted, then clearly these lobes may be accepted as their distal articles, for which SNODGRASS (1957) has proposed the term **telomeres** (equivalent to the 'dististyli' of many authors). Their formation as lobes of the developing lateral papillae of the genital disc described by BLACK (1966) fully accords with this interpretation. In some Cyclorrhapha additional lobes are differentiated from the periandrium, or the telomeres are divided. I do not agree with STEYSKAL's (1957a) view that such conditions indicate that the 'epandrium' is composed of fused 9th and 10th terga. In those groups in which discrete 9th and 10th terga are present (for instance the Rhagionidae), these are usually simple plates which do not bear articulated lobes. Only in certain Asilidae are more or less articulated lobes of the epandrium present (for which the name 'surstyli' may appropriately be retained). However KARL's (1959) analysis indicates that the presence of surstyli cannot be ascribed to the groundplan of the Asilidae, and such conditions therefore do not indicate the character sequence leading to the conditions shown by Cyclorrhapha. It is interesting that BRUNDIN (1966: 80) concluded that in 'the basic design of the Nematocerous Diptera' (by which he must mean the groundplan of the Diptera as a whole, since the Nematocera are not a monophyletic group), the telomeres were double and articulated separately with the basimeres. Thus, even if the presence of double telomeres is a groundplan condition for the Cyclorrhapha, rather than secondary in this group as HENNIG (1936a) maintained, no special difficulty need arise in interpreting this condition. However I think HENNIG's interpretation more probable, as double or divided telomeres are only found in a few families of Cyclorrhapha.

SNODGRASS (1935) stated that 'movable claspers that can be identified with the harpagones, or styli of the gonopods, are absent in muscoid Diptera, but the 9th tergum commonly bears on its lower posterior angles a pair of long lobes, which may be flexible at their base but are not provided with muscles'. SNODGRASS' opinion was probably based on the morphology of *Pollenia* (Tachinidae *s.l.*). In that family movement of the telomeres is produced by contraction of muscles inserted on the processus longi, not on the telomeres themselves, as has been confirmed by SALZER's (1968) detailed studies on *Calliphora*. However the condition shown by the Tachinidae *s.l.* is clearly apomorphous and not a general condition of the Schizophora ('muscoid Diptera'), as SNODGRASS assumed. Certainly in some groups of Cyclorrhapha muscles are inserted on the base of the telomeres, for instance muscle '3' in HENNIG's (1936a) description of the musculature of *Calycopteryx* (Micropezidae). RIVOSECCHI (1958) has described muscles inserted on the telomeres in *Musca* (Muscidae). I have also noted muscles inserted on the base of the telomeres in serial sections of a specimen of a *Syrphus* species (Syrphidae).

In most Cyclorrhapha the integument is strongly infolded between the sides of the periandrium above the base of the aedeagus, thus forming what may be called the **periandrial fold**. Sclerites linking the hypandrium and telomeres are often present in the upper wall of this fold. For a single such sclerite I propose the new term **interparameral sclerite**. The use of the term '10th sternite' for such a sclerite is clearly unwarranted, since the 10th segment is part of the proctiger, which lies above the telomeres. In some groups the sclerotization of the upper wall of the periandrial fold consists of a pair of lateral rod-like sclerites, known as the **processus longi** or **bacilliform sclerites**. It is possible that the processus longi of some Calyptratae arose as apodemes from the base of the telomeres, since they bear muscles apparently homologous with those inserted on the base of the telomeres in other groups (see preceding paragraph). The lower wall of the periandrial fold above the aedeagal apodeme is usually membranous, but sclerotization is developed here in a few groups (e.g. Micropezoinea).

The cyclorrhaphous **hypandrium** in my view consists solely of the 9th sternum and is fully homologous with the hypandrium of the Empididae (Orthogenya). In conformity with my conclusion that the periandrium is formed from the basimeres, I cannot accept suggestions that these may have been involved in the formation of the hypandrium. CHILLCOTT's (1958) homology of the hypandrium with the basimeres (which he called 'gonocoxites') was based on direct comparison with the condition in the Bombyliidae, where the basimeres and telomeres are ventrally situated and a true epandrium retained. While CHILLCOT's assumption that the conditions of the genitalia shown by the Bombyliidae are relatively plesiomorphous for the Brachycera seems substantially correct, he failed to consider the conditions shown by the Empididae, whose study is in my opinion essential to an understanding of the character sequences leading to the organization of the genitalia in Cyclorrhapha. GOODING & WEINTRAUB (1960) tentatively suggested that the hypandrium may include elements derived from the preceding segments, because of the presence of a partial transverse suture in *Hypoderma*. However neither my comparative studies nor BLACK'S ontogenetic studies provide evidence for such a fusion. HENNIG (1936a) also suggested that the hypandrium was formed by fusion of two sclerites, but subsequently abandoned this view; VAN EMDEN & HENNIG (1956) accepted the hypandrium as the 9th sternum.

The anterior end of the hypandrium is in some species produced into a rod-like process called the **hypandrial apodeme**. Posteriorly the hypandrium usually divides into a pair of **hypandrial arms**. In some groups these arms bear more or less vertically directed extensions posteriorly (called the vertical sections of the hypandrial arms in my descriptions).

My interpretation of the origin and homology of the hypandrium, periandrium and telomeres (which for brevity may be referred to as the 'periandrium hypothesis') may at first seem improbable, because it breaks with tradition. But it has much greater explanatory power in terms of the available evidence than the traditional view that the parameres (or 'gonopods') have been lost or fused with the hypandrium, and that the large dorso-lateral sclerite represents an epandrium from which secondary appendages ('surstyli') are differentiated. The following points of comparison seem significant:

(1) According to the periandrium hypothesis the clasping mechanism in the Cyclorrhapha does not differ substantially from that of many nematocerous families, and both the sclerites and some of the muscles involved may be homologized widely throughout the order Diptera. According to the traditional interpretation the sternal clasping mechanism (parameres) of the Nematocera and most non-cyclorrhaphous Brachycera has been lost and functionally replaced by secondary structures of tergal origin.

(2) The periandrium hypothesis implies a gradual process of dorsal expansion of the basimeres, which is exemplified by the conditions shown by certain Empididae. The complete loss of the parameres which the traditional interpretation implies has not been supported by a convincing sequence of character change.

In the light of the above considerations I maintain that the periandrium hypothesis is much more consistent with the known facts and provides a simpler interpretation of them than the traditional homologies.

PREGONITES AND POSTGONITES

In some Cyclorrhapha (see figs. 60 and 70) there are two pairs of lobes near the base of the aedeagus. Following CRAMPTON and others I call these the **pregonites** and **postgonites**. The homology of these structures has long been a source of difficulty. Either or both have often been called 'parameres', a usage which SNODGRASS (1957: 48) justifiably rejects. The postgonites are clearly a form of paraphysis, as there defined by SNODGRASS. These are sensory lobes arising from near the base of the aedaegus; they are probably homologous with lobes present in certain Empididae and other 'lower Brachycera', but the extent to which they can be homologized with similar structures in nematocerous families and other holometabolous orders has not been clarified. According to BLACK's (1966) study of *Eucalliphora* the pregonites (which she calls 'anterior parameres') develop as lobes from the same papillae as the postgonites (anterior papillae 2). They thus represent additional paraphyses, not appendages of the hypandrium as some authors have suggested. Similar articulated pregonites are widespread among the Calyptratae, but the extent to which homologous structures occur in other groups of Cyclorrhapha is unclear. In some groups the structures which are called pregonites (or the equivalent) may in fact be processes or lobes of the hypandrium rather than paraphyses. My use of the term pregonites for groups other than Calyptratae is provisional, and does not indicate any firm opinion on homology. It has frequently been suggested that the pregonites of Calyptratae may be the distal articles of the parameres (the 'styli of the gonopods' of VAN EMDEN & HENNIG 1956), but this view cannot be reconciled with BLACK's account of their ontogeny.

AEDEAGUS

The intromittent organ itself I call the **aedeagus**, and regard the terms 'phallus' and 'penis' as synonyms. The aedeagus provides many characters which are important for the classification of the Schizophora. In many Orthogenya the aedeagus is long and slender, more or less uniformly sclerotized, upcurved distally. I postulate below that an aedeagus of this type can be ascribed to the groundplan of the Eremoneura (see section 4). Among the Cyclorrhapha this type of aedeagus is shown by many Platypezidae (Platypezidea) and Lonchaeoidea (Schizophora) (figs. 21 and 23). In some Schizophora a complex mechanism has evolved for swinging the aedeagus about its articulation with the aedeagal apodeme into an anteriorly directed rest position. The groups which show this apomorphous condition are classified below in the Nothyboidea and Muscoidea. SALZER (1968) has described in detail the functioning of this mechanism in *Calliphora erythrocephala* Meigen (Tachinidae *s.l.*) (figs. 52 and 53), in which the arc of movement is about 160°. In other superfamilies of Schizophora (Lonchaeoidea, Lauxanioidea and Drosophiloidea) and in all non-schizophorous Diptera, only a limited degree of movement of the aedeagus in relation to the aedeagal apodeme is possible.

In those groups of Schizophora in which this swinging mechanism is present (Nothyboidea and Muscoidea), the walls of the aedeagus are differentiated into sclerites and membranous areas. The basal sclerite on which muscles are inserted is called the **phallophore** (= basiphallus). This is a large cylindrical structure in some groups, but in others consists of only a narrow basal ring or partial ring of sclerotization. The term 'theca' recommended by VAN EMDEN & HENNIG (1956) was originally applied by WESCHÉ (1906) throughout the Diptera to structures of several different origins, and is clearly inappropriate in this context since it literally means a 'sheath'. Some authors have suggested that the term aedeagus or phallus should be restricted to parts distal to the phallophore. I follow VAN EMDEN & HENNIG in rejecting this usage, because only the whole intromittent organ can be widely homologized; the distinction between a phallophore and 'aedeagus' in the above sense cannot be applied satisfactorily in groups other than Muscoidea and Nothyboidea. METCALF (1921) considered the hypandrium and aedeagal apodeme of Syrphidae as parts of the 'penis', and some of the terms used for the hypandrium by other authors on Syrphidae ('inner copulatory organ', 'phallosome', 'penis sheath') are also open to the misunderstanding that this is part of the intromittent organ. Such terminology is in my opinion misleading and should not be used. I apply the term **paraphalli** to lateral sclerites in the walls of the aedeagus of Muscoidea (see my characterization of that group below). In some groups of Muscoidea the phallophore bears a characteristic sclerotized fold or process which extends posteriorly from the base of the aedeagus. This is known as an **epiphallus** (= spinus). The epiphallus moves with the aedeagus when this is swung about its articulation with the aedeagal apodeme into the copulatory position. It is likely that an epiphallus has been evolved independently in different groups of Muscoidea.

AEDEAGAL APODEME AND EJACULATORY APODEME

The **aedeagal apodeme** (or **phallapodeme**) is usually a conspicuous structure articulated with the base of the aedeagus; it bears strong musculature involved in the extrusion and retraction of the aedeagus. Ontogenetically the aedeagal apodeme develops as an ingrowth of the integument at the base of the aedeagus (SCHRÄDER 1927, BLACK 1966). HENNIG'S (1958: 539-540) statement that the aedeagal apodeme 'has a similar morphological value' to the epiphallus, because both structures represent sclerotized folds of the integument, is correct as regards their ontogeny, but these structures are not functionally equivalent. I suggest that in the groundplans of the Syrphidea and Schizophora the aedeagal apodeme was rod-like (bacilliform), attached to the body wall only where it articulates with the base of the aedeagus. This condition is shown by all Syrphidea, as far as I am aware, and by many groups of Schizophora. The presence of any sclerotized link or fusion between the aedeagal apodeme and the hypandrium or body wall between the hypandrial arms, is treated as an apomorphous condition in my analysis. Such conditions occur widely among some groups of Schizophora, and have probably been evolved independently on many occasions.

The **ejaculatory apodeme** (fig. 8) is a completely internal apodeme, arising from the wall of the ejaculatory duct. It serves as the attachment for muscles which force seminal fluid from the **ejaculatory bulb** (an expanded chamber of the ejaculatory duct) through the narrow terminal part of the duct. Although an internal structure, this apodeme is conveniently considered together with the external genitalia, as it is ectodermal in origin and can be readily studied in macerated preparations. Such an apodeme occurs in all Syrphidea and Schizophora, except where secondarily reduced.

Further investigation is needed to clarify the homologies of the aedeagal apodeme and ejaculatory apodeme of Cyclorrhapha with structures found in other Diptera. HENNIG (1936a) suggested that both these apodemes may have arisen by splitting of a previously uniform structure equivalent to the 'ejaculatory apodeme' of Orthogenya, but no convincing evidence supports this hypothesis. An alternative explanation is that the 'ejaculatory apodeme' of Orthogenya is homologous with the aedeagal apodeme of Cyclorrhapha, and that the ejaculatory apodeme of some Cyclorrhapha (Syrphidea and Schizophora) is neomorphous. Two criteria can be adduced in support of the interpretation that the aedeagal apodeme of Cyclorrhapha is homologous with the 'ejaculatory apodeme' of Orthogenya: similarity of position and similarity of musculature. Both structures articulate with the base of the aedeagus, and hence occupy a similar position in relation to other hypopygial structures; and according to TREHEN's (1960, 1962) studies the muscles of the 'ejaculatory apodeme' of Empidinae all insert on the hypandrium or at the base of the aedeagus, as is the case with the cyclorrhaphous aedeagal apodeme. However, this interpretation conflicts with another valid criterion of homology, continuity of function; for it involves postulating that the ejaculatory function has been lost by one structure and taken over by a neomorphous structure. I think any conclusion must await detailed comparative information on the structure of the Platypezidea and Hypocera. In two species of Platypezidae, *Polyporivora polypori* (WILLARD) and *Plesioclythia agarici* (WILLARD), I have found sclerotized areas and musculature on the terminal part of the ejaculatory duct; but the conditions are very different in each case, and it is not clear whether either is relevant to the evolution of the characteristic ejaculatory bulb and apodeme of Syrphidea and Schizophora.

PROCTIGER

The **proctiger**, containing elements derived from the 10th to 12th segments of primitive insects, is usually much reduced in the Cyclorrhapha. The 10th tergum seen in Rhagionidae (fig. 5), like the 9th tergum, seems completely lost. The only large sclerotized structures are the **cerci**. Some authors have maintained that these are not cerci but should be called 'paraprocts' or 'parapodial plates'. However, HERTING (1957) has pointed out that in the groundplan of the Brachycera these structures are biarticled in the female, which supports the view they are true cerci. The homology of the male and female structures has been demonstrated by MILANI & RIVOSECCHI (1955), who have described a convincing series of intermediate conditions from sexually abnormal specimens of *Musca domestica* L. The **10th sternum** (on the ventral side of the proctiger) is vestigial or absent in many Cyclorrhapha, but is distinct in the hypoceran *Ironomyia* figured by J. F. MCALPINE (1967) (as '? sternite 11').

The meaning of such terms as 'dorsal', 'ventral', 'left' and 'right' is potentially ambiguous because parts of the male postabdomen of Cyclorrhapha are rotated and deflexed (as explained in detail in section 3.2). I disregard the effects of rotation in applying these terms, so that, for instance, the large inverted 8th sternum is described as occupying a 'dorsal' position. However deflexion of the hypopygium must be taken into account in interpreting my use of such terms as 'dorsal' and 'ventral'. Thus the cerci and the centre-line of the periandrium are always considered to indicate the 'dorsal' side of the hypopygium, irrespective of the degree of deflexion.

3.2. The 'hypopygium circumversum' condition

The subject of hypopygial circumversion (rotation through 360°) in male cyclorrhaphous Diptera has attracted much interest and comment over the years, although few authors have attempted to discuss all the extensive pertinent literature. Perhaps the most balanced previous review is that in LINDNER'S (1949) handbook in 'Die Fliegen der paläarktischen Region'. Very recently the subject has acquired a new dimension as the result of discoveries made by E. L. KESSEL of the University of San Francisco in the course of his studies on Platypezidae (KESSEL & MAGGIONCALDA 1968a, KESSEL 1968b). These discoveries provide the first important new factual information on this subject since the 'thirties.

Confusion has long persisted in the literature because of the widespread currency of theories which assume some direct causal relationship between hypopygial circumversion and the asymmetrical conditions of the postabdomen shown by the Muscoidea (in the sense defined below). The most influential source of such theories was G. C. CRAMPTON. HENNIG (1958) criticized some aspects of CRAMPTON'S views, but several other authors have continued to accept them in their entirety. In view of this continuing controversy, I think that progress will be best served if I include here a critique of the views of CRAMPTON and G. H. HARDY, who approached this question with similar assumptions.

THE EVIDENCE FOR CIRCUMVERSION

FEUERBORN (1922) proposed the term '*hypopygium circumversum*' in the sense of a rotational circumversion of the hypopygium. He did not observe the rotation, but postulated its occurrence inside the puparium of *Calliphora* to explain the looping of the ejaculatory duct over the hind gut in this genus (to which BRÜEL (1897) had first drawn attention). FEUERBORN's conclusion was fully confirmed by the work of his pupil SCHRÄDER (1927), who demonstrated that in *Calliphora* a 360° clockwise rotation of the hypopygium took place on roughly the fifth day after pupation. SCHRÄDER thought that the duration of the process was at most 24 hours. He claimed that a growth process was responsible for the rotation, since he could see no muscles in his serial sections.

GLEICHAUF (1936) published a detailed study of the development of *Drosophila melanogaster* (MEIGEN). He concluded that circumversion began on the second day after pupation, and was completed at latest by the end of the third day. The direction of the rotation was clockwise, as in *Calliphora*.

Further observations on the process of circumversion within the puparium are found in recent works by MILANI & RIVOSECCHI (1955), referring to *Musca domestica* L., and GOODING & WEINTRAUB (1960), referring to *Hypoderma* spp. These accounts are less detailed than the earlier works, but clearly indicate the occurrence of rotation within the puparium, as described for *Calliphora* and *Drosophila*.

The process of circumversion has certain characteristic effects upon the internal structure of the mature adult. First, the ejaculatory duct is looped over the hind gut, passing upwards on the left side of the insect, then crossing above the gut, and then passing downwards to the aedeagus on the insect's right side (fig. 9). Both SCHRÄDER and GLEICHAUF demonstrated that this looping resulted directly from the rotation of the hypopygium within the puparium, thus vindicating FEUERBORN's (1922) hypothesis. Secondly, the postabdominal nervous system shows a double chiastoneury (crossing over) of the *nervi terminales*, as SALZER (1968) has described in much detail. Thirdly, the longitudinal tracheal trunks are crossed posterior to the last pair of spiracles (fig. 12).

As far as is known, the process of circumversion is completed within the puparium in all Schizophora. But KESSEL & MAGGIONCALDA (1968a) and KESSEL (1968b) have reported that in certain Platypezidae only the first 180 degrees takes place within the puparium; the additional 180 degrees takes place in the teneral state immediately following emergence. The extent to which muscular action is involved in the first part of rotation within the puparium has not been investigated; but it is evidently involved at least in the part of rotation completed after emergence. KESSEL and MAGGIONCALDA state that all the movements involved in the post-emergence part of circumversion in platypezids occur to the right of the main axis of the body; in the most teneral flies examined (of the genus *Plesioclythia*), the postabdomen projected

'...sharply out to the right in the horizontal plane and a little toward the rear with reference to the longitudinal axis of the body. The seventh and particularly the eighth abdominal segments are somewhat attenuated to form a stalk which bears the hypopygium at its end. The hypopygium is inverted, with the hypandrium and its parameres above, and the epandrium and its surstyles below, showing the circumversion has progressed 180 degrees... Immediately following emergence, the second half of circumversion takes place as the stalked abdomen continues to move in its arc, rotating another 180 degrees as it passes downward and posteriad to finally reach the position of its beginning at the midline. These movements complete the 360 degrees of circumversion and the hypopygium is right-side-up once more' (KESSEL & MAGGIONCALDA 1968a: 82).

A similar description of the process in *Paraplatypeza coraxa* (KESSEL) is given by KESSEL (1968b: 246), with a photograph of a newly emerged living male.

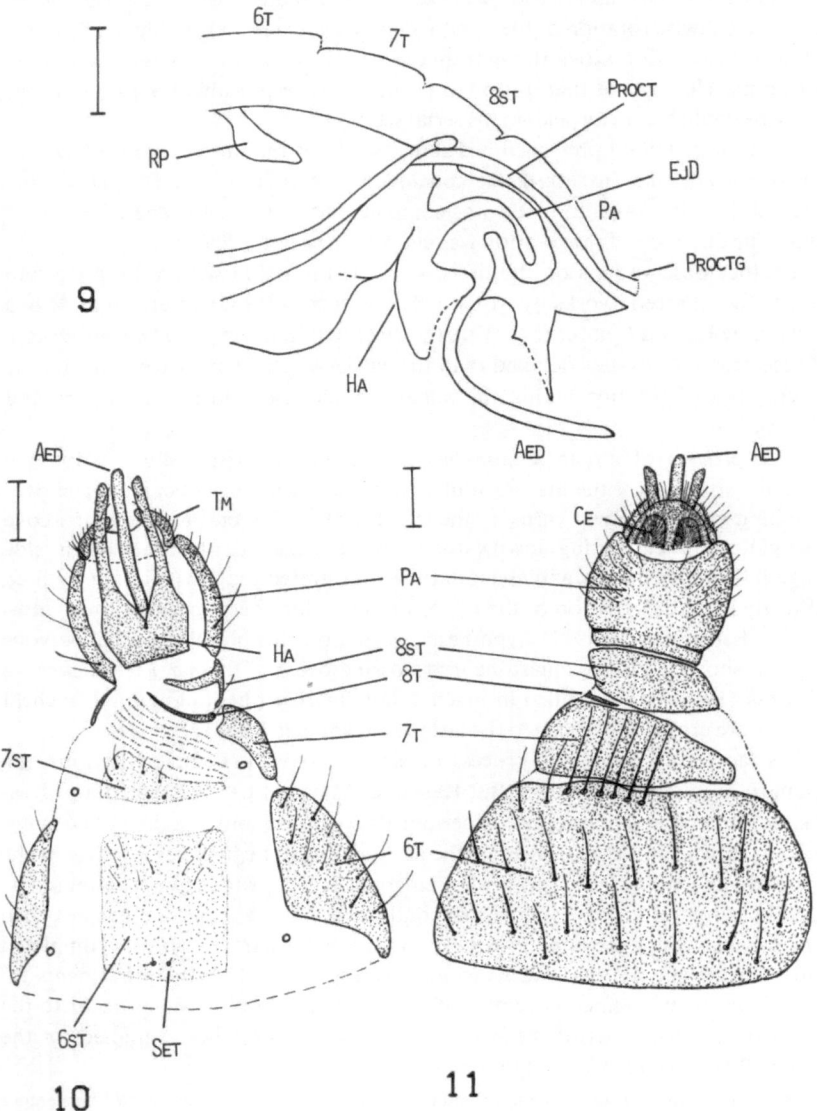

Figs. 9 - 11. 9. *Plesioclythia agarici* (WILLARD) (♂) (Platypezidae), schematic representation of the ejaculatory duct and hind gut, with an outline of the body wall as seen in sagittal section along the mid-line. 10. *Plesioclythia agarici* (WILLARD), postabdomen in ventral view of mature male, showing a *hypopygium circumversum* condition. 11. *Plesioclythia agarici* (WILLARD), postabdomen in dorsal view of mature male, showing a *hypopygium circumversum* condition.

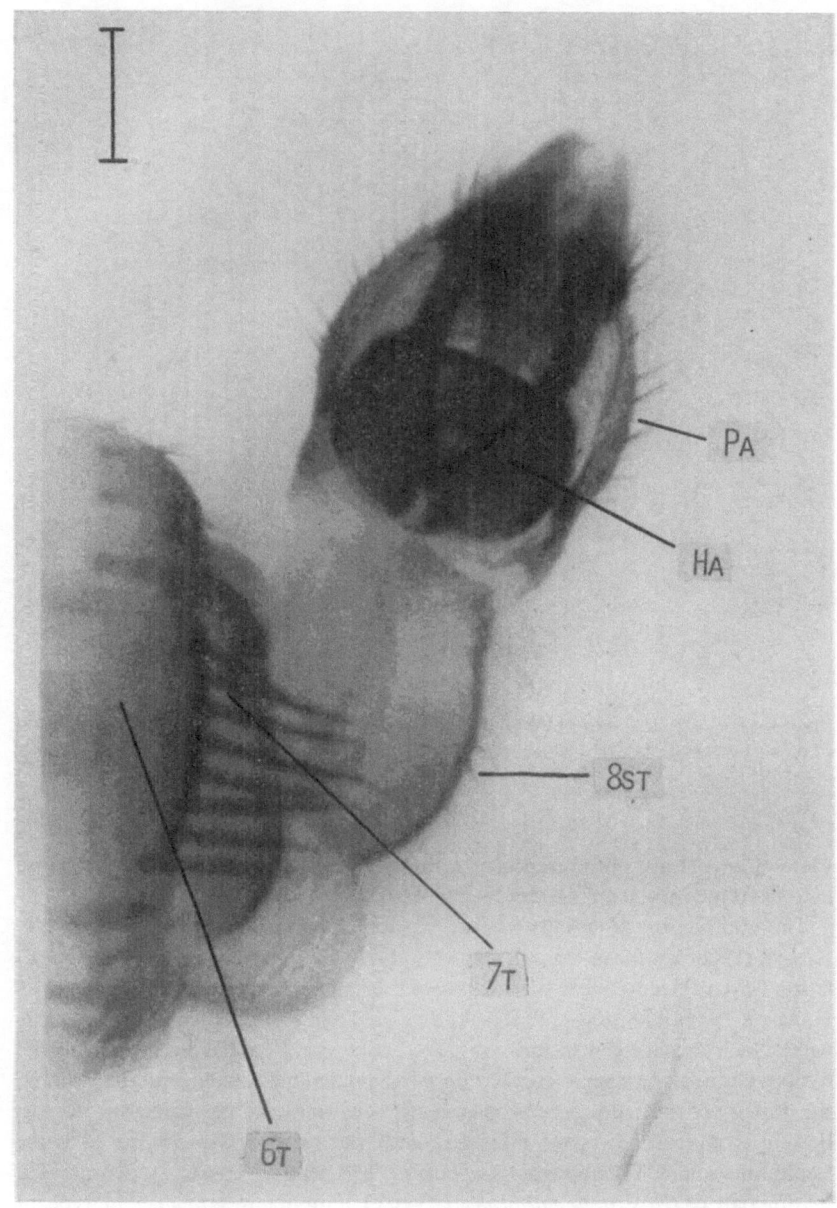

Plate I. *Plesioclythia agarici* (WILLARD) (Platypezidae), postabdomen in dorsal view of newly emerged male, showing a *hypopygium inversum* condition (photograph by DR. D. A. McB. CRAIG).

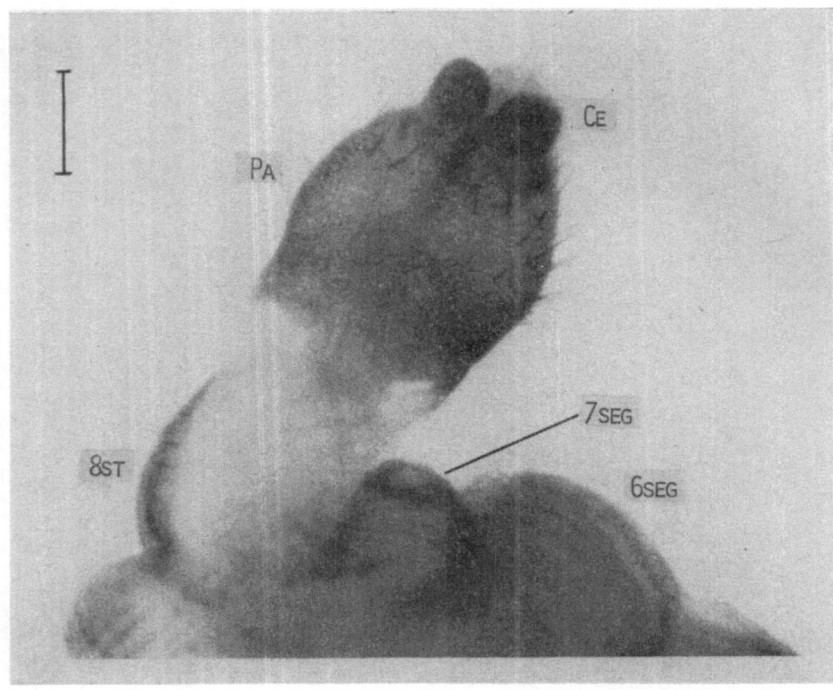

Plate II. *Plesioclythia agarici* (WILLARD) (Platypezidae), postabdomen in ventral view of newly emerged male (the same individual as shown in plate I) (photograph by DR. D. A. McB. CRAIG).

Plates I and II are photographs of a newly emerged specimen of *Plesioclythia agarici* (WILLARD), from material which DR. KESSEL kindly sent me.

The completion of circumversion in such Platypezidae is not irreversible. KESSEL (1968b) reports that all Platypezidae which he has observed in copulation have utilized a linear position (or rather 'opposed position', if one follows HARDY's 1944 definitions), that is 'tail-to-tail and right-side-up for both partners'. Such a mating position can only be expected in flies with an inverted hypopygium, since inverse correlation of the male and female genitalia (that is, the dorsum of the aedeagus contiguous with the venter of the vagina) appears to be a mechanical necessity in insects, with the probable exception of some Lepidoptera and Trichoptera (RICHARDS 1927, HARDY 1944). A mount of a mating pair of *Grossoseta californica* (KESSEL), a unique mount since platypezid pairs usually separate on capture, confirms that the male hypopygium is in an inverse position. KESSEL has concluded that the hypopygium was unwound (anticlockwise) through 180 degrees from the circumverse to the inverse position. I have seen the mount in question and agree with this interpretation. Opposed mating positions are, for obvious reasons, normally assumed after copulation has begun in some other position. KESSEL has not yet observed how

44

copulation in platypezids is initiated, but suggests that the male vertical pose (a 'superimposed position' in HARDY's terminology) is first assumed, with the male hypopygium still in the circumverse position. However this needs to be confirmed by observation, as there are other possibilities (see HARDY 1944).

The process of reversible circumversion now established for the Platypezidae may clearly be regarded as a less modified condition than that shown in the Schizophora, where the entire process occurs within the puparium and is irreversible. The change which has occurred is the acceleration of an ontogenetic process, so that it is completed at an earlier stage in relation to the overall ontogeny of the animal. It has unfortunately not been made clear whether the insect is a pupa or pharate adult when rotation takes place in Schizophora.

While the number of species for which circumversion has been demonstrated directly in ontogenetic studies is very low, the occurrence of circumversion can also be inferred from looping of the ejaculatory duct over the hind gut.

Since looping of the ejaculatory duct over the hind gut has now been confirmed for a representative range of species, I am confident that the *hypopygium circumversum* condition is universal among male Cyclorrhapha, and that the more plesiomorphous reversible expression of this character shown by the Platypezidae occurred in the last common ancestor of the group. Table I sets out in detail the evidence for making such a generalization.

The information on the distribution of circumversion is still inadequate in some respects. In particular, no detailed information is available on the process of circumversion and mating positions in the Lonchopteridae. The only relevant information is HENNIG's (1958: 535) brief statement that he thought he saw looping of the ejaculatory duct over the hind gut in *Lonchoptera* ('*Musidora*'). In view of the postulated sister-group relationship between the Lonchopteridae and other Cyclorrhapha (see section 5), investigations on circumversion, mating positions and related questions in this family might yield interesting results from an evolutionary point of view. The other major gap in the information concerns the timing of the process of circumversion in the Hypocera and Syrphidea. I suspect that in the Syrphidea the process is completed at an early stage within the puparium, as in Schizophora, since the lack of extensive intersegmental membranes in the postabdomen of Syrphidea makes the possibility of rotation after the sclerites have been formed seem unlikely. But there have been no ontogenetic studies which clarify this point.

Loss of circumversion has not been demonstrated in normal males of any species of Cyclorrhapha, although abnormal specimens with only partial rotation of the hypopygium are known in *Drosophila* (extensive literature) and *Musca* (MILANI & RIVOSECCHI 1955). In the micropezid *Calycopteryx moseleyi* EATON from Kerguelen Island the rotation has been reduced to about 320° (see HENNIG 1936a); this is doubtless a secondary modification, since the normal

45

TABLE I

List of genera for which the *hypopygium circumversum* condition has been established by ontogenetic studies, or by examination of the course of the ejaculatory duct in relation to the hind gut.

Acroptera
  Lonchopteridae – *Lonchoptera* (HENNIG 1958: 535)
Atriata
  Platypezidea
    Platypezidae – *Calotarsa** (KESSEL & MAGGIONCALDA 1968a),
        Plesioclythia* (KESSEL & MAGGIONCALDA 1968a), *Polyporivora***
  Hypocera
    Phoridae – *Anevrina***
  Syrphidea
    Pipunculidae – *Alloneura***
    Syrphidae – *Syrphus sens. lat.***, genus not stated (VAN EMDEN 1951)
  Schizophora
    Lauxaniidae – *Sapromyza* (HAHN 1929), *Lauxania***
    Chamaemyiidae – *Chamaemyia***, *Leucopis***
    Drosophilidae – *Drosophila* (GLEICHAUF 1936, MILLER in DEMEREC 1950,
        etc.)
    Ephydridae – *Scatophila* (BOLWIG 1940), *Psilopa***
    Psilidae – *Loxocera***
    Micropezidae – *Micropeza* (HENNIG 1934)
    Sciomyzidae – *Pherbellia***
    Sepsidae – *Nemopoda***
    Sphaeroceridae – *Leptocera sens. lat.***
    Trixoscelididae – *Zagonia***
    Anthomyzidae – *Anthomyza***
    Clusiidae – *Clusia***
    Agromyzidae – *Melanagromyza** (IPE 1967), *Phytomyza***
    Chloropidae – *Chlorops***
    Conopidae – *Thecophora***
    Tephritidae – *Dacus, Chaetostomella* and *Urophora* (= *Euribia*) (HENNIG
        1936a), *Tephritis* (RIVOSECCHI 1957)
    Anthomyiidae – genus not stated (CHILLCOTT 1958)
    Muscidae – *Musca* (HEWITT 1907, MILANI & RIVOSECCHI 1955, RIVOSECCHI
        1958), *Stomoxys* (TULLOCH 1906)
    Hippoboscidae – *Raymondia* (JOBLING 1951), 'Nycteribiidae' (genus not
        stated) (THEODOR & MOSCONA 1954a)
    Glossinidae – *Glossina* (MINCHIN 1905, ZUMPT 1936)
    Tachinidae – *Calliphora* (BRÜEL 1897, SCHRÄDER 1927, RICHARDS 1927,
        SALZER 1968), *Phormia* (CRAMPTON 1944a), *Ernestia* (PETZOLD
        1927), *Hypoderma* (MOTE 1929, GOODING & WEINTRAUB 1960)

* The single asterisk indicates that the published record has been confirmed by examination of serial sections of the abdomen in my possession.
** The double asterisk indicates a new record based on serial sections of the abdomen in my possession.

360° rotation occurs in other Micropezidae. SKAIFE's (1921) account of the internal reproductive system of *Braula* (Braulidae) does not include reference to looping of the ejaculatory duct over the hind gut, which is surprising as there seems no doubt that *Braula* is a true schizophoran. However SKAIFE's description is very brief, and I suspect that his observations were incomplete. Unfortunately I was unable to obtain fresh material of *Braula* to check this point. FERRIS (in DEMEREC 1950: 418) published some remarks which cast doubt on the occurrence of rotation in *Drosophila*, and several later authors have cited these. However, FERRIS seems to have been unaware that circumversion in *Drosophila* had been conclusively demonstrated by GLEICHAUF (1936), since he makes no reference to that paper.

DIRECTION OF ROTATION

Rotation normally occurs in a clockwise direction (as seen from behind) in all species investigated.

MILANI & RIVOSECCHI (1954, 1955) described a 'countercoiled' mutation in *Musca domestica* L. in which the hypopygium is rotated in an anticlockwise direction, and referred to the existence of a similar mutation in *Drosophila* (MILANI & RIVOSECCHI 1955: 347). These authors concluded that in *Musca* the mutation was highly disadvantageous because of imperfect integration with the rest of the genome, and occurred only at frequencies in the order of magnitude of 1:10,000 in natural populations.

WHICH SEGMENTS ARE INVOLVED IN ROTATION?

Circumversion clearly involves the whole hypopygium, that is the 9th (genital) segment and the proctiger, including the hypandrium, which ontogenetic studies have shown to be wholly derived from the 9th segment. But difficulties have arisen in judging to what extent segments preceding the hypopygium are involved. Some authors (notably CRAMPTON, ACZÉL and G. H. HARDY) drew conclusions from the strongly asymmetrical structure of the postabdomen in some groups, in particular the Syrphidea and Muscoidea. But this reasoning is clearly unwarranted, as the asymmetries in question are not present in the groundplan of the Cyclorrhapha and have not been shown by ontogenetic studies to be the result of rotation. My views in this respect accord fully with those expressed by HENNIG (1958) in criticizing the work of CRAMPTON. In the groundplan of the Schizophora the 6th and 7th abdominal segments were more

47

or less unmodified, with terga and sterna in normal dorsal and ventral positions (as also in the Platypezidae); the *hypopygium circumversum* condition must therefore have evolved long before the asymmetrical conditions of the sclerites of the 6th and 7th segments with which it has been causally linked by CRAMP-TON, HARDY and others. In view of the above considerations, I reject all conclusions based on inferences from the arrangement of the sclerites of the mature adult, and proceed first by considering whether there is ontogenetic evidence for rotation of segments preceding the hypopygium in the Schizophora. Relevant information is given by SCHRÄDER (1927) and GLEICHAUF (1936). After considering the evidence for Schizophora (on which the views of earlier authors were largely based), I then support my argument with the new information now available on circumversion in the Platypezidae (KESSEL & MAGGIONCALDA 1968a, KESSEL 1968b).

The point of rotation is well known for many '*hypopygium inversum*' forms belonging to nematocerous groups. But the point of rotation in the Cyclorrhapha cannot be reliably inferred from the conditions shown by such forms, since none can be synapomorphous in this respect with ancestors of the Cyclorrhapha. The most closely related groups to the Cyclorrhapha (the Orthogenya and other groups of Brachycera) consist, with few exceptions, of species without hypopygial rotation. FEUERBORN (1922) and several subsequent authors seem to have assumed that the 180° rotation shown by certain nematocerous groups was an ancestral condition leading to the 360° rotation of Cyclorrhapha. But this view cannot be reconciled with modern conclusions on the phylogeny of the Diptera.

SCHRÄDER (1927) demonstrated that the 6th sternum ('Basalring') in *Calliphora* takes no part in the rotation. It became visible before the start of the process and remained in unchanged position throughout. Unfortunately SCHRÄDER was unable to reach a firm conclusion on the extent to which rotation occurred in the area immediately preceding the hypopygium, which in *Calliphora* bears a large sclerite formed by partial fusion of elements derived from the 7th and 8th segments (called '7th tergite' by SCHRÄDER). The sclerotization of this area did not become distinct until rotation was almost completed. However his observations suggested that some degree of rotation probably occurred here.

GLEICHAUF (1936) concluded that in *Drosophila* the sclerites preceding the hypopygium were not rotated. This conclusion confirms SCHRÄDER's finding that the 6th segment is not rotated, and also rules out the possibility of rotation of the 7th segment (whose tergum is fused with the 6th tergum in male Drosophilidae). However his observations could not clarify to what extent the 8th segment might be rotated, as the sclerites of this segment are much reduced or absent in male Drosophilidae. When GLEICHAUF writes of rotation between the 7th and 8th segments, he means between the 7th segment and the hypopygium. He thought that the basal phragma of the periandrium represented the 8th tergum, but this is now known to be incorrect (see my treatment of the Drosophiloidea in section 6.2).

I think it must be accepted that GLEICHAUF's observations definitely rule out any involvement of the 6th and 7th segments in rotation, despite the speculation to the contrary by later authors, such as HARDY (1944). The view that the 7th segment is substantially rotated ('lateroverted', that is turned through 90°,

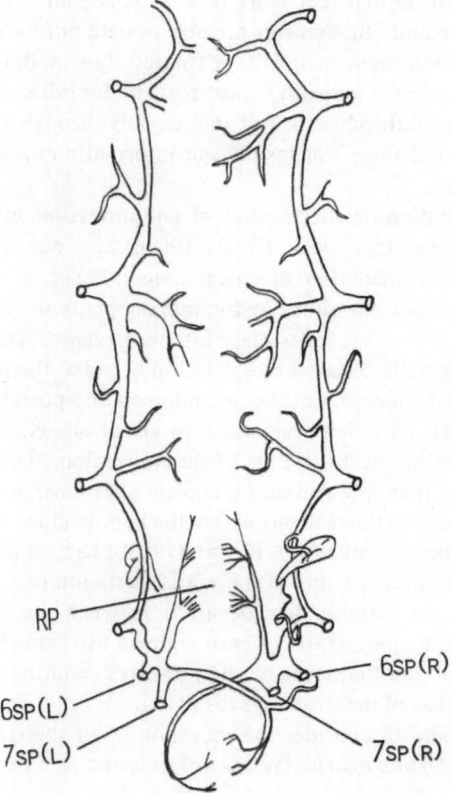

Fig. 12. *Musca domestica* L. (Muscidae), tracheation of male abdomen (after RIVO-
SECCHI 1958) (dorsal view).

according to CRAMPTON 1942) has not been supported by ontogenetic evidence.
Furthermore, the pattern of sclerotization on which CRAMPTON's and other
similar views have been based is an apomorphous condition confined to part of
the Schizophora (the Muscoidea in the sense proposed in this work), and not a
condition of the whole of the Schizophora, as CRAMPTON seemed to assume.

MUNRO (1947) suggested that in *Calliphora* the spiracles of the 7th segment
'will have reversed their positions, and the right spiracle of the synsternite will
have come to lie next to the sixth spiracle of the left side'. KIM & COOK (1966)
suggested that one of the postabdominal spiracles is carried to the other side of
the body in Sphaeroceridae. But these suggestions are contrary to the evidence.
RIVOSECCHI'S (1958: 480) figure of the tracheal system of *Musca domestica* L.
(copied here as fig. 12) shows clearly that the lateral tracheal trunks are only
crossed posterior to the spiracles of the 7th segment. The same is the case with
every other cyclorrhaphan whose tracheation has been examined. There are no
grounds for postulating that any spiracles have crossed from one side of the
body to the other.

To summarize, the ontogenetic work of SCHRÄDER and of GLEICHAUF clearly showed that the 6th and 7th segments are not rotated. SCHRÄDER's observations indicated that the 8th segment might be rotated, but he did not reach a firm conclusion. The argument between CRAMPTON and his critics about whether the 8th segment is rotated through the full 360° or only through 180° was unresolvable at the time, since there was insufficient information upon which the question could be judged.

The new information on the process of circumversion in the Platypezidae (KESSEL & MAGGIONCALDA 1968a, KESSEL 1968b) does not conflict with any of the conclusions on segmental involvement stated above; but it clarifies several points, since in Platypezidae all the abdominal segments are discrete and part of the process of circumversion takes place after emergence. KESSEL & MAGGIONCALDA (1968a) originally believed that rotation began at the junction of the 6th and 7th segments (the junction of the preabdomen and postabdomen according to their definitions). This view was based on visual observation under the dissecting microscope of the final 180° of circumversion, which occurs rapidly (requiring no more than a few minutes) shortly after emergence of the fly from its puparium. Following the realization that the hypopygium is 'unwound' to the inverse position during copulation, KESSEL (1968b) has somewhat modified his earlier view on the basis of study of the available mount of a pair *in copula*. He now reports that no rotation has occurred between the sixth and seventh segments, but that the part of the seventh segment after the 7th tergum is 'considerably distorted' and 'flattened by the pressures resulting from the progressively greater rotation of the structures distal to it'.

It is pertinent here to consider the morphology of the terminal abdominal segments in *Plesioclythia agarici* (WILLARD), selected as a representative of the Platypezidae (figs. 10 and 11). Dorsally there is a complete series of eight sclerites before the periandrium, and ventral sterna of normal appearance continue as far as the 7th segment. These sterna bear *sensilla trichodea*, thus confirming WHEELER's (1960) conclusion that the sensilla form a homonomous ('serially homologous') series as far as the 7th sternum. The 8th segment is largely membranous ventrally, but bears a band-like sclerite which extends from the venter around the right side of the insect, where it lies inside the apex of the 7th tergum.

It is clear from the structure of newly emerged males (plates I and II) that the definitive external structures of the 7th segment, namely the tergum, the sternum and the spiracles, are not rotated. The 7th segment is only involved in rotation to the extent that its membranous areas (especially of the venter) take up some part of the turn between the sclerites of the 7th and 8th segments. In *Plesioclythia* the ventral membranes after the 7th sternum appear involved in rotation to some extent, while the lateral membranes (where the 7th spiracles are situated) are not. These two areas of membrane are delimited in the specimen photographed (plate II): the swollen central area is continuous with the membrane of the 8th segment, and no clear intersegmental boundary is evident. Most of the first 90° of the observed post-emergence rotation of 180° occurs between the sclerites of the 7th and 8th segments. In newly emerged specimens (plates I and

II) the large sclerite of the 8th segment lies on the left side of the insect, and only moves through 90° to its normal dorsal position as the hypopygium is moved through 180° (from the inverse to the circumverse position). The mount of the pair of *Grossoseta in copula* shows that the same differential rotation occurs when the hypopygium is unwound to the inverse position during copulation. The remaining 90° of the post-emergence rotation of 180° must therefore occur between the sclerites of the 8th segment and the hypopygium.

This observed differential rotation of the 8th segment indicates that this segment is not rotated through the full 360°. If the same differential rotation occurs within the puparium as that observed after emergence (which seems to me the most probable hypothesis), then the 8th segment must be inverted when the hypopygium is in its circumverse rest position, as CRAMPTON thought. Since I believe that the irreversible '*hypopygium circumversum*' condition of the Schizophora is a further development of the condition shown by the Platypez-idae, I therefore homologize the large dorsal sclerite of the 8th segment present in some groups of Schizophora with the 8th sternum, and consider the band-like ventral sclerite (retained only in a few Schizophora) to be a reduced 8th tergum.

To summarize, I conclude that rotation in male Cyclorrhapha occurs between the hypopygium and the sclerites of the 8th abdominal segment, and between the sclerites of the 7th and 8th segments (with a slight involvement of part of the ventral membrane of the 7th segment); the sclerites and spiracles of the 7th segment are not involved in rotation. Differential rotation between the 8th segment and the hypopygium can be observed in adult platypezids, in which the last 180° of hypopygial rotation takes place after emergence from the puparium and is reversible. Ontogenetic studies on Schizophora (where the whole process of circumversion takes place within the puparium) give no reason to suppose that the segmental involvement differs from that observed in Platypezidae. The opinion of some authors that the 6th and 7th segments are rotated to some degree was based on the assumption that the asymmetrical conditions of the sclerites of these segments shown by certain groups were caused by rotation; this assumption is not supported by ontogenetic evidence. While it is not impossible that in some groups with a highly modified postabdomen, parti-cularly the Syrphidea and Muscoidea, the process of circumversion has been modified to involve to a limited degree segments preceding the 8th, I think it unwarranted to assume that this is the case in the absence of evidence. Asym-metrical sclerites may develop *in situ*, and this is always the simplest explanation in my opinion. In accordance with the principle of parsimony we should not assume that any unusual ontogenetic phenomenon such as rotation has oc-curred, unless there is positive evidence to this effect.

THE VIEWS OF G. C. CRAMPTON

CRAMPTON (1941, 1942, 1944a) propounded a theory in which various asym-metrical conditions of the postabdomen in Cyclorrhapha were explained as the direct result of circumversion. This theory still commands widespread support, in spite of the criticisms of VAN EMDEN & HENNIG (1956) and HENNIG (1958). CRAMPTON's method was to place the conditions found in various families of Cyclorrhapha in an evolutionary series, leading from the 'borderline' families Lonchopteridae and Phoridae, through the Syrphidea (Syrphidae and Pipunculidae), through various 'acalyptrate' families, culminating in the conditions found in various Calyptratae. He claimed that the 7th segment in the Cyclorrhapha becomes 'lateroverted' (turned through 90°) and the 8th segment inverted (turned through 180°) through involvement in the process of rotation.

I support HENNIG's (1958) criticism of CRAMPTON's theory, for the following reasons:

(1) CRAMPTON's placing of conditions shown by certain Syrphidea in an evolutionary sequence leading to the conditions shown by the Calyptratae seems to me untenable, because the Syrphidea show in their groundplan apomorphous conditions of the postabdomen which are not present in the groundplan of the Schizophora (see also under section 5). According to my present analysis the diverse arrangements of postabdominal sclerites shown by various groups of Schizophora have been derived from a more or less symmetrical condition of the type shown by the Platypezidae. If this view is correct, the condition shown by the Syrphidea must be considered an independent modification of the structure shown by Platypezidae, not a stage in a character sequence involving ancestors of the Schizophora.

(2) CRAMPTON assumed that the asymmetrical arrangement of the postabdominal sterna shown by the Muscoidea (in the sense of this paper) was a general condition of the Schizophora. My analysis indicates that this is incorrect, since there are a few groups of Schizophora in which both the 6th and 7th sterna are symmetrically developed in ventral position.

(3) No ontogenetic evidence supports CRAMPTON's view that the 7th abdominal segment is 'lateroverted' (rotated through 90°), as discussed previously in this section.

In one important respect I think that CRAMPTON was right: this is that the 8th segment is inverted and bears its sternum in dorsal position. However, at the time CRAMPTON wrote there was no convincing evidence to prove that he was right, and it is understandable that many authors, including HENNIG, were sceptical.

After publication of his theory CRAMPTON attempted a wider study of the terminalia of the so-called 'Acalyptratae', but never published his results in detail. He produced a 'purely tentative arrangement' of some families of 'Acalypterates' (CRAMPTON 1944b), but this paper contained only the briefest characterization and has had little influence on subsequent workers. It was strongly criticized by HENNIG (1958). CRAMPTON classified the 'Acalypteratae' (in which he included the Syrphidae, Pipunculidae and Platypezidae, as well as the customary range of schizophoran families) into two divisions, the 'Syrphomorpha' and 'Platypezomorpha', on the basis of the sclerites of the 6th abdominal segment. Such a classification cuts right across all other evidence; it was a logical consequence of his assumption that the conditions of the terminalia found in the Syrphidae and Pipunculidae were a stage leading to the evolution of the asymmetrical conditions of the groups which I classify as Muscoidea. Groups whose structure conflicted with this assumption because they showed an obviously more plesiomorphous condition of the 6th segment, were assumed to belong to a different evolutionary line, and were therefore classified in the highly heterogenous assemblage 'Platypezomorpha'.

THE VIEWS OF G. H. HARDY

HARDY (1944) published a useful analysis of copulating positions in Diptera, but unfortunately his speculation in that paper regarding the origin of the 'hypopygium circumversum' condition is misleading. HARDY seems to have been unaware of SCHRÄDER's (1927) and GLEICHAUF's (1936) ontogenetic studies which conclusively proved the correctness of FEUERBORN's (1922) hypothesis of rotational circumversion in the Cyclorrhapha; instead he elaborated a theory which purported to show that FEUERBORN's conclusion was a misunderstanding, and that the observed looping of the ejaculatory duct over the hind gut in *Calliphora* indicated a 'hypopygium inversum' condition. The condition shown in Syrphidae was considered to be a stage from which that of the Calyptratae was derived by movement of the anus. In support of this remarkable suggestion he argued that 'authors had overlooked the fact that the aedeagus had remained inverted in an apparently erect hypopygium'.

The evolutionary series constructed by HARDY is clearly contrary to the available evidence. His suggestion that the condition found in Syrphidea had originated by 'curving round' of a *hypopygium inversum* is contradicted by the evidence now available that the *hypopygium circumversum* condition (in its partly reversible expression) was already present in the groundplan of the Cyclorrhapha. I have explained above (in the discussion of CRAMPTON's views) why I do not think that the condition found in Syrphidea is similar to that of any ancestor of the Schizophora. HARDY's generalization that the aedeagus is 'inverted' (directed anteriorly rather than posteriorly) in Schizophora (called 'Muscoidea' by HARDY) will not bear close examination. His remarks on an 'inverted' aedeagus presumably refer to the anteriorly directed rest position assumed by the aedeagus in certain groups of Schizophora (Nothyboidea and Muscoidea in the sense of this work). Most members of these groups are able to swing the aedeagus through a wide arc against the aedeagal apodeme by muscular action. During copulation the aedeagus is swung out, so that it becomes directed posteriorly or posteroventrally. It is clearly this copulatory position which is comparable with the position of the aedeagus in other groups of Diptera. In those groups of Schizophora in which this swinging mechanism has not evolved and the aedeagus retains a more or less fixed orientation (Lonchaeoidea, Lauxanioidea and Drosophiloidea), it is directed posteriorly or posteroventrally. The evolutionary changes which have occurred thus involve the orientation of the aedeagus, not movement of the anus.

MUNRO (1947) also published criticisms of HARDY's paper, but these were made from a standpoint different from mine above, since MUNRO accepted CRAMPTON's theory.

ORIGIN AND FUNCTIONAL SIGNIFICANCE OF CIRCUMVERSION

Since the process of hypopygial circumversion through 360° returns the aedeagus and anus to their original positions, it is scarcely possible to explain the evolution of this process on the basis of comparison of the *hypopygium circumversum* condition directly with the normal unrotated hypopygium. The only authors who have speculated on this basis are KESSEL & MAGGIONCALDA (1968a: 85), who suggest that circumversion may possibly bring certain advantages through greater flexibility of the postabdomen. But such an explanation hardly seems adequate to me, as the degree of flexibility of the postabdomen is more dependent on the musculature and the extent of membranous areas than on circumversion as such. ZAKA-UR-RAB (1963) appears to have despaired of finding an explanation, since he states that torsion is an 'accidental development in the phylogeny of male Diptera'.

Faced with this dilemma, many authors postulated an intermediate *hypopygium inversum* stage in the evolution of the cyclorrhaphous *hypopygium circumversum*. This suggestion was soundly based, since it is difficult to believe that such a major organizational change as circumversion could have been achieved in one step (apart from the rotation as such, major changes in musculature are involved). Unfortunately, confusion arose because the *hypopygium inversum* condition has evolved independently in different groups of Diptera, including many families of 'Nematocera' and a few genera of 'lower Brachycera' (included in the Asilidae and Bombyliidae). None of these *hypopygium inversum* conditions can be ascribed to synapomorphy with ancestors of the Cyclorrhapha, since most members of the Orthogenya, the group most closely related to the Cyclorrhapha, show a normal unrotated hypopygium. So do the great majority of other 'lower Brachycera'. Thus there is no group showing the *hypopygium inversum* condition which has yet been demonstrated as synapomorphous in this respect with any ancestor of the Cyclorrhapha. KESSEL & MAGGIONCALDA's (1968a) suggestion that the empidid genus *Microphorus* may represent such a group requires much more investigation (see section 4). Nevertheless, the hypothesis that there was a *hypopygium inversum* stage in the phylogeny of the Cyclorrhapha is practically inevitable in view of the recent discovery that Platypezidae copulate with the male hypopygium in the inverse position. I set out below what seems to me the simplest possible evolutionary sequence for the evolution of the *hypopygium circumversum* condition.

1. Hypopygium normal, not rotated.
[2. Hypopygium rotated through 180° to the inverse position during copulation (rotation reversible).]

[3. Hypopygium rotated to the inverse position after emergence; copulation in the inverse position (rotation irreversible).]

[4. Hypopygium rotated to the inverse position within the puparium; copulation in the inverse position (rotation irreversible).]

5. Hypopygium rotated to the inverse position within the puparium; then rotated through a further 180° to the circumverse rest position after emergence, but unwound to the inverse position during copulation (rotation partly reversible).

6. Hypopygium rotated through the full 360° to the circumverse position within the puparium; copulation in the circumverse position (rotation irreversible).

Stage 1 in the above sequence is the condition inferred for the groundplan of the Eremoneura, as retained by the Orthogenya. Stages 2, 3, and 4 (in square brackets) are still hypothetical as far as the ancestry of the Cyclorrhapha is concerned (although conditions equivalent to stages 2 and 3 are known in some groups of Diptera, as previously discussed). Stage 5 is the condition shown by all Platypezidae which have been studied. Stage 6 is the condition shown by the Schizophora and probably also by the Syrphidea. It will be seen from the above sequence that acceleration of the process of rotation in relation to the ontogeny of the insect must be assumed to have occurred in at least two stages. The *hypopygium inversum* condition is assumed to have been reversible when first evolved, as has now been demonstrated for the *hypopygium circumversum*.

The evolution of the *hypopygium inversum* condition must clearly have been linked with a change of mating position. The platypezids which have been studied mate on the wing in an opposed (tail-to-tail) position (KESSEL 1968b). It is evident from the condition of reversible circumversion shown by these platypezids that the circumverse position was first evolved not as a mating position but as a rest position, which allows the external genitalia to be folded under the abdomen and thereby protected. Later the process became irreversible, and the male vertical (or 'superimposed') position adopted for copulation. This is the usual mating position of the Syrphidea and Schizophora. The opposed position with one of the partners upside-down figured by HENNIG (1966b) for Conopidae can be assumed from the male vertical position without any rotation of the postabdomen.

SALZER (1968) has suggested a different sequence of mating positions leading to the evolution of the *hypopygium circumversum* condition. However he was writing before the new information on the copulation of Platypezidae became available, and this clearly necessitates revision of his views. I doubt whether he was right in postulating steps of 90°, since in those Diptera which copulate in the air rotation always involves all-or-nothing movements of about 180°.

CIRCUMVERSION, DEFLEXION AND ASYMMETRY

It is evident from my review of the literature that many taxonomists have been

55

thoroughly confused over the relationships between circumversion, deflexion and asymmetry. This is hardly surprising, since such confusion was inherent in the discussion of these questions by such authors as CRAMPTON, G. H. HARDY and ACZÉL. Circumversion of a segment means that the segment has been rotated through 360° about the longitudinal axis of the insect. Deflexion of a segment means that the segment has been turned downwards, and I regard the longitudinal axis also as turned downwards. Asymmetry in this context means bilateral asymmetry, that is unequal development of a structure on either side of the centre-line of the insect. These words all denote different concepts.

As has been pointed out in the previous discussion, circumversion has not been demonstrated to produce asymmetry in the development of the post-abdominal sclerites. In the Platypezidae circumversion occurs in two stages, 180° of the rotation occurring between the sclerites of the 7th and 8th segments, and the additional 180° between the sclerites of the 8th segment and the hypopygium. Rotation through any angle other than 180° or a multiple thereof would produce bilateral asymmetry of an originally symmetrical structure, but rotation through 180° does not. There is no convincing evidence that the asymmetry of the sclerites of the 6th and 7th segments in some groups of Cyclorrhapha is caused by anything other than asymmetrical development *in situ*. The Cyclorrhapha include many groups with a fully or nearly symmetrical post-abdomen (not only the Platypezidae, but also some groups of Schizophora, such as the Drosophiloidea), and these are just as much '*hypopygium circumversum*' forms as those with a strongly asymmetrical postabdomen.

In the case of the Cyclorrhapha there seems to be a consistent correlation between circumversion and deflexion. KESSEL's observations on the Platypezidae indicate that the hypopygium 'passes downward' as the final 180° of rotation is completed. Deflexion and the completion of circumversion seem to be manifestations of a single process in this case. Nevertheless, we should recognize a conceptual difference between deflexion and circumversion. We would still describe a rotational movement of 360° as circumversion, even if no deflexion occurred. And there are many insects with deflexed terminalia which are not rotated. What has been observed in the Platypezidae is a complex movement which we analyse into two components, a rotational component and deflexion. Failure to appreciate the conceptual difference between rotation and deflexion has led to some confusion in the literature on Syrphidae, which I illustrate from the work of ZUMPT & HEINZ (1949). ZUMPT and HEINZ state that in *Eristalis* 'we are dealing with a *hypopygium inversum*', thus apparently contradicting the view (which I hold to be correct) that all Cyclorrhapha possess a *hypopygium circumversum*. However, if ZUMPT & HEINZ's arguments are followed closely, it will become apparent that they have confused rotational movement and deflexion. The hypopygium of *Eristalis* is 'inverse' in the sense that it is so strongly deflexed that it points anteriorly and its 'dorsal' side has become ventral. But the phrase '*hypopygium inversum*' was proposed by FEUERBORN (1922) for rotational inversion, as shown by some Psychodidae, Dixidae, Tipulidae and Culicidae, and does not refer to deflexion. *Eristalis* does not have a '*hypopygium inversum*' in FEUERBORN's sense, but a '*hypopygium circumversum*'. To prevent

56

misunderstanding some word other than inversion should be used to describe the strongly deflexed position of the hypopygium in Syrphidae. STEYSKAL's (1957a) word 'reflexion' seems appropriate. The hypopygium of Syrphidae may thus be described as reflexed or, if we wish to extend FEUERBORN's Latin terminology, as a '*hypopygium circumversum et reflexum*'.

STEYSKAL (1957a) rightly points out the distinction between circumversion and reflexion. But I do not understand his reference to a 'third type of movement, which we may call *strophe*'. In my opinion the structure of the postabdomen of all male Cyclorrhapha can be explained as the result of a combined rotational and deflexional movement followed by asymmetrical development of some sclerites *in situ* in certain groups; I see no need to postulate a third type of movement.

4. THE RELATIONSHIP OF THE CYCLORRHAPHA TO OTHER EREMONEURA

The following three sections (4-6) deal with classification. A summary of the classification proposed is given in section 7.

I have not undertaken a new investigation of the relationship of the Cyclorrhapha to other Eremoneura, but include this brief review of the available information as orientation for my treatment of the Cyclorrhapha. I use the names Eremoneura LAMEERE (1906) and Orthogenya BRAUER (1883) in preference to HENNIG's names Muscomorpha and Empidiformia, both on grounds of priority and because at least the use of the suffix '-morpha' leads to formal difficulties, as it is also used to indicate groups of higher rank in the classification (such as Culicomorpha and Biblionomorpha). In general the use of neutral names without connotations about rank does not seem objectionable for taxa above superfamily, and may be positively advantageous because revisions of the rank of taxa do not necessitate consequent names changes (see further the discussion by HENNIG 1968b).

CHARACTERIZATION OF THE EREMONEURA

The characterization of the Eremoneura (= Muscomorpha HENNIG) was discussed by HENNIG (1952a, 1954), who classified the Cyclorrhapha and Orthogenya (= Empidiformia) in this group. This classification agrees exactly with that proposed by LAMEERE (1906). HENNIG referred to the following groundplan conditions which are apomorphous with respect to the groundplan of the Brachycera, as indicating that the Eremoneura probably represent a monophyletic group.
(1) Hypopharyngeal skeleton of larvae V-shaped.
(2) Media with only 3 branches (m_4 fused with m_3).
(3) Anal cell closed apically; veins cu_{1b} and la with common terminal section (the 'anal vein').

In the light of my reinterpretation of BÄHRMANN's (1960) data on the male genitalia of Empididae and my own data on the genitalia of Cyclorrhapha, I am able to add the following, all of which appear to be autapomorphous groundplan conditions of the Eremoneura.

58

(4) Basimeres (♂) expanded dorsally, forming large lateral plates which are only narrowly separated on dorsum of genital segment.

This characterization depends on the validity of the periandrium hypothesis discussed in section 3.1. I postulate that the possession of separated basimeres, as in the Empidinae, is a more plesiomorphous condition than the possession of a periandrium formed by fusion of the basimeres across the dorsum of the genital segment.

(5) Epandrium (♂) either lost or fused with cerci.

BÄHRMANN's (1960) statement that the simplest form of epandrium in the Empididae consists of two large lobes which are fused basally on the dorsum and extend laterally towards the ventral surface of the hypopygium, refers to the type of sclerite which I call a periandrium. In those Empididae in which the basimeres remain separate (Hemerodromiinae, *Empis*, *Hilara*, *Rhamphomyia*), BÄHRMANN reports that the epandrium is usually cleft, and that the cerci are either absent or more or less fused with the epandrium. Since a discrete epandrium and cerci are apparently never present in these insects, I suggest an alternative explanation; that the true epandrium (9th tergum) has been lost in all Eremoneura, and that the sclerites in question are all referable to the proctiger, representing modified cerci and/or 10th tergum. The presence of the 10th tergum in the groundplan of the Brachycera is indicated by the presence of this sclerite in male Rhagionidae (fig. 5).

(6) Aedeagus (♂) slender, upcurved distally.

BÄHRMANN (1960) did not give a precise estimate of what he thought might be the groundplan condition of the aedeagus for Empididae, beyond the general statement that 'der Aedoeagus besitzt eine schlauchförmige Gestalt von unterschiedlicher Länge und ist mitunter stark gebogen'. A rather slender, upcurved aedeagus was probably present in the groundplan of the Cyclorrhapha, as shown for instance by many Platypezidae and Lonchaeoidea (figs. 9, 21 and 23). The aedeagus of the Dolichopodidae and some groups of Empididae (such as *Empis* and *Rhamphomyia*) is also of this type. From this distribution I infer that a slender upcurved aedeagus was present in the groundplan of the Eremoneura. I list this condition of the aedeagus as an autapomorphous groundplan condition of the Eremoneura, since as far as I am aware such a condition is not shown in the groundplan of other groups of Brachycera. However the information available is not complete.

The Eremoneura, as defined above, do not include *Hilaromorpha*, which BÄHRMANN (1960) has transferred from the Empididae to the Bombyliidae. The structure of the male genitalia of *Hilaromorpha* figured by BÄHRMANN fully supports his exclusion of this genus from the Empididae. *Hilaromorpha* shows a true epandrium in dorsal position (as in the Bombyliidae and others).

As far as I am aware, the only family of Brachycera outside the Eremoneura in which no true epandrium seems to be present is the Scenopinidae. The figures in KELSEY's (1969) revision of this family indicate that the so-called epandrium consists of a pair of lateral plates which are contiguous or slightly separated on the dorsum of the genital segment. It is possible that these plates represent laterally situated basimeres (as in the groundplan of the Eremoneura), rather than a cleft epandrium. However, the larvae of the Scenopinidae do not show the modifications characteristic of the Eremoneura, but are very similar to the larvae of the Therevidae (HENNIG 1952a). If the Scenopinidae are in fact more closely related to the Therevidae than to the Eremoneura (as is currently believed), then the similarity between the genitalia of the Scenopinidae and Eremoneura can only be ascribed to convergence or parallelism. On the other hand, if the similarities between the Therevidae and Scenopinidae are due to symplesiomorphy, the possibility that the Scenopinidae are the sister-group of the Eremoneura cannot be exluded. I offer no opinion on this question, but draw attention to it for consideration in future studies.

PROTEMPIDIDAE

Until recently no information on the age of the Eremoneura was available, but USSATCHOV (1968) has described as *Protempis* (Protempididae) a relevant fossil species from Jurassic deposits in Kazakhstan. The wing venation of *Protempis* corresponds exactly with what can be inferred about the groundplan condition of the wing venation for Eremoneura. The Protempididae may thus represent the stem-group of the Eremoneura, in accordance with the definition of 'stem-group' proposed by HENNIG (1965b). The full suite of plesiomorphous conditions shown by *Protempis* are not shown in combination by any recent species, as far as I am aware, although a few recent species classified in the Empididae have wing venation little removed from that of *Protempis*.

CHARACTERIZATION OF THE CYCLORRHAPHA

The Cyclorrhapha (= Musciformia HENNIG) are a subordinate monophyletic group of the Eremoneura. The earliest known fossil Cyclorrhapha are from the Cretaceous Period (see J. F. MCALPINE and MARTIN 1969 and J. F. MCALPINE 1970). The apomorphous conditions of the Eremoneura are as follows.
(1) Larval head capsule reduced: individual parts of cephalic and pharyngeal skeleton fused into uniform 'cephalopharyngeal skeleton': pupa enclosed within puparium formed by contraction and hardening of integument of 3rd larval instar.

The morphology of the larvae of Cyclorrhapha was reviewed by HENNIG (1952a). The name Cyclorrhapha, first proposed by BRAUER, refers to the circular ecdysial suture around the first abdominal segment of the larva and puparium of most Schizophora (except Crypto-chetidae). However HENNIG indicated that the position of the ecdysial sutures on the puparia of other groups of Cyclorrhapha varies, and the groundplan condition for the group as a whole has not been established.

(2) Radial sector two-branched; r_{4+5} not forked (HENNIG 1954).

This condition is also shown by various groups of Empididae and such reduction may in some cases represent synapomorphy with the Cyclorrhapha. The second venational character given by HENNIG (1954), loss of r_3, does not in my opinion differentiate the groundplan of the Cyclorrhapha from the groundplan of the Eremoneura, since the presence of an apparent r_3 as a cross-vein in *Dolichocephala* (Empididae) is most probably secondary (apomorphous). Such a vein is not normally present in other Brachycera except in members of groups which show a clear tendency to develop additional cross-veins (in particular the Nemestrinidae and some Bombyliidae).

60

(3) Hypopygium ($\male$) rotated to inverse position within puparium, then rotated through further 180° to circumverse rest position soon after emergence; 8th abdominal segment rotated through half angle of hypopygial rotation (through 90° within the puparium, then through further 90° to inverse position after emergence).

This characterization is the result of the analysis presented in section 3.2. The postulated groundplan condition is retained by the Platypezidae.

(4) 8th tergum ($\male$) (normally in ventral position in mature adult) reduced to narrow band (fig. 10).

Reduction of the 8th tergum also occurs in *Atelestus* and in some other genera of Empididae (see BÄHRMANN 1960, fig. 1). Since the Empididae are probably paraphyletic (see below), the possibility that such reduction may represent synapomorphy with the Cyclorrhapha should be borne in mind. The reduced 8th tergum in these Empididae is dorsal, since their post-abdomen is not rotated.

ATELESTUS

The delimitation of the Cyclorrhapha has long been largely settled, and the only remaining dispute in recent literature concerns the classification of the genus *Atelestus* (= *Platycnema*). Most authors have included this genus in the Empididae, but others (notably KESSEL 1960 and KRYSTOPH 1961) have referred it to the cyclorrhaphous family Platypezidae. Clearly there is substance in the arguments of KESSEL and KRYSTOPH that *Atelestus* is strongly divergent from most other Empididae and shows certain resemblances to the Platypezidae. Another resemblance, besides those stated by these authors, lies in the structure of the male genital segment (figs. 6 and 7). This bears a large symmetrical periandrium, agreeing remarkably with the condition shown by the Cyclorrhapha. Nevertheless, I do not think that *Atelestus* should be included in the Platypezidae, because it shows at least one feature which seems to me irreconcilable with such a classification: this is that the 8th abdominal segment of the male bears a narrow band-like sclerite dorsally and a large sclerite ventrally. This is the reverse of the groundplan condition for the Cyclorrhapha (retained by the Platypezidae), in which the inverted 8th segment bears its large sternum in a dorsal position. The obvious inference from the condition of the 8th segment shown by *Atelestus* is that its terminalia are not rotated: at any rate it seems reasonable to assume this unless contrary evidence can be found, for instance from dissection of fresh material (which was not available to me). If *Atelestus* does not possess rotated terminalia, then the genus cannot be referred to the Platypezidae (or any other cyclorrhaphous family), and the possibility must be considered that its resemblances to the Platypezidae are in fact resemblances to the groundplan of the Cyclorrhapha as a whole. Since the Platypezidae show the least modified adult morphology of all groups of Cyclorrhapha in many respects, such an interpretation raises no conceptual difficulty. The question of how *Atelestus* should be classified probably cannot be settled at present, since

61

the phylogenetic relationships of the groups currently included in the Empididae require clarification. Pending such clarification I think it best to retain the conventional classification of the genus in the Empididae for purposes of nomenclature.

The Orthogenya

BRAUER (1883) proposed the group Orthogenya to include the Empididae and Dolichopodidae, that is all Eremoneura excluded from the Cyclorrhapha. Identical group concepts were also proposed by LAMEERE (1906) ('Orthorrhapha') and HENNIG (1952a, 1954) ('Empidiformia'). HENNIG's discussions of this group have been concerned mainly with demonstrating its affinity with the Cyclorrhapha. The question of whether the Orthogenya are monophyletic or paraphyletic (containing more than one lineage of non-cyclorrhaphous Eremoneura), has not been settled. ACZÉL (1954) and KESSEL & MAGGIONCALDA (1968a) have implied that the Orthogenya are not monophyletic, because they suggest that particular groups of Orthogenya are more closely related to the Cyclorrhapha. I do not think that this question can be settled from existing analyses, but briefly review these authors' views with the aim of clarifying the issues involved.

ACZÉL (1954) considered that the Dolichopodidae should be grouped with the Cyclorrhapha, and consequently proposed a new subdivision of the Brachycera into two 'Divisions', the Orthopyga and Campylopyga. He included in the Campylopyga the Dolichopodidae and Cyclorrhapha, and characterized the group as follows:

'Male postabdomen without exception folded beneath last tergite of the preabdomen and circumverted. Antennae inserted below the prefrontal suture and consisting of three segments, scape, pedicel and postpedicel only.'

The value of the form of the antennae in indicating the relationships of the Dolichopodidae is uncertain, since there is no clear-cut distinction between the types of antennae shown by the Cyclorrhapha, Dolichopodidae and some groups of Empididae. Further investigation of the character sequence involved is needed. The part of ACZÉL's characterization which I am able to assess definitively is the first sentence, which implies that the Dolichopodidae possess a *hypopygium circumversum* and are hence monophyletic with the Cyclorrhapha. Until recently some doubt has remained as to whether this is correct. HARDY (1953) stated that in Dolichopodidae the alimentary and genital tracts are 'in one plane', and that 'this difference in anatomical detail shows that the Dolichopodidae are not in the direct evolutionary line that leads to Cyclorrhapha'. However D. K. McALPINE (1960) stated that 'in *Sciapus* the relative twisting of the hind gut and vas deferens is indicative of circumversion'. BÄHRMANN (1966) has clarified these apparent contradictions. He reports that the genus *Dolichopus* contains species without looping of the ejaculatory duct over the hind gut, as well as species in which such looping is almost complete. He concludes that the irregularity of this looping is evidence against the assumption of synapomorphy between the Dolichopodidae and Cyclorrhapha in respect of hypopygial rotation. Thus the only demonstrated similarity between the hypopygia of the Dolichopodidae and Cyclorrhapha is that in both groups the hypopygium is deflexed. This does not provide sufficient grounds for inferring that the Dolichopodidae are monophyletic with the Cyclorrhapha, when the detailed structure of the hypopygium is very different in the two groups. The hypopygium in the Dolichopodidae is enclosed by a highly modified 'genital capsule', which is connected with the preceding segment only through a narrow opening sistuated asymmetrically on its left side (BÄHRMANN 1966). This capsule is more or less uniformly sclerotized, and its homology is in doubt. However, its muscular connections with the aedeagal apodeme suggest that it is largely of hypandrial origin (see BÄHRMANN 1966: 70). Possibly the nearest relatives of the Dolichopodidae should be sought amongst those groups referred to the Empididae in which the hypandrium is enlarged at the expense of the basimeres (as for instance in *Heleodromia*). BÄHRMANN (1960) discusses the possibility of such a relationship, but reaches no definite conclusion. Thus, although the available analyses of the relationships of the Dolichopodidae are not conclusive, it seems unlikely that ACZÉL's Campylopyga is monophyletic, and I therefore follow HENNIG, BÄHRMANN and others in rejecting Aczél's classificatory proposals.

KESSEL & MAGGIONCALDA (1968a) have advanced a different hypothesis about the relationship of the Cyclorrhapha with particular groups of Empididae, in elaboration of views expressed by KESSEL in earlier papers. These authors consider that the Cyclorrhapha are phylogenetically related to certain genera of Empididae ('that branch of the Hybotinae which leads to the Platypezidae'). The genera stated to belong to this 'branch' are *Meghyperus*, *Ocydromia*, *Leptopeza*, *Bicellaria*, *Euthyneura*, *Oedalea*, *Syndyas*, *Trichina*, *Hybos* and *Microphorus*. The order in which the authors comment on these genera (as given in my previous sentence) corresponds with the degree of apparent rotation of the postabdomen. *Meghyperus*, whose hypopygium is fully symmetrical, is considered as 'the perfect representative of the ancestral line forming the base of that branch of the Hybotinae which leads to the Platypezidae'; *Ocydromia* and *Leptopeza* are then mentioned as 'other primitive genera'. The authors then mention *Bicellaria*, *Euthyneura*, and *Oedalea* as examples of genera 'in which there is already a slight twist of the postabdomen towards the right. At still a higher level, and in such genera as *Syndyas*, *Trichina* and *Hybos*, this twist has become a full 90-degree rotation with reference to the main axis of the abdomen, and this rotation has become the rule in that hybotine line which leads towards Platypezidae and Tachydromiinae'. Finally, *Microphorus* is discussed, 'a genus of Hybotinae above *Hybos* and also representing the empidid stock leading to Platypezidae. In this form the postabdomen is stalked and bears the hypopygium in inverted position, the relationships of these male sexual structures resembling those of Platypezidae at the time of emergence'. These statements imply that the Orthogenya are not monophyletic, since a particular group of Empididae is considered to be 'ancestral to the platypezids'. The latter phrase should not, however, be taken as a denial of the monophyly of the Cyclorrhapha, since the authors explicitly state (page 103) that they are convinced that the Cyclorrhapha are properly considered to be a monophyletic group.

KESSEL & MAGGIONCALDA's argument is mainly based on placing conditions of the hypopygium shown by Recent forms in a series which they believe to represent stages in the evolution of the cyclorrhaphous hypopygium. In attempting to judge to what extent their conclusions are warranted, I now consider their argument in the light of the two recent comparative morphological studies of the Empididae by BÄHRMANN (1960) and KRYSTOPH (1961).

I must first remark that the Hybotinae in the sense followed by KESSEL and MAGGIONCALDA are no doubt paraphyletic. Both BÄHRMANN and KRYSTOPH use the name Hybotinae in a more restricted sense, and conclude that the Hybotinae (in their restricted sense), the Ocydromiinae and the Tachydromiinae together form a monophyletic group. This group is well characterized by apomorphous conditions of the mouthparts (KRYSTOPH 1961). The tormae are free from the clypeus in their lower two-thirds, not articulated with the labrum, but standing separate from it on the margins of the cibarium; the maxillary laciniae have been lost, and the palpi are separated from the maxillae (borne on palpifers).

Of the genera stated by KESSEL and MAGGIONCALDA to belong to their 'branch leading to the Platypezidae', all belong to the Hybotinae/Ocydromiinae/Tachydromiinae group of KRYSTOPH, with the exception of the last genus *Microphorus* (discussed further below). However, the markedly apomorphous features of the mouthparts shown by this group clearly exclude the possibility that the Cyclorrhapha were derived from it (in the sense of having a more recent common ancestry with particular genera within the group than these have with each other). In many Cyclorrhapha well-developed maxillary lacicinae are retained, and a palpifer of the type stated is not developed; the tormae are usually not separated from the labrum in the manner stated. Such conditions are plesiomorphous in comparison with the conditions shown by the Hybotinae/Ocydromiinae/Tachydromiinae group, and indicate that the Cyclorrhapha could not have been derived from this group, although the possibility of some close relationship between the Cyclorrhapha and the group as a whole is not thereby excluded. The 'rotated' hypopygia shown by certain genera within this group cannot be ascribed to synapomorphy with conditions shown by ancestors of the Cyclorrhapha. In many genera of the group the hypopygium seems unrotated; and apparent rotation, where it occurs, is usually associated with asymmetrical modifications of the periandrium (which is almost invariably symmetrical in Cyclorrhapha). I am doubtful whether any true rotation in fact occurs, since the change in the orientation of the genitalia may be the result of asymmetrical development *in situ*. I think that the sequence of hypopygial modifications shown by members of the Hybotinae/Ocydromiinae/Tachydromiinae group proceeded independently of the evolution of the *hypopygium circumversum* condition of the Cyclorrhapha. If there was a common ancestor of the Cyclorrhapha and that group, this probably had a symmetrical unrotated hypopygium, as in *Atelestus* (figs. 6 and 7).

The genus *Microphorus* is probably related to the Hybotinae/Ocydromiinae/

Tachydromiinae group according to KRYSTOPH (1961), but retains a more plesiomorphous maxilla; the laciniae are well developed and the palpi in normal position, borne on the maxillae (without any intervening palpifer). Similar plesiomorphous conditions must be assumed for ancestors of the Cyclorrhapha (but are not evidence of phyletic relationship, which can only be inferred from common possession of apomorphous conditions). The common apomorphous condition of the mouthparts which led KRYSTOPH to infer a relationship between *Microphorus* and the Hybotinae/Ocydromiinae/Tachydromiinae group, is the structure of the tormae. The condition of the tormae in the groundplan of the Cyclorrhapha is thus highly relevant to the suggestion of an affinity between the Cyclorrhapha and *Microphorus*, but I am not at present able to judge this question. FREY's (1921) classical comparative study of the mouthparts was unfortunately confined to the Schizophora; there is insufficient comparative information on the mouthparts of the other subgroups of the Cyclorrhapha. It certainly seems possible that the condition of the postabdomen shown by *Microphorus* represents (through synapomorphy) a stage in the evolution of the conditions shown by the Cyclorrhapha, since the inverted position and rightward direction of the 8th segment and the hypopygium in this genus resembles the condition shown by newly emerged platypezid males (as KESSEL and MAGGIONCALDA have stated). But further detailed comparative studies are needed before firm conclusions can be drawn.

The conclusion which I draw from the above discussion is that, while there are good reasons for doubting whether the Orthogenya are monophyletic, the authors who have raised this question have not made sufficiently extensive studies to settle the matter.

THE VIEWS OF HENNIG (1970)

Shortly before completing the manuscript of this work I received a new paper by HENNIG (1970), in which he reviews some of the material which I have covered (independently) in the preceding discussion. HENNIG also affirms that the Hybotinae/Ocydromiinae/Tachydromiinae group is monophyletic, and discusses other characters in addition to those which I have mentioned. He suggests that this group can probably be classified with the Microphorinae and Atelestinae in a 'subfamily-group' Ocydromioinea, while all other Empididae can be referred to another group of the same rank, the Empidoinea.*

This seems a useful working classification for the present, although doubt must remain about whether these 'subfamily-groups' are monophyletic until the relationships of the Dolichopodidae and Cyclorrhapha have been clarified. The groups which HENNIG includes in the Ocydromioinea are those which show a periandrium, and I think that the possibility that the whole or part of this group

* HENNIG's use of the suffix '-oinea' for 'subfamily-groups' is inconsistent with the proposal of VON KÉLER (1963) to use this suffix for a category between family and superfamily (see section 6.1). But this discrepancy is likely to be removed in future revisions since the present wide limits of the 'Empididae' are untenable.

65

is more closely related to the Cyclorrhapha than to other 'Empididae' remains open (see previous discussion).

In this paper HENNIG expresses doubt about the monophyly of the Eremoneura on the basis of a character not previously considered, the structure of the antennal arista. He points out that a three-articled arista (with two small basal articles) must be ascribed to the groundplan of the Cyclorrhapha, but the arista has only two articles in those recent Orthogenya which he examined (including *Atelestus*). This character seems to him to indicate the possibility that the Orthogenya should be included in the Asilomorpha (which have a two-articled arista), as in ROHDENDORF's (1964) classification, rather than considered more closely related to the Cyclorrhapha.

I do not share HENNIG's newly raised doubts about the monophyly of the Eremoneura. The evidence for the monophyly of the Eremoneura has become very substantial. To the resemblances in wing venation and larval morphology discussed in HENNIG's previous works, I can now add the similar structure of the male genital segment. In particular, the genus *Atelestus* shows certain strong similarities to some groups of Cyclorrhapha (such as the Platypezidae), which cannot be summarily dismissed as due to convergence. The hypothesis that the similarities between *Atelestus* and the Platypezidae are retained from a common ancestor of the Ocydromioinea and Cyclorrhapha has much to commend it. If HENNIG's suggestion that the Cyclorrhapha may be the sister-group of the Asilomorpha (including Orthogenya) is correct, then we must infer the existence of an independent lineage leading to the Cyclorrhapha at least since the Jurassic Period. Any fossils belonging to this lineage would be recognizable by a series of antennal modifications leading to the formation of a three-articled arista. But nothing in the assemblages of Jurassic Diptera so far described suggests the existence of such an independent lineage. In default of such contrary evidence, I think that the three-articled arista condition was probably derived from the two-articled condition, not through an independent series of modifications. This interpretation is fully compatible with the evidence of other characters.

66

5. THE MAJOR SUBORDINATE GROUPS OF THE CYCLORRHAPHA

The characterization of the Cyclorrhapha has been discussed in the previous section. HENNIG (1952a) followed DE MEIJERE (1900) in dividing the Cyclorrhapha into two sister-groups: the Acroptera (= Anatriata) containing the single family Lonchopteridae, and the Atriata containing all other Cyclorrhapha (fig. 13). I also accept this division, but prefer to use BRAUER's (1883) name Acroptera to HENNIG's Anatriata (= Anatria DE MEIJERE) on grounds of priority. Many authors still follow a traditional division of the Cyclorrhapha into the Aschiza and Schizophora. However only the latter group is monophyletic. The 'Aschiza' constitute a residual paraphyletic assemblage of all Cyclorrhapha which are excluded from the Schizophora because they lack a ptilinum. To my knowledge it has never been claimed that the Aschiza show any apomorphous conditions in common. Such a group has no place in a phylogenetic classification in the sense followed in this work.

The justification for the division of the Cyclorrhapha into the Acroptera and Atriata rests largely on larval characters. In the Acroptera a small remnant of the larval head capsule is visible in dorsal view (DE MEIJERE 1900, HENNIG 1952a), a relatively plesiomorphous condition in comparison with the highly apomorphous 'headless' condition shown by all other cyclorrhaphous larvae, in which no free remnant of the head capsule is visible externally. Adult lonchopterids are of uniform appearance, characterized by several apomorphous conditions, for instance in the shape and venation of the wing (see HENNIG 1954). Thus there appears to be an alternance of apomorphous and plesiomorphous conditions between the Acroptera and Atriata, which is the basis for inferring that these groups are sister-groups in the sense proposed by HENNIG (1950, 1966a).

The name Atriata refers to the formation of an 'atrium' before the primary mouth opening of the larva (see HENNIG 1948b). The 'headless' condition of the larva is clearly one of considerable complexity (in the sense in which I use this term in section 2.3), and thus provides firm grounds for inferring that the Atriata are monophyletic. The groups of Atriata which show the least modified adult morphology are the Platypezidae and Ironomyiidae. When the adult morphology of members of these families is considered, it is scarcely possible to find autapomorphous conditions of the Atriata, except perhaps for reduction of the costa on the posterior margin of the wing (given as an apomorphous condition of this group by HENNIG 1954). However the difference in this respect is

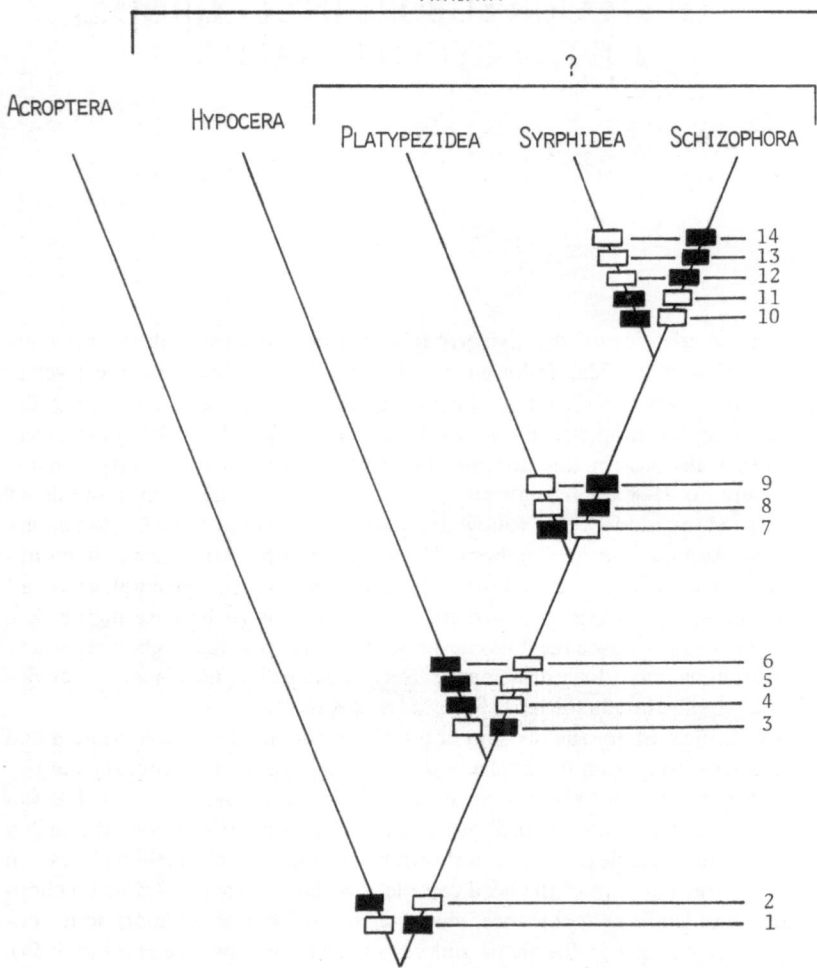

Fig. 13. Systematic division of the Cyclorrhapha, in accordance with the argumentation scheme of phylogenetic systematics.

Apomorphous conditions (black rectangles)	Plesiomorphous conditions (white rectangles)
1. Larval head capsule reduced: atrium formed	Small head capsule retained: atrium absent
2. Shape and venation of wing modified	Wing as in groundplan of Cyclorrhapha
3. m_1 and m_2 forked beyond discal cell or at least at its distal corner	Distal section of m_2 arising from discal cell
4. Subcosta partly fused with r_1	Subcosta complete
5. Anal cell shortened	Anal cell elongate
6. Only a single dorsal sclerite between 6th segment and hypopygium (♂)	All sclerites of 7th and 8th segments retained (♂)

7. Hind tarsi expanded	Hind tarsi not expanded
8. Hypopygial circumversion completed within puparium ($\male$)	Final 180° of hypopygial circumversion occurring after emergence from puparium ($\male$)
9. Ejaculatory apodeme free from body wall ($\male$)	(not clarified)
10. Frons without macrochaetae	Frons with well differentiated macrochaetae
11. 6th and 7th abdominal segments asymmetrically developed on left side, with reduced terga; 8th sternum enlarged and asymmetrical; hypopygium strongly deflexed, directed anteriorly ($\male$)	Postabdomen symmetrical or only weakly asymmetrical; 6th and 7th terga not reduced; 8th sternum not enlarged; hypopygium less strongly deflexed ($\male$)
12. Ptilinum and temporary musculature present: 1st abdominal segment with adventitious suture	Ptilinum and temporary musculature absent: adventitious suture absent
13. m_{1+2} not forked	m_{1+2} forked
14. Anal cell shortened	Anal cell elongate

slight, since in *Opetia* a trace of sclerotization around the posterior margin of the wing is still apparent (HENNIG 1954). In respect of all other characters so far analysed, the groundplan condition of the adult stage of the Atriata seems not to differ from the relevant groundplan condition of the Cyclorrhapha as a whole.

The structure of the Platypezidae has already been discussed in section 3.2, in the light of KESSEL's recent discovery that in this group part of the hypopygial rotation in the male does not take place until after emergence from the puparium. This condition corresponds to an earlier stage in the evolution of the irreversible '*hypopygium circumversum*' condition shown by the Schizophora, in which circumversion is completed within the puparium. The segmental sclerotization of the male postabdomen of some Platypezidae, such as *Plesioclythia* (figs. 10 and 11), is the least modified of that shown by any cyclorrhaphan known to me. The 6th and 7th segments bear well developed terga and sterna in the normal dorsal and ventral positions. The 7th sternum is more or less symmetrical, or appears slightly asymmetrical due to weaker sclerotization of its right side (as in fig. 10). The 8th segment bears a large sternum, which is dorsally situated (inverted) when the hypopygium is in its circumverted rest position; the 8th tergum is reduced to a narrow band, which in *Plesioclythia* (fig. 10) extends from the venter around the right side of the insect (in the rest position). The 8th segment is rotated through half the angle of hypopygial rotation, which results in the sternum occupying a lateral position on the left side at emergence and during copulation, when the hypopygium is inverted (rotated through only 180°). The arrangement of postabdominal sclerites shown by the Platypezidae corresponds closely with what can be inferred about the groundplan condition for the Schizophora, and the conditions shown by other groups of Atriata are derivable from such a condition. For this reason I am convinced that one of the keys to an understanding of the morphological changes which have occurred in the male postabdomen of various groups of Atriata lies in detailed study of the

Platypezidae. However, another consequence of the retention of many plesio-morphous conditions by the Platypezidae is that it is difficult to interpret their relationships to other Atriata because of a paucity of apomorphous conditions, at least as known from existing analyses. HENNIG (1952a, 1954) accepted the classification of this family in the Hypocera (following BRAUER 1883), but was not able to advance any clear-cut apomorphous conditions to demonstrate the validity of this classification. His comment that all the families of Hypocera show a 'tendency to reduction of the discal cell' (HENNIG 1954: 360) is uncon-vincing. If the character sequence involved in the reduction of the discal cell in true Hypocera is compared with the condition shown by the Platypezidae, the latter condition does not appear to belong to the same sequence. In *Irono-myia* (an undoubted hypoceran) the distal section of vein m_2 arises from the discal cell (J. F. MCALPINE 1967, fig. 3), as in the postulated groundplan of the Eremoneura (corresponding to the conditions shown by the Jurassic Protempid-idae and some recent groups of Empididae). In Platypezidae the discal cell is elongate and bounded distally only by the posterior cross-vein (m-m); the fork of veins m_1 and m_2 lies beyond the discal cell, or at least at its distal corner (in *Platypezina*). This apomorphous condition of the median field of the wing is hardly different from the groundplan condition of the Syrphidea, and the con-dition shown by the Schizophora probably represents a further modification of the same character sequence (differing from the groundplan of the Platypezidae and Syrphidea in respect of loss of the fork of m_{1+2}). Thus the evidence of the wing venation does not support the classification of the Platypezidae in the Hypocera, but suggests that this may be incorrect. If it is later confirmed that the Platypezidae, Syrphidea and Schizophora are monophyletic, SCHINER'S (1864) name Orthocera is available for such a group.

J. F. MCALPINE & MARTIN (1966) also accepted the classification of the Platypezidae in the Hypocera as a starting point for their studies. In that paper they reconstructed some aspects of the groundplan of the Hypocera ('the original phoroid fly') by comparing conditions shown by *Ironomyia* (Ironomyiidae) and the platypezid genera *Platypezina*, *Opetia*, *Microsania* and *Melanderomyia*. However, if the Platypezidae are not referable to the Hypocera, then the most recent common ancestor of the Platypezidae and Ironomyiidae may well have been the common ancestor of the Atriata as a whole, and MCALPINE & MAR-TIN's characterization of the 'original phoroid fly' may consequently refer to the groundplan of the Atriata. I do not find in their data any strong support for the view that the Platypezidae are more closely related to the undoubted hypoceran families (Ironomyiidae, Sciadoceridae and Phoridae) than to the Syrphidea or Schizophora. Clearly the relationship of the Platypezidae to other Atriata is in need of further study. In section 7 I provisionally list the Platypezidae as con-stituting one of the major subordinate groups of the Atriata, the Platypezidea (a name already proposed by ROHDENDORF 1964). This classification leaves open the question of whether the group is most closely related to the Hypocera, Syrphidea or Schizophora. A sound basis for further studies of the Platypezidae has been provided by KESSEL & MAGGIONCALDA's (1968a) revision. The presence of expanded hind tarsi at least in the male (clearly an apomorphous condition) suggests that at least the majority of the genera accepted as Platypezidae by these authors constitute a monophyletic group. However *Atelestus* (= *Platycnema*) should not in my opinion be referred to this family (see section 4). KESSEL and MAGGIONCALDA have also accepted as Platypezidae two other problematical genera, *Opetia* and *Melanderomyia*, which they consider to be 'primitive' Platypezidae. Both genera are unknown to me. Neither shows the tarsal modification characteristic of the other genera referred to the Platypezidae. Since none of the other diagnostic characters of the Platypezidae have yet been shown to be autapomorphous, the possibility remains that these genera may be misplaced or may represent a group or groups meriting separate family status. Detailed morphological studies on *Opetia* and *Melanderomyia* might thus advance our understanding of the evolution of the Atriata. But until further information is available, I provisionally accept the classification of these genera in the Platypezidae. The larvae of *Melanderomyia* are known to be fungivorous, as are all the known larvae of undoubted Platypezidae, but no description is available.

As a consequence of my exclusion of the Platypezidae, I restrict the limits of the Hypocera (= Phoridea) to the Ironomyiidae, Sciadoceridae and Phoridae (including Termitoxeniinae). The characterization of this group has been clarified by HENNIG (1954, 1964), J. F. McALPINE & MARTIN (1966) and J. F. McALPINE (1967). These studies indicate that the Hypocera are characterized in their groundplan by the following apomorphous conditions with respect to the groundplan of the Atriata.

(1)   Apex of second antennal article deeply inserted into base of third.
(2)   Subcosta partly fused with vein r_1.
(3)   Anal cell shortened.
(4)   Male postabdomen with only one dorsal sclerite between 6th segment and hypopygium.

The groundplan condition of the wing venation in Hypocera is probably that shown by the Tasmanian *Ironomyia* (see J. F. McALPINE 1967). The structure of the male postabdomen of *Ironomyia* (McALPINE 1967, figs. 9 and 10) also appears more plesiomorphous than that of the Phoridae, in which the protandrial sclerites are still more reduced. My statement of the groundplan for the Hypocera (character 4 above) consequently refers to the condition shown by *Ironomyia*. This characterization may possibly need revision, when information becomes available on the structure of the male postabdomen of recent Sciadoceridae. The available descriptions of the latter family do not give details of the postabdomen and genitalia.

The works of HENNIG (1964), J. F. McALPINE & MARTIN (1966) and J. F. McALPINE (1967) have considerably advanced our understanding of the evolution of the Hypocera. McALPINE and MARTIN's proposal of a new family Ironomyiidae based on the Tasmanian *Ironomyia nigromaculata* White seems justified by the information presented, which suggests that this species is the sister-group of all other known Hypocera. Following these authors I recognize three families of the Hypocera in the restricted sense here followed: the Ironomyiidae, Sciadoceridae and Phoridae. McALPINE & MARTIN's key summarizes the characterization of these families. The Termitoxeniinae (hermaphroditic forms living in termite colonies) are a subordinate group of the Phoridae in the phylogenetic system. ROHDENDORF'S (1964) proposal of infraordinal rank ('Termitoxeniomorpha') for this group reflects his phenetic principle of classification in accordance with which highly divergent groups are given high rank in the system. He does not interpret the phylogenetic relationships of the group differently, for he states it to be a 'very altered descendant of the Phoridae'.

The third major subordinate group of the Atriata here recognized is the Syrphidea. This group is conventionally divided into two families, the Pipunculidae and Syrphidae, whose close relationship to each other has long been accepted. HENNIG (1954) characterized the group on the basis of wing venation, indicating the following apomorphous conditions with respect to the groundplan of the Atriata:

(1) m_1 and m_2 arising with common stem from discal cell; and

(2) m_2 not reaching wing margin.

However this characterization does not demonstrate the monophyly of the Syrphidea, since the wing venation of the Schizophora may be a further modification of the groundplan condition of the Syrphidea. Conditions which constitute more conclusive evidence of the monophyly of the group (all autapomorphous except number 7) are as follows.

(3)   Frons without conspicuous macrochaetae.

(4)   6th and 7th abdominal segments ($\male$) asymmetrically developed on left side, with terga relatively small in comparison with the large sterna.

(5)   8th sternum ($\male$) enlarged and strongly asymmetrical, occupying terminal position on abdomen.

(6)   Hypopygium ($\male$) very strongly deflexed and directed more or less anteriorly, adpressed against right side of protandrium when at rest.

(7)   Ejaculatory apodeme free from body wall ($\male$) (fig. 8).

The structure of the male postabdomen of the Syrphidea was correctly interpreted in METCALF'S (1921) now classical paper, except that his numbering of segments requires revision. METCALF accepted a theory then current that a segment had been lost at the base of the abdomen in Diptera. This theory now seems defunct, and consequently all METCALF'S segment numbers should be reduced by one. Unfortunately, much confusion has arisen in the literature on the Syrphidea because some subsequent workers have not accepted METCALF'S homologies, due apparently to an assumption that the larger sclerite of any segment must be the tergum. ACZÉL'S (1948) homologies of the postabdominal sclerites of the Pipunculidae clearly illustrate the untenable consequences to which rigid adherence to this assumption can lead. ACZÉL stated that terga are usually more resistant to reduction than sterna in the Diptera, because the sterna of the preabdominal segments are usually smaller than the terga. Applying this principle to the structure of the male abdomen of the Pipunculidae, he counted the first six abdominal terga in the normal manner, but then numbered the 6th to 8th sterna as the '7th tergite', '8th tergite' and 'epandrium' or '9th tergite'. He consequently considered the true genital segment to be the 10th visible segment, which is scarcely tenable because workers on other groups of

Diptera have never reported more than eight visible segments before the genital segment. ZUMPT & HEINZ (1949) also confused the postabdominal terga and sterna, and their work was justifiably criticized by HARDY (1950) and VAN EMDEN (1951) on this account. But METCALF (1921) had given figures of the tracheation of the postabdomen which demonstrated that his interpretation of the terga and sterna of the 6th and 7th segments was correct. The large sclerite of the 8th segment (called 'urite nine' by METCALF because he was not sure whether it was a tergum or a sternum) should clearly be homologized with the inverted 8th sternum on the basis of comparison with the structure of other Cyclorrhapha. METCALF did not mention any sclerite which might represent the 8th tergum, but I have noted a small sclerotized band which possibly represents a remnant of this sclerite in *Xylota* (Syrphidae). In respect of the condition of the ejaculatory apodeme the Syrphidea are probably synapomorphous with the Schizophora (see the discussion in section 3.1).

The fourth major subordinate group of the Atriata is the Schizophora, which are treated in detail in the next section of this work. The Schizophora are characterized primarily by the development of a new and complex organ, the ptilinum, associated with temporary musculature in the first four abdominal segments. The temporary muscles pump haemolymph into the ptilinum; after retraction of the ptilinum the temporary abdominal muscles, as well as the muscles of the ptilinum itself, are broken down. This complex condition involves such major structural and physiological modifications that it is scarcely possible to doubt that the Schizophora are monophyletic. STRICKLAND (1953) has published a comparative study of the structure of the ptilinum. The name Schizophora (proposed by BECHER 1882) refers to the ptilinal suture which remains on the frons and face of the mature fly as a result of retraction of the ptilinum. Additional apomorphous groundplan conditions of the Schizophora with respect to the groundplan of the Atriata are as follows.

(2)   Vein m_{1+2} not forked.

(3)   Anal cell shortened.

In some members of a few schizophorous families, notably the Conopidae and Micropezidae, the anal cell is elongate, reaching to near the wing margin. But HENNIG (1958) concluded that the elongate anal cell has probably evolved secondarily in these groups, with uncertainty remaining only in respect of the Conopidae. However, if my reference of the Conopidae to the Tephritoinea is accepted (see section 6.2), then the elongate anal cell shown by some members of this family must also be interpreted as secondary.

(4)   Two pairs of vertical bristles (*vti* and *vte*) present.

Most Schizophora are characterized by well differentiated frontal and vertical bristles, including typically 1-3 orbital bristles, a pair of anteriorly directed ocellar bristles, two pairs of vertical bristles (of which the *vti* are directed inwards and the *vte* outwards) and a pair of postvertical brisles. Some of these may be homologous with bristles shown by other Cyclorrhapha, such as the Platypezidae and Hypocera; but at least the presence of two pairs of vertical brisles can be considered an autapomorphous condition of the Schizophora, since no more than a single pair are shown by other cyclorrhaphans. Whether the presence of well differentiated postvertical brisles is also an autapomorphous groundplan condition of the Schizophora is not clear, as these bristles are scarcely differentiated in members of the Lonchaeoidea. HENNIG (1958) has discussed the chaetotaxy of the head in the Schizophora in some detail.

(5) Hypopygium (♂) irreversibly rotated through 360° to circumverse position within puparium.

I have reviewed the subject of hypopygial rotation in section 3.2. There I concluded that the completion of circumversion within the puparium is a more apomorphous condition than the condition shown by the Platypezidae, in which the last 180° of rotation is facultative and occurs only after emergence from the puparium. Unfortunately, detailed observations on the process of circumversion in the Syrphidea and Hypocera are not yet available, so that it is not clear whether these groups are synapomorphous with the Schizophora in respect of this character.

(6) First abdominal segment with 'adventitious suture' extending posterodorsally from its anterior margin (YOUNG 1921).

I suspect that the adventitious suture is a ridge serving as attachment for some of the temporary muscles which pump haemolymph into the ptilinum. But I have not been able to find any definite information on this point.

(7) Ejaculatory apodeme (♂) free from body wall.

See the discussion in section 3.1. The same condition is shown by the Syrphidea.

74

6. REVIEW OF THE CLASSIFICATION OF
THE SCHIZOPHORA

6.1. Preliminary remarks

The classification of the Schizophora has long been recognized as presenting formidable difficulties, not the least of which is the sheer number and diversity of the species and species-groups included. Conventionally the group has been divided into two 'sections', the Acalyptratae and Calyptratae. The current delimitation of these 'sections' was worked out by GIRSCHNER (1893), whose work was a major advance in the field of classification. However, only the Calyptratae in GIRSCHNER's revised sense constitutes a probable monophyletic group. The 'Acalyptratae' are a residual group (paraphyletic in HENNIG's sense) which contains all Schizophora excluded from the Calyptratae. Such groups do not belong to the phylogenetic system. I doubt whether even adherents of other principles of classification would find the group 'Acalyptratae' satisfactory, if they examined it critically. It seems of little value for purposes of identification, since I have not yet seen a key in which members of all groups of Schizophora are likely to be taken under the appropriate alternative in the couplet where the 'Acalyptratae' and Calyptratae are separated. One of the consequences of my rejecting the group 'Acalyptratae' is that the group Calyptratae loses relative rank because of subordination to a wider group (Muscoidea in the new sense proposed below). This does not necessitate a change of name, since the name Calyptratae carried no connotation of rank. This relative downgrading of the Calyptratae was already implied in HENNIG's (1958) treatment.

HENNIG (in press) refers to the work of DE MEIJERE (1902) on the prothoracic spiracles of dipterous pupae as possibly providing grounds for supposing that the 'Acalyptratae' are monophyletic. DE MEIJERE reported that in many Calyptratae the prothoracic spiracular horn of the pupa pierces the wall of the puparium, and evaluated this condition as plesiomorphous ('primitivste') for the Schizophora. I agree with HENNIG that this evaluation seems justified by the data presented. However, I do not see how the 'Acalyptratae' can be characterized by synapomorphous reduction of the spiracular horn ('aüsseres Tüpfelstigma'). DE MEIJERE reported that the spiracular horn also pierces the puparium in some Heleomyzidae (*Leria* and *Heleomyza*), and the possibility that such a condition occurs more widely among 'Acalyptratae' cannot be excluded, as representatives of only a few families were examined.

HENNIG's (1958) work was the starting point of my present study. In that work HENNIG reviewed previous proposals for the classification of the Schizophora, and presented a revision based on his own analysis of certain character sequences (mainly involving the chaetotaxy of the head and the wing venation). HENNIG's classification has a provisional appearance. Some of his proposed superfamilies were stated to be tentative and weakly based, and many families were not placed in any superfamily but treated as 'families of uncertain relationships'. HENNIG's acknowledgement of the inconclusive results of his analysis in some areas of the system is commendable, since he largely resisted the temptation of proposing groups on too speculative a basis for the sake of achieving a more satisfying formal presentation of the system. Some of the uncertainty which attends the results of HENNIG's analysis is due to the nature of the character sequences chosen. The modifications of wing venation shown by the Schizophora mostly involve relatively minor changes which have occurred independently in many different groups. Similar changes in the chaetotaxy of the head have also occurred independently in different groups, and the frequency of individuals with abnormal frontal chaetotaxy in many species clearly suggests that some modifications of this chaetotaxy (such as changes in the number of orbital bristles) may involve little genetic complexity. Because of difficulties in separating true synapomorphies from the numerous homoplastic similarities shown in these character sequences, HENNIG was only able to propose a few groups above the family level which he considered to be firmly based. The need to widen the scope of HENNIG's analysis by incorporating additional characters is clear. In this study I have incorporated characters of the male postabdomen and genitalia into the analysis. HENNIG (1958) used characters of the male postabdomen to a limited extent, but some of his characterization is not sufficiently explicit. In particular, his statements that only a single dorsal sclerite is present between the preabdomen and the hypopygium in some families do not explain the homology of the sclerite and the nature of the character sequence which has led to this condition. These comments on the limitations of HENNIG's work are not intended as criticisms. For the Schizophora are such a large and diverse group that restriction of the characters analysed is necessary, if a comparative study is to be completed within a reasonable length of time. For this same reason my own investigations have been largely restricted to a limited range of characters (involving the structure of the male postabdomen and external genitalia). Undoubtedly there are other characters which, if analysed comprehensively, could be used to improve the classification. The structure of the female postabdomen and reproductive system is an obvious example.

The above remarks on restriction of the characters which I have analysed do not mean that I propose a classification based solely on these characters. All groupings suggested by my new data have been tested for compatibility with other character sequences already analysed in the literature (both in HENNIG's works and elsewhere), and in principle I admit no restriction of the range and kind of characters which should be considered. If I do not discuss certain characters used in taxonomic descriptions, this is because they have not been sufficiently analysed from an evolutionary point of view.

My interpretations of the characters analysed by HENNIG (1958) follow, with few exceptions, those given in that work. It is possible that HENNIG's interpretation of two characters, the development of the costa and postvertical bristles, should be reversed in some cases; for the suggested apomorphous conditions (costa broken; postverticals convergent) are at least as widely distributed as conditions which HENNIG considers usually plesiomorphous (costa unbroken; postverticals divergent). This question is complicated because these character states must on any reasonable interpretation be postulated to have changed independently in several different lineages, possibly with reverse changes in some cases. On balance I have decided not to reappraise extensively characters apart from the male postabdomen and genitalia. I think it preferable to see how well my interpretation of this character complex stands the test of criticism, before using it as a yardstick to test the compatibility of other workers' interpretations of other characters.

In addition to the groundplan characterization I give descriptions of the male postabdomen and genitalia of each group accorded family rank. I think it useful to present this information, because the treatment of this character complex in the available monographs of particular families is, with few exceptions, incomplete. The preparation of these descriptions was not easy. Some families include subordinate groups with highly modified morphology, with the result that a range of alternatives has to be presented in descriptions which do not carry implications about the direction of character change. In other cases difficulties arose because of gaps in the available information. I often found that, while some of the characters were described in the literature for a wide range of species of a family, the states of other characters could only be established for the few representatives available to me for study. These difficulties should be borne in mind when using my descriptions. They should be considered as provisional drafts, whose details need to be confirmed or revised by more extensive studies of each family.

The assignment of formal rank in the system here presented is provisional, because the available historical information is too little to apply the criterion of age of origin to the ranking of particular groups. The Jurassic deposits from Kazakhstan studied by ROHDENDORF and his collaboratores have not yielded any Cyclorrhapha, but the material described from them includes *Protempis* (Protempididae), which is possibly referable to the stem-group of the Eremoneura (see section 4). If this is so, then these deposits preceded the age of origin of the Schizophora. The earliest fossil Schizophora so far reported are an undescribed specimen in Canadian amber believed to be of Late Cretaceous age (listed as a chloropid by J. F. McALPINE & MARTIN 1969) and some puparia tentatively referred to the Calliphoridae (J. F. McALPINE 1970) from the Edmonton Formation (Alberta) of latest Cretaceous age. These meagre finds provide no basis for applying HENNIG's (1966a: 186) proposal that in classifying insects the boundary between the Upper and Lower Cretaceous can be taken as the boundary between the family level and lower levels. Consequently it is impossible to say that authors who have treated the Schizophora as consisting largely of a single family 'Muscidae', such as LAMEERE (1906), are wrong in any formal sense. At present we cannot apply any guidelines to settle this question. My approach in these circumstances has been, in general, to disturb minimally the ranks conventionally assigned to taxa. As a result some discrepancy has been accepted between the sequence of subordination indicated by my analysis and the categorical ranks formally assigned. This discrepancy manifests itself in my arrangement of some families in 'family-groups', which would be ranked only as families according to the sequence of subordination (see section 7). The levels of taxa in the sequence of subordination are indicated by the letters A, B, C and D before the number assigned to each taxon. I propose to introduce into the classification of the Schizophora a category between superfamily and family, indicated by the suffix '-oinea'. This follows the proposal of VON KÉLER (1963: 636), who uses the name 'suprafamilia' (German: 'Oberfamilie') for such a category. However, the prefix 'supra-' is merely a variant of 'super-' and does not seem sufficiently distinctive. I propose therefore to substitute the prefix 'prae-' ('pre-' in English), so that the category name becomes prefamily (or praefamilia in Latin). If my proposal gains acceptance, similar changes should be made to the Latin and English names of other categories for which VON KÉLER uses the prefix 'supra-'. According to my revision of VON KÉLER's proposals the complete list of family-group categories is as follows: superfamily, prefamily, family, subfamily, infrafamily. Between the family-group and order-group VON KÉLER interposes the phalanx-group, to which I refer the taxa Eremoneura, Cyclorrhapha and Schizophora (see section 7). The use of so many intermediate categories between family and order will doubtless be deprecated by some. But in relation to the numbers of species classified, the numbers of categories used is not unusually high. Of course the full list of categories given by VON KÉLER should not be regarded as mandatory, for the use of so many categories would be superfluous in classifying less diverse groups of organisms.

ROHDENDORF (1964) has classified the Diptera on different principles from those followed in this work. The taxa in his system are based directly on conformity with particular functional-morphological types, and are not intended to represent monophyletic groups in all cases. ROHDENDORF's classification should therefore be considered a morphological-phenetic classification, and should not be confused with the phylogenetic system which I discuss in this work. Consistent with his principles of classification ROHDENDORF accords high formal rank to certain highly modified groups, notably the Termitoxeniinae ('infraorder Termitoxeniomorpha'), the Braulidae ('infraorder Braulomorpha') and the groups which I include in the Hippoboscidae family-group ('infraorder Nycteribiomorpha', 'infraorder Streblomorpha', 'superfamily Glossinidea', 'superfamily Hippoboscidea'). In a phylogenetic system these groups should be given lower rank because of their position in a sequence of subordination which is determined, at least in a relative sense, by the time dimension (see section 2.1). Both treatments are formally correct within the framework of the type of classification followed, and the striking differences which result from applying different principles of classification do not necessarily indicate differing opinions about either the phylogenetic relationships or morphological divergence of the groups concerned. However the subdivision of the 'infraorder Myiomorpha' (including most Schizophora) in ROHDENDORF's classification seems to me unduly arbitrary, and many of the included superfamilies are insufficiently characterized. I doubt whether some of these proposed superfamilies would stand the test of more critical analysis even in terms of ROHDENDORF's own principles of classification. For instance, his 'superfamily Borboridea', containing the Agromyzidae, Milichiidae, Sphaeroceridae (= Borboridae) and Cryptochetidae, is a highly heterogenous group, both from the standpoint of phenetic and phylogenetic classification. Because ROHDENDORF does not classify in accordance with the principles of phylogenetic systematics, many of his group names do not refer to groups in the phylogenetic system, and are consequently not used in the classification presented in this work.

The classification presented here is a classification of existing groups, and I have not attempted to review the fossil forms. HENNIG's (1965b, 1967, 1969b) fine works on the Baltic amber fauna provide the main body of useful information on fossil Schizophora. Much of the material preserved in other media is in too poor condition to be of much value for evolutionary studies, although this has not deterred some authors from naming numerous specimens of doubtful significance.

At a late stage in the preparation of this work I received two important papers dealing with the classification of the Schizophora, and revised my manuscript to take account of these. The first was SPEIGHT's (1969) work on the prothoracic morphology of 'acalypterates'. SPEIGHT offered only a few firm conclusions on the relationships of the families, because he found that few of them could be characterized by apomorphous conditions of the prosternum. His work is marred in some places by confused argumentation and is consequently criticized in the paper by HENNIG discussed in the next paragraph. But on the whole it is a helpful contribution to our knowledge of the Schizophora, and the range of material studied (2150 species) is unusually extensive.

I also received at a late stage of this study a manuscript copy of a paper by HENNIG entitled 'Neue Untersuchungen über die Familien der Diptera Schizophora'. This is a supplement to his 1958 paper, and includes some important new information and discussion on Drosophiloidea and Anthomyzoinea. HENNIG's manuscript was ready for press at the time I sent him a preliminary draft of my manuscript for comment. He decided that the best procedure would be to let his paper go to press without substantial alteration, while making the manuscript available to me so that I could take account of it in completing my manuscript. At the time of writing (March 1971) HENNIG's paper is still unpublished.*

MORPHOLOGICAL FOUNDATIONS OF SUPERFAMILY CLASSIFICATION

In the section which follows (6.2) I present a classification of the families of Schizophora in five superfamilies: Lonchaeoidea, Lauxanioidea, Drosophiloidea, Nothyboidea and Muscoidea. Some of the characters on which this classification is based are indicated on figure 14.

* HENNIG's paper has since been published as Stuttgarter Beiträge zur Naturkunde, no 226 (1971). The only part of it not included in the manuscript sent me was the addendum (Nachtrag) on pages 70-73.

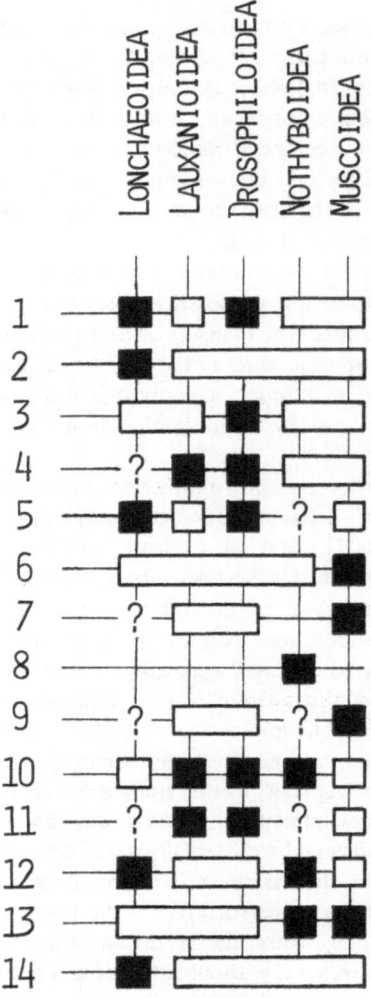

Fig. 14. Distribution of some character states among the superfamilies of Schizophora.

Apomorphous conditions   Plesiomorphous conditions
   (black rectangles)   (white rectangles)

Head
  1. 2nd antennal article with cleft   2nd antennal article without cleft
  2. Only 1 *ors* present   At least 2 *ors* present
  3. Proclinate *ors* present   Proclinate *ors* absent
  4. Postverticals convergent   Postverticals divergent
Wing
  5. Costa broken at end of sc   Costa unbroken
Male postabdomen
  6. 6th sternum more strongly develop-   6th sternum symmetrical
     ed on left side

80

7. 7th sternum asymmetrical, extending dorsally on left side	7th sternum symmetrical
8. 7th sternum lost	(as 7)
9. 7th tergum reduced to lateroventral vestige	7th tergum well developed in dorsal position
10. 8th tergum vestige lost	8th tergum vestige present
11. 8th sternum reduced or lost	8th sternum large
12. Aedeagal apodeme linked to hypandrium or body wall between hypandrial arms	Aedeagal apodeme free from hypandrium and body wall
13. Aedeagus swung through wide arc against aedeagal apodeme	Aedeagus without well developed swinging mechanism
Female postabdomen	
14. 7th tergum and sternum fused; 8th segment elongate	7th tergum and sternum discrete; 8th segment short

I will explain the grounds for my proposing a revised superfamily classification first by discussing the Drosophiloidea and Muscoidea, leaving aside for the moment the other three superfamilies since ontogenetic information on them is not yet available. The Muscoidea and Drosophiloidea show in their groundplans gross structural differences in the male postabdomen, which are analysable into an alternance of apomorphous and plesiomorphous conditions. In the groundplan of the Muscoidea the sclerites of the 8th segment are developed as in the Platypezidae (figs. 10-11), with a large inverted sternum in dorsal position and reduced ventral tergum. In this respect the Muscoidea are relatively plesiomorphous, since in the Drosophiloidea the sclerites of the 8th segment are vestigial or absent in the male. However, the Drosophiloidea are relatively plesiomorphous in respect of the sclerites of the 7th segment, which in the groundplan of the group are fully or nearly symmetrical plates in dorsal and ventral positions. In Muscoidea the 7th tergum is never developed as a symmetrical dorsal plate, but is asymmetrically reduced to a vestige situated lateroventrally on the right side in the groundplan of the group. Because of this alternance of relatively plesiomorphous and apomorphous conditions, I think it would be unwarranted to maintain that either type of structure was derived from the other; more probably both types are independent modifications of the type of structure shown by the Platypezidae. As a result of new information now available on the development of the genital discs in Muscoidea (DÜBENDORFER 1970), I can reinforce this conclusion with ontogenetic evidence. It now seems likely that the differences in postabdominal structure between the Drosophiloidea and Muscoidea rest on differences in the ontogenetic process which can be traced back to the blastoderm stage of the embryo, when the imaginal presumptive areas are determined (see ANDERSON 1966). In order to make this point clear, I set out the relevant data in table II. I use *Mycetophila* as the starting point of the comparison, since this is probably the most closely related to the Brachycera (of which the Cyclorrhapha are a subordinate group) of the non-brachycerous groups for which relevant information is available (see HENNIG 1968b). The larva of *Mycetophila* resembles that of the Cyclorrhapha in showing a series of imaginal discs on all abdominal segments (ABUL NASR 1950), but there are three pairs of discs on

81

TABLE II
Development of imaginal discs in Bibionomorpha and Brachycera.

Imaginal discs on larval 8th and 9th abdominal segments *(Mycetophila)* or 8th abdominal segment (Cyclorrhapha)		♂♂	♀♀
Mycetophila (Bibionomorpha, Mycetophilidae) (after ABUL NASR 1950)*	8th larval segment — ventral discs	not established	Vagina, sperma-thecae and com-mon oviduct
	lateral discs	not established	not established
	dorsal discs	not established	not established
	9th larval segment — ventral median discs	9th sternum, parameres, aedeagus, para-physes and ejaculatory ducts	caecus (parovaria)
	lateral discs	10th sternum (hypoproct)	10th sternum
	dorsal discs	cerci	cerci
Tephritidae and Calyptratae (Cyclorrhapha, Muscoidea) (after ANDERSON 1966, BLACK 1966 and DÜBEN-DORFER 1970)	lateral genital discs	8th sternum (? also 8th tergum, when present)	8th tergum and 8th sternum, vagina, sperma-thecae and com-mon oviduct
	median genital disc	all external structures of genital (9th) segment, ejacu-latory duct, accessory glands (paragonia), proctiger and proctodaeum (hind gut)	parovaria, proctiger and proctodaeum (hind gut)
Drosophila (Cyclorrhapha, Drosophiloidea) (many refer-ences)	median genital disc	all external structures of genital (9th) segment, ejacu-latory duct, accessory glands (para-gonia), proct-iger and proctodaeum (hind gut)	vaginal plates (8th sternum), vagina, sperma-thecae, common oviduct, paro-varia, proctiger and procto-daeum (hind gut)

each segment, compared with only two pairs in the cyclorrhaphous Tephritidae (ANDERSON 1966).

If the position and development of the imaginal discs in the Muscoidea is compared with that in *Mycetophila*, it seems likely that the lateral genital discs of Muscoidea are homologous (at least in part) with the ventral discs of the 8th abdominal segment in *Mycetophila*. However, the lateral genital disc of Muscoidea generates structures in addition to those generated by its equivalent in *Mycetophila*, so that an increase of the field which the discs are capable of generating or coalescence of previously separate discs must be postulated. The same applies to the median genital disc of Muscoidea, which generates the proctiger (including the cerci) and proctodaeum (hind gut), in addition to structures homologous with those generated by the ventral median discs of *Mycetophila*. In *Mycetophila* the sclerites of the proctiger develop from two additional pairs of discs, while the proctodaeum is not reported to develop from any disc. But it is important to note that the genital arch (8th tergum) in female *Drosophila* is not generated by the genital disc. I think it improbable that the median genital disc of *Drosophila* was derived through coalescence of discs from the condition shown by the Muscoidea, for in the Muscoidea the structures generated by the lateral genital discs include the 8th tergum. More probably the conditions shown by the Muscoidea and Drosophiloidea were not either derived from the other, but are independent modifications of some earlier condition, in conformity with the conclusion reached above from considerations of comparative morphology.

Where then do the Lonchaeoidea, Lauxanioidea and Nothyboidea fit into this picture? For the present I leave open the question of whether each of these groups is more closely related to the Drosophiloidea or to the Muscoidea. The arrangement of the male postabdominal sclerites in the Lauxanioidea and Nothyboidea may agree with that in the groundplan of the Drosophiloidea (with the 7th tergum retained in dorsal position, but with the sclerites of the 8th segment lost). However the homologies of the sclerites concerned have not been critically demonstrated, and I therefore do not think that the evidence is sufficient for formally grouping them with the Drosophiloidea. A particular problem arises in interpreting the relationships of the Nothyboidea. This group shows in its groundplan the characteristic mechanism for swinging the aedeagus through a wide arc against the aedeagal apodeme to an anteriorly directed rest position, as in the Muscoidea. If the Nothyboidea are more closely related to the Drosophiloidea and Lauxanioidea than to the Muscoidea, then possibly the

Footnote to table II:
* There are certain differences between ABUL NASR's (1950) terminology and that which I use in this work. The structures which ABUL NASR calls 'gonapophyses' are what I call parameres: he applies the term' parameres' to some kind of paraphyses (possibly homologous with the postgonites of the Cyclorrhapha). I do not accept his homology of the cerci with the '10th tergum'. He describes biarticled cerci in the female of *Mycetophila*, as also in the groundplan of the Brachycera (see HERTING 1957). ABUL NASR did not clarify the origin of the 9th tergum in either sex. Therefore his suggestion (his page 381) that the imaginal terga are derived from discs (or buds) is not proven. I interpret ABUL NASR's figures as indicating that the true 10th tergum is absent in both sexes of *Mycetophila*. Since this sclerite is retained in the groundplan of the Brachycera (in both sexes), it is possible that there are some differences between the pattern of imaginal discs shown by *Mycetophila* and that shown in the groundplan of the Brachycera. However, in the absence of relevant information on any group of Brachycera whose postabdominal morphology is close to the groundplan condition, I can only use the information on *Mycetophila* for purposes of comparison.

swinging mechanism has been secondarily lost in the Drosophiloidea and Lauxanioidea. On present information I doubt whether the latter hypothesis is correct, since the aedeagus of some Lauxanioidea (especially Chamaemyioinea) is upcurved and uniformly sclerotized, as I postulate for the groundplan of the Schizophora (compare the types of aedeagus shown by the Lonchaeoidea and Platypezidae). However, this question is not likely to be settled until comparative information is available on the imaginal discs and their development in each of the groups of Schizophora recognized as superfamilies in this work. I regard the extension of studies on the imaginal discs to a wider range of groups than hitherto as of the highest importance for clarifying the evolution of the Schizophora, and indeed of the Brachycera as a whole.

Both in *Mycetophila* and in those Cyclorrhapha which have been studied the arrangement of imaginal discs is initially the same in both sexes. This gives grounds to expect that structural differences between the females of Muscoidea and Drosophiloidea will be found to result from the different ontogenetic processes. So far little detailed comparative information on the structure of the female postabdomen is available for groups other than the Calyptratae. Wider comparative studies in this field would be of much interest.

6.2. Treatment of subordinate groups to the family level

A. 1. Superfamily Lonchaeoidea

My proposal of this group breaks with tradition. I include in the Lonchaeoidea two families, the Lonchaeidae and Cryptochetidae, which have usually been dissociated and placed in different family groups.

The Lonchaeidae have usually been grouped with the 'Pallopteridae', which are referred in this paper to the Tephritidae *s.l.* (Muscoidea, Tephritoinea). HENNIG (1948a) argued that such treatment was erroneous, but reverted to grouping the Lonchaeidae with the Pallopteridae in his 1958 paper. The resemblances between Pallopteridae and Lonchaeidae suggested as synapomorphous by HENNIG (1958) are: (1) the formation of an ovipositor sheath (♀) by the completely sclerotized 7th segment, (2) 6th sternum (♀) with anteriorly directed apodeme, (3) the presence of only a single pair of upper fronto-orbital bristles *(ors)*, and (4) aedeagus (♂) tubular. However, there are serious obstacles to the inclusion of the Lonchaeidae in the Tephritoinea. The aedeagus of the Lonchaeidae can hardly be considered synapomorphous with that of the Tephritoinea, since it is a rigid, uniformly sclerotized structure, not flexible and extruded by pressure of body fluid as in the latter group. The development of the male postabdominal sclerites in the groundplan of the Lonchaeidae seems to indicate that the family is not referable to the Muscoidea. HENNIG's (1958) assumption that the postabdominal morphology of *Protearomyia obscura* (WALKER) (which

84

has strongly asymmetrical 6th and 7th sterna similar to the condition shown by most Muscoidea) was relatively plesiomorphous for the Lonchaeidae, is not supported by the more detailed analyses subsequently available (J. F. McAL-PINE 1962b, MORGE 1963 and 1967). It now seems clear that the groundplan of the family should include a rather large, symmetrical 6th sternum and a weakly asymmetrical 7th sternum, which (unlike in Muscoidea) was not fused with the inverted 8th sternum. A large unmodified 6th sternum is shown by all Dasiopinae. and some species of that group also retain a discrete 6th tergum. Clearly the conditions of the 6th segment shown by the Dasiopinae are relatively plesiomorphous in comparison with those shown by the Lonchaeinae *s.l.*, as stated by MORGE (1963). The development of the 7th sternum is variable in both subfamilies, and some degree of asymmetry may probably be ascribed to the groundplan of the family. However this asymmetry is weak in many groups, as indicated for instance by McALPINE's (1962b) figures of *Dasiops relicta* Mc-ALPINE (fig. 32) and *Lonchaea subpolita* MALLOCH (fig. 38). There is no indication in the structure of Lonchaeidae of the asymmetrical reduction of the 7th tergum characteristic of the Muscoidea, where this sclerite is represented only by a vestige on the right side (near the 7th right spiracle). Probably McALPINE and MORGE are right in assuming that the 7th tergum is represented by the last distinct dorsal sclerite, which is symmetrical or, if asymmetrical, more strongly developed on the left side. The fate of the sclerites of the 8th segment requires clarification. A small ventral sclerite which probably represents the 8th tergum vestige is often present (called 'secondary sclerite' by McALPINE 1962b). No discrete 8th sternum (inverted) has been reported in Lonchaeidae, and Mc-ALPINE (1962b: 33) interprets the structure of certain groups of *Dasiops* as indicating that this has been fused with the 7th tergum.

The evidence of the structure of the male postabdomen and genitalia, as outlined above, provides no firm grounds for including the Lonchaeidae within the Muscoidea, let alone in a close relationship with *Palloptera* (Tephritoinea). Males of the latter genus are characterized by a flexible aedeagus which is extruded by pressure of body fluid, and by loss of both pairs of postabdominal spiracles. This combination of apomorphous conditions indicates that *Palloptera* belongs to the Tephritidae family-group (Muscoidea, Tephritoinea). Male Lonchaeidae do not show these conditions, but always retain a rigid aedeagus and the complete complement of seven pairs of spiracles.

From the above considerations I think that the hypothesis of a sister-group relationship between the Loncaheidae and 'Pallopteridae' should be definitely rejected, as the evidence of synapomorphy in the structure of the male postabdomen between *Palloptera* and other members of the Tephritidae family-group is overwhelming. If, as seems probable, the presence of an anteriorly directed apodeme on the female 6th sternum is not a groundplan condition of the Tephritidae family-group, then the presence of such an apodeme in *Palloptera* can only be homoplastic with the similar condition shown by the Lonchaeidae. A modified version of the above hypothesis has been presented by Mc-ALPINE (1962b: 69), whose dendrogram indicates the Lonchaeidae as the sister-group of the Tephritidae family-group as a whole (= Otitoidea + Pallopteroidea). This view depends on the interpretation that certain similarities in the structure of the 7th and 8th segments of the female postabdomen in Lonchaeidae and Tephritoinea are synapomorphies. Even this view conflicts seriously with the evidence of the male postabdomen and genitalia, which provides no basis for classifying the Lonchaeidae in the Muscoidea (as previously discussed). Detailed comparative studies of the structure of the female postabdomen among the families of Schizophora are unfortunately not yet available, and it is possible that such studies will show that the resemblances between the Lonchaeidae and Tephritoinea are not so close as has been supposed. I have not attempted detailed comparison in the course of this study, but it may be useful for me to draw attention to one structural difference. In Lonchaeidae the sclerites of the 7th segment extend as flexible rods on the posterior part of the segment which is invaginated when the 8th segment is retracted. In the extended position these rods are seen to extend to the apex of the 7th segment, where they meet the sclerites of the 8th segment (see character 4 below). In all the preparations of Tephritoinea which I have examined the invaginated area is membranous, so that in extended position the sclerites of the 7th and 8th segments are seen to be well separated by membrane. At any rate I postulate that the resemblances in the structure of the female postabdomen between the Lonchaeidae and the Tephritidae *s.l.* rest on convergence, because of the incompatibility of their interpretation as synapomorphies with the evidence of other characters. If any authors continue to maintain that the structural similarities of the female postabdomen of the Lonchaeidae and Tephritidae *s.l.* (or any part of this group) are due to synapomorphy, then they should explain how this interpretation can be reconciled with the evidence of the male postabdomen and genitalia.

HENNIG's (1958: 607) remarks on the possession of elongate spermathecae by *Palloptera* and the Piophilidae *s.l.* seem to imply that the Lonchaeidae share this apomorphous condition with these groups. However it is doubtful whether elongate spermathecae can be ascribed to the groundplan of the Lonchaeidae or of the Tephritidae family-group, since both groups contain genera with orbicular spermathecae (*Dasiops* in the case of the Lonchaeidae, as indicated by McALPINE 1962b).

The only other family which can probably be associated with the Lonchaeidae is the Cryptochetidae, a small group of highly modified flies whose larvae develop as internal parasitoids of Coccidae (Hemiptera). HENNIG (1958) treated the Cryptochetidae as a family of uncertain relationship. He suggested that their affinity probably lay with the Drosophiloidea, although admitting that the evidence was not really convincing. However he now definitely rejects this interpretation (HENNIG, in press). I can find in the literature no strongly held opinion on the relationship of the Cryptochetidae to other Schizophora.

The larvae of Cryptochetidae exhibit a wealth of autapomorphous conditions, but the morphological gaps between them and all other known larvae of Cyclorrhapha are so great that one can scarcely judge the relationship of the group from characters of the larvae and puparium. HENNIG (1952a: 126) has drawn attention to an interesting similarity between the Cryptochetidae and

Platypezidae in respect of the position of the ecdysial sutures on the puparium. It is not clear whether this represents a plesiomorphous condition in the Cryptochetidae or has evolved secondarily.

A case can be made on the basis of adult morphology for grouping the Cryptochetidae with the Lonchaeidae, with whom they share certain distinctive apomorphous conditions, notably involving the structure of the antennae and the development of a dense coat of setulae over most of the body. Both the Lonchaeidae and Cryptochetidae exhibit a similar form of aedeagus (upcurved, with uniform sclerotization, without any swinging mechanism), a plesiomorphous condition for the Schizophora (see also section 3.1) which seems to exclude the possibility of a close relationship between these families and the Drosophiloidea. A similar form of aedeagus is shown by some Lauxanioidea, but clear synapomorphies between that group and the Lonchaeoidea seem lacking. HENNIG's (1948a) view that the Lonchaeidae were closely related to the Chamaemyiidae (Lauxanioidea) was subsequently abandoned by him (HENNIG 1958). Another probably symplesiomorphous condition of the Lonchaeidae and Cryptochetidae is their retention of uniformly dark body colour with dark halteres. In this respect they resemble the Pipunculidae (Syrphidea), Platypezidae and Hypocera. HENNIG (1967) originally stated that the Baltic amber species *Morgea mcalpinei* HENNIG had pale halteres, but his remarks on a second specimen (HENNIG 1969b) indicate that the halteres of this species were probably infuscated in life.

The Lonchaeoidea in the sense here proposed may be characterized by the following apomorphous conditions with respect to the groundplan of the Schizophora.

(1) 2nd antennal article with longitudinal cleft or suture; 3rd article downwardly directed with subbasal arista (see HENNIG 1965b: 178 and J. F. McALPINE 1962b: 18).

Similar conditions occur through convergence in the Calyptratae, some Tephritidae *s.l.* (Tephritoinea), *Loxocera* (Psilidae) and most Drosophiloidea. While Recent Cryptochetidae have highly modified antennae (entirely lacking the arista), the Baltic amber fossil *Phanerochaetum* described by HENNIG (1965b) has similar antennae to those of some Lonchaeidae.

(2) Only one upper fronto-orbital bristle (*ors*) present, situated on short vertical plates.

I list this character as apomorphous following HENNIG (1958). It seems clear from HENNIG's analysis that at least two pairs of orbital bristles were present in the groundplan of the other superfamilies of Schizophora, and that the presence of only a single pair in some members of these superfamilies should be considered apomorphous. Only in the Lonchaeoidea are there no species which show any indication of a second pair of orbitals. It is difficult to judge whether the presence of only a single pair of orbitals in the Lonchaeidae is primary, that is inherited from the groundplan of the Schizophora, or is an autapomorphous condition of the Lonchaeoidea. On balance I accept HENNIG's evaluation of this condition as apomorphous as the more probable interpretation, since two pairs of fronto-orbitals are present in members of the non-schizophorous groups Platypezidae and Hypocera.

The frons in Lonchaeoidea is covered by an extensive coat of setulae. Postvertical bristles are poorly differentiated from the surrounding setulae, and their homology with the postverticals of other Schizophora requires clarification. J. F. McALPINE (1962b) refers to two pairs of postverticals, one convergent and one divergent, in some Lonchaeidae. It is not clear whether such a condition should be considered apomorphous or plesiomorphous with respect to the groundplan of the Schizophora.

(3) Costa broken at end of subcosta.
(4) 7th abdominal segment (♀) with its tergum and sternum broadly fused anteriorly, forming an oviscape (basal cone of ovipositor); the tergum and sternum extend as pairs of flexible rods on the posterior half of this segment, which forms an eversible sheath for the 8th segment, when the latter is retracted within the oviscape: 8th segment elongate ('the main shaft of the ovipositor', J. F. McAlpine 1962b), with its tergum and sternum both modified to form pairs of slender rods.

Both McAlpine and Morge have interpreted the eversible sheath in Lonchaeidae as representing the 8th segment, and the shaft of the ovipositor as representing the 9th segment. I think this incorrect, as the female gonopore is situated on the posterior margin of the 8th segment in Diptera. The probable explanation of this discrepancy in segment numbering is that the eversible sheath does not constitute a separate segment.

The above characterization refers to the condition shown by the Lonchaeidae. In Cryptochetidae the 7th and 8th segments are membranous. Thorpe (1934) describes the 7th segment as 'membranous with incompletely defined tergum and sternum and... covered with minute backwardly projecting spines'. The evolution of internal parasitism by the Cryptochetidae has doubtless been accompanied by profound changes in the structure of the ovipositor. It seems reasonable to suppose that the condition in Cryptochetidae may have been derived from the condition in Lonchaeidae through reduced sclerotization and modification of the hairs which the 7th segment bears in Lonchaeidae to form the 'spines'.

(5) Dorsal sclerites of 7th and 8th abdominal segments reduced (♂) (figs. 18 and 22).

In Lonchaeidae a narrow dorsal pregenital sclerite lies between the 6th tergum and the periandrium. This sclerite is free in *Dasiops*, but is partly fused with the 6th and 7th sterna in other genera (J. F. McAlpine 1962b). It is not clear whether this sclerite represents the 7th tergum or a fusion of the 7th tergum and 8th sternum, as implied by McAlpine (1962b). In Cryptochetidae no such sclerite is present, the last dorsal sclerite before the genital segment being the large 6th tergum.

(6) Aedeagal apodeme (♂) linked with hypandrium (figs. 21 and 23).

There is some variation in the form of the aedeagal apodeme and hypandrium among the Lonchaeoidea, and I am not able to define the groundplan condition precisely. However, some form of fusion or connection between the hypandrium and aedeagal apodeme is common to all members of this group. In Cryptochetidae the connection takes the form of a long process ('Medianfortsatz der Tragplatte', Hennig 1937b). In all Lonchaeidae the aedeagal apodeme is partly fused with the hypandrium (at most with its anterior end free in *Dasiops*).

Combined with their characteristic apomorphous features (as listed above), the Lonchaeoidea show several plesiomorphous conditions which suggest that they early became separated from the other groups of Schizophora. In the males of many Lonchaeidae a complete series of ventral postabdominal sclerites is retained (fig. 20), including the 7th sternum (which retains *sensilla trichodea* like the preceding sterna) and the inverted 8th tergum (called 'secondary sclerite' by McAlpine 1962b). Outside the Lonchaeidae the only Schizophora which retain the latter sclerite are certain Muscoidea. The type of aedeagus shown by the

88

Lonchaeoidea seems hardly different from that shown by the Platypezidae, and I consider it to be close to the groundplan condition for the Schizophora. In both Lonchaeoidea and Platypezidae the aedeagus is normally uniformly sclerotized, expanded at its base but becoming slender and upcurved along most of its length. The aedeagus is directed more or less posteriorly in relation to the longitudinal axis of the insect, and cannot be 'swung' against the aedeagal apodeme to any significant degree. In some Lonchaeidae the aedeagus is divided into two sections through a break in the sclerotization, but this seems to be an apomorphous condition (as stated by MORGE 1963), not attributable to the groundplan of the family. The extremely short aedeagus of some groups of *Dasiops* is probably also apomorphous. Consistent with my interpretation of the early separation of the Lonchaeoidea from other Schizophora, I accept the uniformly dark coloration and dark halteres of this group (as in Pipunculidae, Platypezidae and Hypocera) as a plesiomorphous condition retained from the groundplan of the Schizophora.

Particularly interesting is the prevalence of synorchesic swarming among male Lonchaeidae. Correlated with this behaviour are certain structural peculiarities such as enlarged male eyes and strong anal lobes on the wings. This subject has been ably reviewed by J. F. MCALPINE & MUNROE (1968c), who conclude that 'the synorchesic swarming habit in Lonchaeidae is believed to be evidence of the primitive nature of this family in relation to most other Acalypteratae'. MC-ALPINE & MUNROE's statement implies that the swarming habit and the associated structural peculiarities were derived directly from ancestors of the Schizophora and not secondarily acquired after diversification of the Schizophora had begun. This interpretation seems to me very probable, and accords well with the early separation of Lonchaeoidea suggested by the morphological evidence. I therefore postulate that the structural features associated with synorchesic swarming in Lonchaeidae, such as enlarged male eyes and large anal lobes on the wing, are groundplan conditions for the Schizophora which have been retained by this family. Among other superfamilies of Schizophora pronounced sexual dimorphism in the size of the eyes occurs only among three groups of Muscoidea, the Calyptratae, the Tanypezoinea and *Neomaorina* (Tephritidae family-group); and also in the lauxaniid genus *Holopticander*, described by HENNIG (1968a). The view that synorchesic male swarming was a plesiotypic behaviour of the Cyclorrhapha is further supported by the prevalence of this behaviour among the Platypezidae, which seem to be the least modified of all Recent Cyclorrhapha (especially in respect of the structure of the male postabdomen and the partial reversibility of the circumversion process). The eyes are always contiguous in male Platypezidae (KESSEL & MAGGIONCALDA 1968a).

B. 1. Family Lonchaeidae

Sources of information

Detailed comparative information on the male postabdomen and genitalia of

Lonchaeidae has been given by J. F. McALPINE (1962b) and MORGE (1963). I studied *Dasiops passifloris* McALPINE and *Lonchaea polita* SAY.

Description of male postabdomen and genitalia (figs. 18-21)

Postabdominal spiracles symmetrically or slightly asymmetrically situated: 6th spiracles in membrane: 7th spiracles variable in position, either both in membrane (e.g. *Dasiops*), or one or both within 7th sternum or within dorsal pregenital sclerite.

Postabdomen weakly or strongly asymmetrical. 6th tergum represented by a small discrete sclerite only in some *Protearomyia* and *Dasiops* species; in other Dasiopinae broadly fused with 5th tergum, or not differentiated from the latter; no trace of 6th tergum reported in Lonchaeinae other than *Protearomyia*. 6th sternum large and symmetrical in Dasiopinae, shorter and often slightly asymmetrical in Lonchaeinae. 7th sternum always asymmetrical, rather weakly so in Dasiopinae, articulated with dorsal pregenital sclerite on left side: a possible composite origin (tergum 7 + sternum 8) for this sclerite is indicated by the free lateral extremities shown by some *Dasiops* species. 8th tergum vestige (called 'secondary sclerite' by McALPINE) usually present (in ventral position).

Periandrium bearing discrete telomeres: the latter more or less enclosed within periandrium (concealed in lateral view) in most Lonchaeinae, but exposed in lateral view in Dasiopinae. Cerci usually discrete, but partly fused in *Protearomyia* and *Chaetolonchaea*. Hypandrial arms fused to form bridge above base of aedeagus: in Lonchaeinae the hypandrium bears a pair of internally directed processes (? pregonites), but these are scarcely developed in Dasiopinae. Aedeagal apodeme short and laterally compressed, fused with hypandrium (at most with its anterior end free in *Dasiops*). Aedeagus of simple tubular structure, uniformly sclerotized throughout or sometimes divided into two sections, very variable in length (short in Dasiopinae but long and upcurved in many Lonchaeinae): at rest remaining posteriorly or posteroventrally directed, not able to be swung through wide arc against aedeagal apodeme. Small postgonites present. Fan-shaped ejaculatory apodeme present.

Characterization and discussion

Many of the conditions shown by the Lonchaeidae are more plesiomorphous than those shown by the highly modified Cryptochetidae. This poses the question of whether the Cryptochetidae may be more closely related to some genera of Lonchaeidae than to others (which would mean that the latter form a paraphyletic group).

The very long aedeagus of the Cryptochetidae could perhaps be considered synapomorphous with the long aedeagi of species of some lonchaeid genera (especially some *Silba* species). However the length of the aedeagus shows much variation between closely related species both among the Lonchaeidae and

90

among the Platypezidae (which I interpret as showing a similar type of aedeagus through symplesiomorphy, as previously discussed). Therefore I consider the interpretation of resemblances in the length of the aedeagus as problematical. In particular I do not think it should be assumed that the very short aedeagus of some groups of *Dasiops* necessarily represents a more plesiomorphous condition than the conditions shown by other genera of Lonchaeidae and by the Cryptochetidae. HENNIG (1967: 10) has expressed a similar opinion on the latter point.

All Lonchaeidae show a more apomorphous condition than the Cryptochetidae in respect of the following character.

(1)   6th tergum (♂) reduced, less than one quarter of length of 5th tergum.

This groundplan condition is shown by some species of *Dasiops* (fig. 18). In most Lonchaeidae the reduced 6th tergum is fused with the 5th tergum or entirely lost. In Cryptochetidae the 6th tergum is large and well developed, a condition which is clearly plesiomorphous in relation to that shown by the Lonchaeidae. The Baltic amber fossil *Phanerochaetum* also appears to have a large 6th tergum (HENNIG 1965b, fig. 263), although HENNIG was unfortunately not able to discern the structure of the postabdomen in detail.

On the evidence of the above character it seems reasonable to accept that the Lonchaeidae probably represent a monophyletic group, the sister-group of the Cryptochetidae.

B. 2.   *Family Cryptochetidae*

Sources of information

My treatment of the male postabdomen and genitalia is based on a preparation of *Cryptochetum nipponense* Tokunaga and the information given by HENNIG (1937b).

Description of male postabdomen and genitalia (figs. 22-23)

Only one pair of abdominal spiracles present (probably the 6th pair).

Postabdomen fully symmetrical. 6th tergum fully as long as 5th tergum. No other postabdominal sclerites present. Last sternum before the genital segment probably the 5th sternum.

Telomeres not discrete, represented by setose lobes of periandrium: cerci discrete. Hypandrium with setose lobes extending posterior to base of aedeagus. Aedeagal apodeme cleft posteriorly, connected with hypandrium anteriorly by long ventral process ('x' on fig. 23). Aedeagus long, slender and upcurved, of simple tubular structure, uniformly sclerotized throughout, at its base passing above level of aedeagal apodeme through the cleft in the latter; at rest remaining posteriorly directed, not able to be swung through wide arc against aedeagal apodeme. Postgonites absent. Ejaculatory apodeme small and slender.

Characterization and discussion

The Cryptochetidae show an extensive suite of autapomorphous conditions, which I review summarily, as follows.
(1) Arista lost: 2nd antennal article without cleft or suture.

This characterization applies to all Recent Cryptochetidae. The Baltic amber fossil *Phanerochaetum* has similar antennae to those of many Lonchaeidae (with subbasal arista and a suture on the 2nd article) (HENNIG 1965b).

(2) Frontal bristles scarcely differentiated (see HENNIG 1965b, fig. 267).

In Cryptochetidae the frons is largely clothed with short setulae. It is scarcely possible to determine the homologues of the fronto-orbital, ocellar, vertical and postvertical bristles of other Schizophora.

(3) Male eyes not enlarged.

The strong sexual dimorphism in the size of the eyes retained by the Lonchaeidae is not reported for the Cryptochetidae. Probably this reflects loss of the synorchesic swarming habit. However I have not found any information on the behaviour of these insects in the field.

(4) Costa with two breaks.

In most Cryptochetidae there is a costal break shortly beyond the humeral cross-vein, in addition to the break at the end of the subcosta. However in *Cryptochetum melanum* GHESQUIÈRE (1942) only the latter break is present. I think that the loss of the second break in this case is probably secondary, since both breaks are present in the Baltic amber fossil *Phanerochaetum*.

(5) Base of m_{3+4} (= 'tb', lower cross-vein) weak or absent (see note by HENNIG 1965b: 173-174).
(6) Anal vein withdrawn close to margin of anal lobe of wing (HENNIG 1958, 1965b).
(7) All abdominal spiracles lost except 6th pair (both sexes).

HENNIG (1958) was only able to establish the presence of a single pair of spiracles in the female abdomen, and I find a similar condition in the male before me (fig. 22). Such an extensive loss of abdominal spiracles occurs among other adult Schizophora only in *Eginia* (Calyptratae). From the position of the retained pair of spiracles in the male before me it is difficult to judge whether they represent the 6th or 7th pair. But since the retained pair of spiracles in the female is clearly the 6th pair, I postulate that probably the same pair is retained in the male.

(8) Female postabdomen with segments 7 and 8 membranous.

As previously stated, I think that the highly modified condition of Cryptochetidae may have been derived from the type of ovipositor shown by Lonchaeidae. THORPE's (1934) account of the structure of the ovipositor of Cryptochetidae is the most detailed available. However his segment numbering was incorrect, as already pointed out by HENNIG (1958).

(9) Two non-sclerotized spermathecae ($\female$) (see STURTEVANT 1926).

(10) All postabdominal sclerites ($\male$) before genital segment lost, except the large 6th tergum (fig. 22).

Whether such an extensive reduction has occurred in all Cryptochetidae requires confirmation, as the structure of the postabdomen has not been described for many species.

(11) Periandrium ($\male$) without differentiated telomeres.

(12) Postgonites lost ($\male$).

(13) Larvae profoundly modified, developing as internal parasitoids in Coccidae (Hemiptera).

Studies on the larvae of Cryptochetidae have revealed many very remarkable modifications in their structure and physiology. A summary is given by HENNIG (1952a).

A. 2. Superfamily Lauxanioidea

My delimitation of the Lauxanioidea follows that of HENNIG (1958), except that I exclude the Periscelididae (here referred to the Nothyboidea). HENNIG (in press) now also excludes the latter family from the Lauxanioidea. The Lauxanioidea in my present sense consists of two well-characterized subordinate groups, here accorded prefamily rank (Lauxanioinea and Chamaemyioinea). I am reasonably confident that each of these groups is monophyletic. But their association in the Lauxanioidea rests on weak characterization and may possibly be incorrect. HENNIG (in press) also alludes to this possibility, stating that only the Lauxaniidae and Celyphidae belong with certainty to the Lauxanioidea. My retention of the group Lauxanioidea in a wide sense is therefore provisional and in need of confirmation (or refutation) through further studies.

If the Lauxanioidea in my present sense are monophyletic, the group may be characterized by the following apomorphous conditions with respect to the groundplan of the Schizophora.

(1) Anal vein shortened, not reaching wing margin (HENNIG 1958).

(2) Postverticals *(pvt)* convergent.

The only group with long divergent postverticals included by HENNIG (1958) in the Lauxanioidea are the Periscelididae, which I refer to the Nothyboidea.

(3) Only one dorsal sclerite (probably 7th tergum) between 6th tergum and periandrium ($\male$) (figs. 26 and 30).

The homology of this 7th dorsal sclerite is not clear. It may represent the 7th tergum, the inverted 8th sternum or a fusion of both (as interpreted by HENNIG 1958). I think the first alternative the most probable, but this question may not be clarified until information is available on the ontogeny of the sclerites or on the morphology of intersexes (see section 6.1).

(4) 8th tergum vestige ($\male$) lost.

93

HENNIG (1938e) labelled the last ventral sclerite befor the genital segment in *Parochthiphila* (Chamaemyiidae) as 'sternite 8', but this is surely the 7th sternum.

The Lauxanioidea exhibit several features in their groundplan which are thought to be plesiomorphous. The costa was unbroken and the subcosta well developed. A complete series of ventral sterna up to the 7th sternum was present in the male, as retained by many Chamaemyiidae (fig. 28). Oral vibrissae were not differentiated. The type of aedeagus was probably similar to that of the Lonchaeoidea and Platypezidae, namely a more or less uniformly sclerotized structure, upcurved along most of its length and always directed more or less posteriorly in relation to the longitudinal axis of the insect. No mechanism has been developed in any Lauxanioidea for swinging the aedeagus through a wide arc against the aedeagal apodeme (as is characteristic of the Muscoidea and Nothyboidea).

B. 3. *Prefamily Lauxanioinea, family Lauxaniidae (including Celyphidae)*

Sources of information

Detailed information on the male postabdomen and genitalia is only available in the literature for a few genera, as given by HAHN (1929), HENNIG (1948a) and TENORIO (1969). I examined preparations of the following: *Celyphus obtectus* DALMAN; *Camptoprosopella borealis* SHEWELL; *Homoneura severini* SHEWELL; *Lauxania cylindricornis* (F.); *Meiosimyza platycephala* (LOEW); *Minettia lupulina* (F.); *Lyciella annulata* (MELANDER); *Lycia rorida* (FALLÉN); *Sapromyza cyclops* MELANDER; *Prorhaphochaeta inusta* (MEIGEN).

Description of male postabdomen and genitalia (figs. 24-27)

6th and 7th abdominal spiracles symmetrically situated: 6th spiracles always in membrane: 7th spiracles in membrane except in forms with 7th tergum and sternum fused to form complete band around postabdomen (*Celyphus, Minettia, Prorhaphochaeta*).

Postabdomen fully symmetrical. 6th tergum large, of similar length to 5th tergum. 7th tergum well developed as dorsal sclerite, but shorter than preceding terga. 6th sternum developed as short bare plate (e.g. *Lauxania*), represented by two fragments (*Celyphus*) or absent (e.g. *Camptoprosopella, Sapromyza*). 7th sternum either absent or fused with 7th tergum to form complete band of sclerotization around postabdomen *(Celyphus, Minettia, Prorhaphochaeta)*. 8th tergum and sternum absent.

Periandrium bearing telomeres of variable shape which are usually discrete: in *Celyphus* periandrium divided dorsally into two halves, but complete in all other genera examined. Transverse interparameral sclerite linking the telomeres

94

sometimes developed (e.g. *Celyphus, Minettia*), but in many genera no inter-parameral sclerotization evident. Cerci small, discrete. Hypandrium usually with long posterior processes (? pregonites), but these are absent in *Prorhapho-chaeta:* in *Celyphus* hypandrium consisting of two halves separate centrally, bearing antler-like processes (? pregonites). Aedeagal apodeme usually present (rodlike and free from hypandrium), but apparently absent in *Sapromyza* and *Prorhaphochaeta*. Aedeagus, when well developed (e.g. *Camptoprosopella*), usually of simple tubular structure, always remaining posteriorly directed when at rest (not able to be swung against aedeagal apodeme through wide arc): however the aedeagus is small or apparently absent in some groups (e.g. *Cely-phus, Homoneura*). Postgonites absent. Small ejaculatory apodeme usually present, but absent in *Celyphus*.

Characterization and discussion

The Lauxaniidae are characterized in their groundplan by the following apomorphous conditions with respect to the postulated groundplan of the Lauxanioidea.

(1) Male accessory glands ('paragonia') repeatedly branched, forming dense tangle (STURTEVANT 1926).

STURTEVANT reports this condition in all the genera of Lauxaniidae which he examined, namely *Calliope, Camptoprosopella, Lauxania, Minettia, Sapromyza* and *Steganolauxania*. He stated that the condition was diagnostic of the Lauxaniidae and not shown by members of any of the other groups (including Chamaemyiidae) which he had examined. STURTEVANT's observations indicate that a major structural modification of the male reproductive system has occurred in Lauxaniidae, and I am convinced by this evidence that the Lauxaniidae are a monophyletic group.

(2) Postgonites lost (♂).

Additional autapomorphous conditions may possibly be found in the structure of the female genital segment, since this shows unusual internal sclerotization and modification of the 8th sternum in the species which I have studied *(Sapromyza, Lauxania)*. Further studies are needed to clarify this point. HENNIG (1958) has suggested as an apomorphous condition of the Lauxaniidae that in the female postabdomen the 7th spiracles lie within the tergum; but this is probably not a groundplan condition of the group, since at least in *Sapromyza* these spiracles lie in membrane (as indicated by HAHN 1929 and confirmed by my own observation).

Plesiomorphous conditions retained in the groundplan of the Lauxaniidae include: (1) a well-developed subcosta which is well separated from r_1 at its apex; (2) a large unmodified 6th tergum in the male; (3) a rod-like aedeagal apodeme (♂), free from the hypandrium (fig. 27); and (4) three spermathecae (♀). It is doubtful whether the condition of the aedeagus in the groundplan of the Lauxaniidae can be characterized as apomorphous in any respect. Although some genera show characteristically modified types of aedeagus (including even complete reduction in *Homoneura laticosta* THOMSON according to HENNIG

1948a), a fairly slender and upcurved aedeagus is shown, for instance, by *Camptoprosopella* (fig. 27).

I propose to include the 'Celyphidae' *(Celyphus* and *Paracelyphus)* within the Lauxaniidae, rather than recognizing this group as a distinct family as did HENNIG (1958). These insects are characterized by several striking modifications, including an enlarged scutellum which shields the wings and abdomen and a tendency to reduction of the chaetotaxy of the head and thorax. However the condition of the postabdomen of both sexes suggests that they may be more closely related to some genera of 'Lauxaniidae' than to others. Among the genera which I have examined, the closest resemblance to *Celyphus* is shown by *Minettia* and *Prorhaphochaeta.* In all three of these genera the 7th sternum is fused with the last dorsal pregenital sclerite. The sclerites of the 7th abdominal segment in the female of *Celyphus* form a complete ring (HENNIG 1958, fig. 163), a condition also possibly due to synapomorphy with some genera of Lauxaniidae. Thus there is a strong possibility that the primary subdivision of the Lauxanioinea (Lauxaniidae *sensu lato*) should not be into 'Lauxaniidae' and Celyphidae, but on some other basis which remains to be clarified. For the present I prefer to use the name Lauxaniidae in a wide sense. The ranking of the Lauxaniidae *s.l.* at the B level (equivalent to a prefamily) in the sequence of subordination presented in this work gives me no grounds for objecting to the assignment of family rank to its major subordinate groups, at least in terms of the current ranking of groups of Schizophora. But the relationships between the included genera have been so little investigated that I see no firm basis for making such a subdivision at this time.

The work of TENORIO (1969) strengthens my opinion that the Celyphidae are a subordinate group of the Lauxaniidae *s.l.* She reports that the genus *Idiocelyphus* shows the modifications characteristic of the Celyphidae to a less pronounced degree than the other genera, and thus 'emphasizes the close relationship of the celyphids to the family Lauxaniidae'.

B. 4. *Prefamily Chamaemyioinea*

I include in this group the families Eurychoromyiidae and Chamaemyiidae. The view that the Chamaemyioinea are monophyletic is supported by the following condition which I interpret as an autapomorphous groundplan condition of the group.

(1) Four spermathecae (♀) (STURTEVANT 1926, HENNIG 1958).

While a reduction in the number of spermathecae from three to two has evidently occurred independently in many lineages of Schizophora, an increase in this number to four has occurred much more rarely. Outside the Chamaemyioinea the presence of four spermathecae has been reported only for two species of *Suillia* (Heleomyzidae) (HENNIG 1958), some genera of Conopidae (STURTEVANT 1925, HENNIG 1966b), *Seioptera* (Tephritidae *s.l.*) (STURTEVANT 1925) and *Salticella* (Sciomyzidae) (STEYSKAL 1965). All these genera are clearly referable to families for which the presence of three or two spermathecae is the ground-plan condition, and the evolution of the four-spermathecae condition was evidently independent in each case. However, the presence of four spermathecae in the Eurychoromyiidae gives grounds for supposing that this group is closely related to the Chamaemyiidae, as suggested by HENNIG (1958), since the condition seems to belong to the groundplan of both families. Other characters of the Eurychoromyiidae are readily compatible with this interpretation, but offer serious obstacles to classifying the group in the Sciomyzoinea (see below).

C. 1. *Family Eurychoromyiidae*

Sources of information

J. F. MCALPINE (1968b) has published a detailed description of the abdomen and genitalia of the male lectotype of *Eurychoromyia mallea* HENDEL, the only known species of this family. I have not seen any specimen.

Description of male postabdomen and genitalia

6th and 7th abdominal spiracles symmetrically situated in membrane near margins of respective terga.

Postabdomen symmetrical except for 6th sternum. 6th tergum large, of similar length to 5th tergum. 6th sternum without setae, 'rather amorphous and apparently highly asymmetrical' (MCALPINE). 7th tergum developed as narrow dorsal sclerite (shorter than preceding terga), with bifurcate lateral extremities. 7th sternum absent. 8th sternum absent, unless perhaps represented by a small strap-like sclerite attached to right anterodorsal margin of periandrium. 8th tergum absent.

Periandrium bearing long discrete telomeres. Cerci discrete. Hypandrium quadrilateral, without lobes. Aedeagal apodeme scarcely differentiated from base of aedeagus. Aedeagus membranous ventrally, heavily sclerotized dorsally, with two anteroventrally curved spines at its apex; remaining posteriorly directed when at rest. Postgonites represented by 'simple, elongate, somewhat dumbbell-shaped sclerite', extending 'from each posterolateral corner of the hypandrium to the posterolateral base of the aedeagus... this structure does not project above the surrounding membrane' (MCALPINE). Small ejaculatory apodeme present.

Characterization and discussion

J. F. MCALPINE (1968b) has compared *Eurychoromyia* with families both of Lauxanioidea and Sciomyzoinea (Muscoidea), in the belief that these groups are closely related. His use of the name 'Sciomyzoidea' in that paper refers to both groups together. On the basis of my present analysis I conclude that these groups are not so closely related as McAlpine thought. The families of Sciomy-zoinea show (at least in their groundplan) the typical muscoid morphology of the postabdomen (with asymmetrical development of the sclerites of the 7th segment) and a well-developed mechanism for swinging the aedeagus, which contains clearly defined sclerites in its walls, against the aedeagal apodeme. These conditions are not shown by *Eurychoromyia*, and I therefore doubt whether the genus should be classified in the Sciomyzoinea or any other group of Muscoidea. On the other hand no such problem of compatibility arises if the genus is classified in the Lauxanioidea. *Eurychoromyia* clearly cannot be includ-ed in the Chamaemyiidae unless the definition of the latter is widened, because in respect of at least two character sequences it exhibits more plesiomorphous conditions than those found in the groundplan of the Chamaemyiidae, as fol-lows: (1) the subcosta is well developed and separated from r_1 at its apex, and (2) the telomeres are large and discrete (not fused with the periandrium). Thus retention of the Eurychoromyiidae as a separate family, in accordance with McAlpine's view, seems to me well justified. The apomorphous conditions shown by Eurychoromyiidae (briefly summarized) with respect to the ground-plan of the Chamaemyioinea are as follows.

(1) 7th abdominal tergum and sternum (♀) fused to form a ring which in-cludes the 7th pair of spiracles (Hennig 1958, fig. 165).

(2) Head transverse, with chaetotaxy much reduced (Hennig 1958, fig. 151).

(3) 6th sternum (♂) asymmetrical (McAlpine 1968b).

(4) 7th sternum (♂) lost.

(5) Aedeagal apodeme (♂) reduced; aedeagus with two anteroventrally curved spines at apex (McAlpine 1968b).

C. 2. Family Chamaemyiidae

Sources of information

Some information on the male postabdomen and genitalia of Chamaemyiidae was given by Hennig (1938e) and J. F. McAlpine (1960). I examined prepara-tions of the following: – *Chamaemyia juncorum* (Fallén); *Parochthiphila spectabilis* (Loew); *Acrometopia wahlbergi* (Zetterstedt).

Description of male postabdomen and genitalia (figs. 28-30)

6th and 7th abdominal spiracles symmetrically situated: 6th pair always in membrane: 7th pair usually in membrane, but at edge of sclerotized area in *Acrometopia* (in which the 7th tergum is partially fused with the 6th sternum).

Postabdomen fully symmetrical (e.g. *Chamaemyia*) or with slightly asymme-trical development of 6th and 7th sterna. 6th tergum discrete, but much shorter than preceding terga. 7th tergum usually present as narrow dorsal sclerite, but absent in *Leucopis*. 6th sternum variably developed, clearly defined and discrete in *Chamaemyia*, fused with 7th tergum on left side in *Acrometopia*, ill-defined in *Parochthiphila*. 7th sternum variably developed, rather large and well defined in *Chamaemyia*, narrower and ill-defined in *Acrometopia* and *Parochthiphila*. 8th tergum and sternum absent.

Periandrium without discrete telomeres: transverse interparameral sclerite present below proctiger (with pair of conspicuous processes in *Chamaemyia*): cerci large and discrete. Hypandrium bearing posterior processes (? pregonites), sometimes fused with base of postgonites (e.g. *Chamaemyia*): hypandrial arms usually fused posteriorly to form bridge above base of aedeagus (but not in *Acrometopia*). Aedeagal apodeme short, linked to hypandrium by ventral process. Aedeagus of simple tubular structure, upcurved, more or less uniformly sclerotized; at rest remaining posteriorly directed, not able to be swung through wide arc against aedeagal apodeme. Postgonites sometimes more or less discrete (e.g. *Parochthiphila* and some *Leucopis* spp.), but sometimes fused basally with hypandrium (e.g. some *Chamaemyia* spp.). Small ejaculatory apodeme present.

Characterization and discussion

I do not include *Cremifania* in the Chamaemyiidae, for reasons which are discussed below in my treatment of this genus (here considered as representing a distinct family of Sciomyzoinea, the Cremifaniidae). Thus the Chamaemyiidae in the sense of this work are equivalent to the Chamaemyiinae of J. F. MC-ALPINE (1963). Both HENNIG and MCALPINE now agree that the genus *Sciochthis* was misplaced in this family. HENNIG (in press) refers it to the 'Pallopteridae' (Tephritidae family-group) on MCALPINE's authority.

The Chamaemyiidae are characterized by the following apomorphous conditions with respect to the groundplan of the Chamaemyioinea.
(1) Vein r_1 approximated to sc, becoming contiguous with this before its apex (HENNIG 1958).
(2) Proscutellum present (given by MCALPINE 1963 as a character of 'Chamaemyiinae').
(3) Telomeres ($\male$) not differentiated from periandrium (fig. 30).

This condition was already stated in different phraseology by HENNIG (1958) and MCALPINE (1963) (for 'Chamaemyiinae').

(4) 6th tergum ($\male$) shortened, less than half as long as 5th tergum (fig. 30).
(5) Aedeagal apodeme ($\male$) linked to hypandrium by ventral process (fig. 29).

HENNIG (1938e) discussed this character. He distinguished two conditions found among the Chamaemyiidae. In *Acrometopia* the apodeme is rod-like, linked to the hypandrium by an anterior ventral process. In the other Chamaemyiidae examined the apodeme is more or less wedge-shaped and it is sometimes difficult to draw any clear distinction between the apodeme proper and the ventral process.

The development of the 7th sternum in some male Chamaemyiidae represents the most plesiomorphous condition of this sclerite found among the Lauxanioidea. In *Chamaemyia*, for instance, this sternum is large and symmetrical (fig. 28).

A. 3. Superfamily Drosophiloidea

My delimitation of the Drosophiloidea follows HENNIG (1958). The included families are: Drosophilidae, Camillidae, Curtonotidae, Campichoetidae and Ephydridae. SPEIGHT (1969) and COLLESS & D. K. MCALPINE (1970) have proposed to widen the limits of this group to include the Chloropidae and various other families. I do not see sufficient justification for these proposals. The structure of the male postabdomen of the Drosophiloidea (*sensu* HENNIG) is very different from that of the Chloropidae and all other families which I refer to the Muscoidea in this work, and HENNIG (in press) presents additional evidence for the monophyly of the Drosophiloidea in his sense from study of the structure of the antennae. The Drosophiloidea in HENNIG's sense seem to me to be one of the most surely grounded monophyletic groups of the Schizophora, characterized in their groundplan by the following apomorphous conditions with respect to the groundplan of the Schizophora.

(1)   Proclinate fronto-orbital bristle (*ors*) present.

   The variation in this character has been discussed in detail by HENNIG (1958 and 1965b: 190).

(2)   2nd antennal article with longitudinal cleft or suture (see discussion by HENNIG, in press).

(3)   Aedeagus (♂) short.

   The type of aedeagus in the groundplan of the Drosophiloidea cannot be defined precisely. However, since nearly all Drosophiloidea (except a few groups of Drosophilidae) show a very short aedeagus, the groundplan condition for the Drosophiloidea was probably apomorphous in this respect in comparison with the elongate condition which I ascribe to the groundplan of the Schizophora (as retained by many Lonchaeoidea and Lauxanioidea). In Drosophiloidea no mechanism has been developed for swinging the aedeagus through a wide arc against the aedeagal apodeme (contrast the Muscoidea and Nothyboidea), and the aedeagus remains posteriorly or posteroventrally directed in the rest position. The available accounts of the hypopygial musculature of *Drosophila* (GLEICHAUF 1936; FERRIS in DEMEREC 1950) do not indicate any muscle equivalent to the '*musculus phallapodemalis basiepiphallicus*' (M 36) of SALZER (1968), which in *Calliphora* swings the aedeagus out of the genital pouch into the copulatory position. The sclerotization of the aedeagus in Drosophiloidea either appears uniform or a pair of lateral sclerites is more or less differentiated. I suggest that the term paraphalli may be applied to such lateral sclerites in the walls of the aedeagus, although it is not clear whether they are homologous with the paraphalli of Muscoidea. Whether or not there is a phallophore in *Drosophila* requires clarification. According to GLEICHAUF's account the aedeagus lacks extrinsic musculature, its retraction being effected by the '*depressores penis*' which insert on the underside of the aedeagal apodeme. However, FERRIS (in DEMEREC 1950) states that these muscles ('retractor muscles of the aedeagus') insert upon the 'base of the aedeagus itself'. This contradiction between FERRIS' and GLEICHAUF's accounts indicates that the hypopygial musculature of *Drosophila* needs further investigation.

(4)   8th sternum (♂) much reduced; 8th tergum lost.

   The identity of all abdominal sclerites as far as the 7th segment in *Drosophila* has been definitely established from study of intersexes (see LAUGÉ 1968). The only remaining sclerites of the male which may belong to the 8th segment are the pair of small dorsolateral sclerites which lie between syntergum (6 + 7) and the periandrium (fig. 33). These can hardly be interpreted as vestiges of the 8th tergum (the most frequent interpretention in the literature on *Drosophila*), since this would be expected in a ventral position in normal males with rotated terminalia. Either they are vestiges of the inverted 8th sternum, or they are neomorphous (secondary) sclerites. The former explanation is here provisionally accepted. These sclerites have so far only been reported for Drosophilidae, where they occur widely in *Chymomyza*, *Scaptomyza* and some groups of *Drosophila*. They are particularly well developed in *Drosophila picta* Zetterstedt (see Tsacas 1969, fig. 1). These sclerites are one of the first male structures to disappear in the morphological sequence from male-type to female-type intersexes which Laugé (1968) has presented. They are retained only in individuals with almost normal male morphology, including extensive rotation, and are not found in individuals in which female structures of the 8th segment develop. The morphology of intersexes is thus compatible with the view that these sclerites belong to the 8th segment, although their homology with any particular sclerite of the female has not been demonstrated conclusively, since no series of intermediate structures has been obtained.

Some authors (such as Gleichauf 1936 and Salles 1948) have interpreted the 'basal phragma' on the anterior margin of the periandrium in *Drosophila* as a fused 8th tergum. But I see no evidence that this is other than an extension of the periandrium. According to Ferris (in Demerec 1950) muscles inserting on the telomeres originate on this phragma. These extrinsic muscles of the telomeres are part of the normal musculature of the periandrium in Cyclorrhapha.

The effect of the reduction of the sclerites of the 8th segment in males of Drosophiloidea has been to increase the sexual dimorphism of the postabdomen, since the male hypopygial structures all belong to the 9th segment, while the female external genitalia belong to the 8th segment. Thus the genital segment of each sex bears no sclerites in the other sex, with the probable exception of the small dorsolateral sclerites of some Drosophilidae (as noted above). Only in intersexes do large sclerites of the 8th and 9th segments occur together in the same individual. The reduction of the 8th segment in male Drosophiloidea may be strongly contrasted with the structure of male Muscoidea; in the latter the 8th segment bears a large inverted sternum, which provides attachment for one of the muscles involved in the copulatory mechanism (see Salzer 1968).

Some of the details of the above argument have so far been confirmed for *Drosophila* only, but the structure of other Drosophiloidea indicates that near or complete loss of the sclerites of the 8th segment has occurred among all males of the group.

(5)   Costa broken at end of subcosta (or near end of r_1 if the subcosta has been reduced).
(6)   Subcosta closely approximated to r_1 distally.

The condition of the subcosta in the Curtonotidae appears to be the most plesiomorphous condition found among the Drosophiloidea, as indicated by Hennig (1958). In other families the subcosta becomes faded distally or fused with vein r_1.

(7)   Anal vein much shortened, not reaching wing margin.

Hennig (1958) concluded that the long apparent 'anal vein' of some Curtonotidae is a secondary sclerotization.

(8)   Postverticals convergent (Hennig 1958).
(9)   Vibrissae present (see Hennig 1958).
(10) Two spermathecae ($\female$).

Two spermathecae have been reported as the normal condition for Drosophilidae (many references), Campichoetidae (Hennig 1958, J. F. McAlpine 1962a) and Curtonotidae (Sturtevant 1926). There are two spermathecal ducts in the Ephydridae (Sturtevant 1926), although the spermathecae themselves are rudimentary in this family.

An unmodified (plesiomorphous) condition of the 7th abdominal segment ($\male$) seems ascribable to the groundplan of the Drosophiloidea, with both the 7th

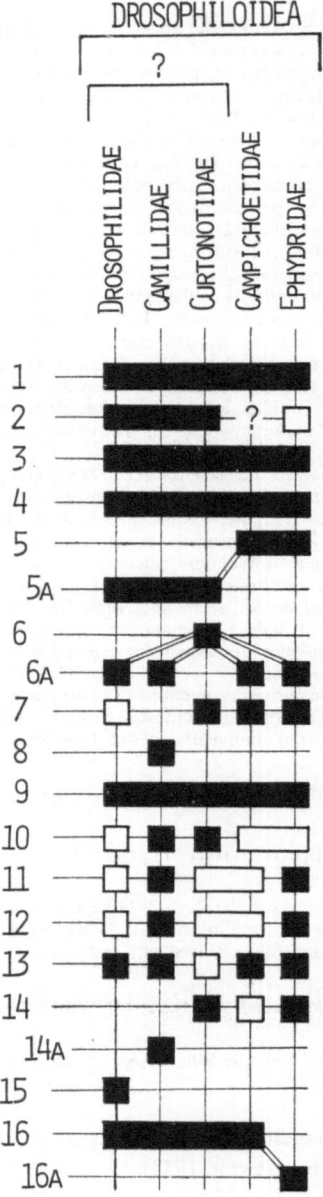

Fig. 15. Distribution of some character states among the families of Drosophiloidea. The black rectangles indicate groundplan conditions interpreted as apomorphous with respect to the groundplan of the Schizophora. Where common possession of such apomorphous conditions is interpreted as due to synapomorphy, a continuous black rectangle is shown against the families concerned. Corresponding plesiomorphous conditions are shown by the white rectangles. Further modifications of a primary apomorphous condition are shown in separate rows, with synapomorphy at the primary

level indicated by linking the rectangles. The list of characters is as follows.

Apomorphous conditions (black rectangles)	Plesiomorphous conditions (white rectangles)

Head
1. 2nd antennal article with cleft
2. 3rd antennal article with long dorso-lateral process at base — 3rd antennal article without long basal process
3. Proclinate *ors* present
4. Postverticals convergent

Wing
5. Costa broken at end of sc or r₁
5a. Costa broken twice
6. Subcosta closely approximated to r₁
6a. Subcosta faded or fused with r₁ distally
7. Anal cross-vein angulate — Anal cross-vein not angulate
8. Anal cross-vein lost — (as 7)
9. Anal vein not reaching wing margin as distinct vein
10. Lower cross-vein lost — Lower cross-vein present

Abdomen
11. 7th spiracles lost (both sexes) — 7th spiracles present in both sexes
12. 6th sternum reduced (♂) — 6th sternum well developed (♂)
13. 7th sternum reduced (♂) — 7th sternum well developed (♂)
14. 6th and 7th terga reduced, much shorter than preceding terga (♂) — 6th and 7th terga well developed (♂)
14a. 6th tergum reduced; 7th tergum lost (♂)
15. 6th and 7th terga fused (♂) — (as 14)
16. 2 spermathecae (♀)
16a. Spermathecae reduced; ventral receptacle sclerotized (♀)

tergum and 7th sternum retained as discrete sclerites in normal dorsal and ventral positions. The 7th tergum is well developed and discrete in Campichoetidae (fig. 37) and also discrete (although small) in the Curtonotidae and *Diastata* (Ephydridae). The 7th sternum is retained as a clearly defined sclerite only in Curtonotidae (fig. 39), but its *sensilla trichodea* are present also in some Drosophilidae (WHEELER 1960). In no Drosophiloidea have I found any trace of the asymmetrical reduction of the sclerites of the 7th segment characteristic of the Muscoidea. This difference in the groundplan structure of the 7th segment is important for my interpretation of how the Schizophora should be divided into superfamilies (see section 6.1).

HENNIG (in press) suggests that the Drosophiloidea may be divided into two sister-groups, the 'Ephydridea' ('Diastatidae' and Ephydridae) and 'Drosophilidea' (Camillidae, Curtonotidae and Drosophilidae). This subdivision may well be valid, but the question is complicated by the existence of the group for which I propose below the name Campichoetidae. In this group the position of the proclinate *ors* is similar to that shown in the groundplan of the Ephydridae (in which I include *Diastata*), but the modifications of the female reproductive system characteristic of the Ephydridae are lacking. Unfortunately HENNIG does not discuss any members of the Campichoetidae when proposing his sub-division. Until this point is clarified I prefer to leave open the question of how the Drosophiloidea should be divided into prefamilies, and present the classific-ation of the group as a series of five families. My delimitation of these families follows that customary in recent literature, except that my transference of *Diastata* to the Ephydridae entails proposing the new name Campichoetidae for the residue of the former 'Diastatidae'. Some of the characters on which the family classification is based are indicated on figure 15.

C. 3. *Family Drosophilidae*

Sources of information

There are numerous papers which treat the male genitalia of various species of *Drosophila*, and I do not give a comprehensive list here. I found the very detailed descriptions of SALLES (1948) particularly helpful, although I do not accept her numbering of some of the sclerites. Comparative data on the character sequen-ces shown by various groups of *Drosophila* was presented by NATER (1953). WHEELER (1960) clarified the homologies of the postabdominal sterna. LAUGÉ's (1968) paper helps clarify the homologies between male and female sclerites, since she describes and illustrates a complete series of intersexes. Finally the basic ontogenetic work of GLEICHAUF (1936) must again be mentioned.

The available information on the male postabdomen and genitalia of genera other than *Drosophila* is far from complete. Two works which should however be mentioned are HACKMAN's (1959) treatment of *Scaptomyza* (whose genitalia are very similar to those of some groups of *Drosophila*), and J. F. MCALPINE's (1968a) revision of *Paracacoxenus* in which he has also included figures of the male genitalia of species of *Cacoxenus*, *Gitonides* and *Acletotoxenus*.

I examined the following in the course of this study: – *Drosophila busckii* COQUILLETT; *Drosophila* sp. cf. *melanogaster* MEIGEN (laboratory strain); *Amiota picta* (COQUILLETT).

Description of male postabdomen and genitalia (figs. 33-34)

6th abdominal spiracles symmetrically situated in membrane below syntergum (6 + 7). 7th spiracles, when present, also symmetrically situated below this sclerite; but absent in at least some genera of Steganinae.

Postabdomen fully symmetrical. 6th and 7th terga fused, forming syntergal plate which is longer than the preceding terga: partial suture along line of fusion between these terga evident in *Paracacoxenus*, but not in Drosophilinae. 6th sternum variably developed, represented by bare or setose plate of similar size to preceding sterna in some groups of *Drosophila*. but often small or completely absent. 7th sternum absent, although its *sensilla trichodea* sometimes present. 8th sternum represented by pair of small lateral fragments in some *Drosophila*, *Chymomyza* and *Scaptomyza* spp., but usually completely absent. 8th tergum absent.

Periandrium narrow, bearing telomeres which are discrete or at least partly articulated. Cerci discrete. Interparameral sclerites absent. Hypandrium usually rounded or quadrate anteriorly, sometimes with pair of posterior processes (pregonites). Aedeagal apodeme variable in form, sometimes rodlike and free from hypandrium (e.g. *Drosophila melanogaster* group), but often linked to hypandrium by some form of anterior ventral process. Aedeagus very variable, always remaining posteriorly or posteroventrally directed when at rest (not able to be swung through wide arc against aedeagal apodeme), usually short (but elongate in *Paracacoxenus*), without complex sclerotization although usually with pair of more or less distinct lateral sclerites. In some groups of *Drosophila* unusual modifications of the external genitalia have occurred, including, for instance, fusion of the aedeagus and aedeagal apodeme, or the formation of an 'aedeagal mantle' above the base of the aedeagus from extensions of the hypandrial arms (see NATER 1953). Small postgonites usually present. Ejaculatory apodeme present (variable in shape).

Characterization and discussion

HENNIG (1965b) commented that it is both astonishing and deplorable, in view of the dominant role that many species of *Drosophila* play in the science of evolutionary genetics, that there is still no modern monograph available which treats this family as a whole from the standpoint of phylogenetic systematics. Good progress has been made with analysis of the phylogeny of *Drosophila* and its close relatives in recent papers by THROCKMORTON (1962, 1966, 1968), but the results of this work have unfortunately not been translated into revision of the genera. The 'genus' *Drosophila* as currently delimited is evidently paraphyletic or polyphyletic, since THROCKMORTON (1968) states that '*Scaptomyza* is an exgroup genus derived from one of the lineages within the genus *Drosophila*, as are most of the other genera of drosophilids that I have seen'. The granting of coordinate rank to 'exgroup lineages' is incompatible with the construction of a system of monophyletic groups, because it necessarily leads to the admission of paraphyletic groups into the system (see HENNIG 1966a).

In the present state of knowledge the characterization and delimitation of the Drosophilidae should be regarded as provisional, since characters of the male postabdomen and genitalia have not been checked in all the genera currently included in the family. Until this has been done, the suspicion will remain that heterogenous genera may be included. With this reservation I think that the Drosophilidae can be characterized by the following apomorphous conditions with respect to the groundplan of the Drosophiloidea.

(1) 3rd antennal article bearing long dorsolateral process which projects within the 2nd article.

105

HENNIG (in press) interprets this condition as due to synapomorphy between the Drosophilidae, Camillidae and Curtonotidae.

(2) 6th and 7th terga (♂) fused (fig. 33).

Studies on the morphology of intersexes in *Drosophila* have shown conclusively that the last complete dorsal sclerite before the periandrium represents the fused 6th and 7th terga. LAUGÉ's (1968) recent treatment confirms the conclusion of earlier workers (such as DOBZHANSKY & BRIDGES 1928) in this respect. J. F. MCALPINE's (1968a) figures indicate the presence of a suture between the fused terga in some species of *Paracacoxenus*.

(3) 7th sternum (♂) reduced, not developed as distinct sclerite although its pair of *sensilla trichodea* may be retained.

WHEELER (1960) drew attention to the presence of the *sensilla trichodea* of the 7th sternum in some Drosophilidae. These lie in membrane posterior to the 6th sternum. WHEELER's finding indicates that the 7th sternum has been reduced *in situ*, not fused with the 6th sternum as assumed by LAUGÉ (1968). LAUGÉ's view seems to have been based on analogy with the form of the terga (see character 2 above). The morphology of the intersexes which she describes does not indicate that any fusion of the sterna has occurred. The only family of Drosophiloidea in which the 7th sternum is developed as a distinct sclerite is the Curtonotidae.

(4) Second costal break present (shortly beyond humeral cross-vein) (HENNIG 1958).
(5) Subcosta fading distally or becoming fused with r_1, not reaching wing margin as distinct vein (HENNIG 1958).

The condition of the wing-base in the groundplan of the Drosophilidae appears more plesiomorphous than the conditions shown by other families of Drosophiloidea. The anal cell and 'anal vein' (cu_{1b} + 1a) are retained in all species, as far as I am aware; and in some groups the basal section of m_4 (= 'tb', lower cross-vein) is also retained.

C. 4. *Family Camillidae*

Sources of information

HENNIG (1958) figured the male postabdomen of *Camilla glabra* (FALLÉN). I examined a preparation of *Camilla atripes* Duda.

Description of male postabdomen and genitalia (figs. 35-36)

6th abdominal spiracles symmetrically situated within 6th tergum: 7th spiracles absent.

Postabdomen fully symmetrical. The largest abdominal segment is the 4th. 5th and 6th terga both much shorter than the large 4th tergum; in *Camilla atripes* Duda 6th tergum split into two fragments. The last well-formed setose sternum is the 5th sternum, but some ill-defined sclerotization after this possibly represents a vestige of the 6th sternum. No further sclerites between 6th segment and hypopygium.

Periandrium narrow, bearing discrete telomeres: interparameral sclerotization weak and ill-defined. Cerci discrete. Aedeagal apodeme rod-like, free from hypandrium. Aedeagus of simple tubular structure, shielded laterally by large postgonites: aedeagus remaining posteriorly directed when at rest, not able to be swung through wide arc against aedeagal apodeme. Ejaculatory apodeme not traced.

Characterization and discussion

The Camillidae show the following apomorphous conditions with respect to the groundplan of the Drosophiloidea.

(1) 3rd antennal article bearing long dorsolateral process which projects within the 2nd article (HENNIG, in press).
(2) 5th and 6th terga (♂) reduced, much shorter than 4th tergum (fig. 35).
(3) 7th tergum and 8th sternum completely lost (♂).
(4) 6th sternum (♂) reduced, represented only by small area of ill-defined sclerotization.
(5) 7th sternum (♂) lost.
(6) 7th abdominal spiracles absent in both sexes.
(7) Postabdomen (♀) largely membranous, with sclerites of 6th and 7th segments completely lost (HENNIG 1958).
(8) Anal vein (cu_{1b} + 1a) and anal cross-vein (cu_{1b}) lost, so that the anal cell is open distally.
(9) Base of m_4 (= 'tb', lower cross-vein) lost (HENNIG 1958).
(10) Second costal break present (shortly beyond humeral cross-vein) (HENNIG 1958).
(11) Subcosta fading distally, not reaching wing margin (HENNIG 1958).

HENNIG (1958) concluded that the Camillidae were the sister-group of the Ephydridae, but in his paper now in press he rejects this opinion and concludes that the Camillidae are more closely related to the Drosophilidae and Curtonotidae. The main reason for his change of opinion is the apparent synapomorphy between the Camillidae, Curtonotidae and Drosophilidae in the structure of the antennae (character 1 above). HENNIG's change of opinion seems to me well justified. *Camilla* does not show the modifications of the female reproductive system characteristic of the Ephydridae (including *Diastata*), but retains two orbicular spermathecae (J. F. MCALPINE, personal communication). Reduction of the anal cell and base of m_4 ('tb') must have occurred in the Camillidae independently of the Ephydridae, since such reduction is not shown by *Diastata*.

Many authors have included *Camilla* in the Drosophilidae, and I see nothing in the characterization given above which excludes this interpretation. However I know of no drosophilids whose abdominal morphology shows any approach to the highly modified conditions shown by both sexes of *Camilla*. In the absence of positive evidence for including *Camilla* in the Drosophilidae, I prefer to follow the classification of the genus in a separate family, thus leaving open the question of whether its closest relatives are the Drosophilidae or Curtonotidae.

C. 5. Family Curtonotidae

Sources of information

HACKMAN (1960) gave figures of the male genitalia of the African genus *Cyrtona*. I studied preparations of *Curtonotum helvum* (LOEW).

Description of male postabdomen and genitalia (figs. 39-40)

6th and 7th abdominal spiracles symmetrically situated in membrane below respective terga.

Postabdomen almost symmetrical, except for slight asymmetry of 6th and 7th sterna. 6th and 7th terga represented by discrete narrow bands of sclerotization in *Curtonotum anus* (MEIGEN): in *C. helvum* (LOEW) only a single band (probably 7th tergum) present. 6th and 7th sterna represented by bare, slightly asymmetrical sclerites which lie below the hypandrium (posterior to the conspicuous angle in the ventral wall of the abdomen produced by the aedeagal apodeme). 8th tergum and sternum absent.

Periandrium bearing discrete, posteriorly directed telomeres. Interparameral sclerites scarcely differentiated. Cerci small, discrete: in *Curtonotum* there is a small posteriorly directed process immediately below the cerci. Hypandrium quadrate anteriorly, bearing slender posterior processes (? pregonites): hypandrial arms fused posteriorly to form bridge above aedeagus. Aedeagal apodeme short, fused with base of aedeagus, linked with anterior end of hypandrium. Aedeagus grossly enlarged, curving downwards anteriorly where it is fused with the aedeagal apodeme. Postgonites absent. Small ejaculatory apodeme present.

Characterization and discussion

The Curtonotidae are characterized by the following apomorphous conditions with respect to the groundplan of the Drosophiloidea.
(1) 3rd antennal article bearing long dorsolateral process which projects within the 2nd article (HENNIG, in press).
(2) Aedeagus (♂) grossly enlarged, curved downwards anteriorly where it is fused with the aedeagal apodeme (fig. 40).

HACKMAN's (1960) figures of the African genus *Cyrtona* indicate a similar condition of this apodeme to that shown by *Curtonotum*.

(3) 6th and 7th terga (♂) reduced, much shorter than preceding terga.

HENNIG (1958, fig. 353) has indicated that in *Curtonotum anus* (MEIGEN) both these terga are retained as discrete sclerites, although they are very short in comparison with the preceding terga. This is probably the groundplan condition for the Curtonotidae. In *C. helvum* (LOEW) only a single dorsal sclerite (probably the 7th tergum) lies between the 5th tergum and the periandrium.

(4) 8th sternum (♂) completely lost.
(5) Second costal break present (shortly beyond humeral cross-vein) (HENNIG 1958).
(6) Base of m_4 (= 'tb', lower cross-vein) lost (HENNIG 1958).
(7) Spermathecae (♀) long and slender (STURTEVANT 1926, HENNIG 1958).

Noteworthy plesiomorphous conditions of the Curtonotidae are: (1) the complete subcosta, and (2) the retention of both the 6th and 7th sterna (♂) (fig. 39). I have not found the 7th sternum in males of any other group of Drosophiloidea.

C. 6. Campichoetidae familia nova (type-genus Campichoeta MACQUART *1835)*

Sources of information

J. F. MCALPINE's (1962a) revision provides a sound basis for further studies of the Campichoetidae. Figures of the male postabdomen and genitalia are included. I studied preparations of *Campichoeta griseola* (ZETTERSTEDT).

Description of male postabdomen and genitalia (figs. 37-38)

6th and 7th abdominal spiracles symmetrically situated in membrane below respective terga.

Postabdomen fully symmetrical. 6th and 7th terga developed as discrete dorsal sclerites, but shorter than preceding terga. 6th sternum usually absent, but present in *Campichoeta obscuripennis* (MEIGEN). 7th sternum absent. 8th tergum and sternum absent.

Periandrium usually with discrete telomeres: but in some species the latter are fused with the periandrium, and they are apparently absent in *C. obscuripennis* (MEIGEN). Cerci discrete. Hypandrium quadrate anteriorly, bearing posterior processes (pregonites). Aedeagal apodeme rod-like, free from hypandrium. Aedeagus short, posteriorly directed when at rest (not able to be swung through wide arc against aedeagal apodeme), of simple tubular structure with only weak differentiation of lateral sclerotization. Postgonites well developed. Small ejaculatory apodeme present.

Characterization and discussion

Hennig (1958) accepted the customary delimitation of a family 'Diastatidae', which included the three genera *Diastata*, *Campichoeta* and *Euthychaeta*. However he was not able to find any autapomorphous condition to demonstrate the monophyly of this group. It seems to me that there is conclusive evidence that *Diastata* is more closely related to the Ephydridae in Hennig's sense than to *Campichoeta*. The characters which indicate this relationship are complex, involving modifications of the female reproductive system and loss of the 7th abdominal spiracles in both sexes (see under the Ephydridae below). I have accordingly proposed to expand the limits of the Ephydridae to include *Diastata*. The genus *Campichoeta* remains as a residual of the old Diastatidae, and a new family name is required for this.

The only genus which I definitely refer to the Campichoetidae is *Campichoeta*. It is likely that *Euthychaeta* also belongs here, but I have not been able to obtain any material to check this. Since the relationships of *Euthychaeta* require confirmation, my present treatment of the Campichoetidae is based solely on the structure of species of *Campichoeta*.

The Campichoetidae are characterized by the following apomorphous conditions with respect to the groundplan of the Drosophiloidea.

(1)   3rd antennal article enlarged, about twice as long as broad.

This is the only probable autapomorphous condition of the Campichoetidae which I am able to suggest. The nature of the connection between the 3rd and 2nd antennal articles has not been established for this group.

(2)   Subcosta becoming fused with vein r_1 distally.
(3)   Anal cross-vein (cu_{1b}) angulate (see Hennig 1958).

This condition is also well shown by *Euthychaeta* and in the groundplan of the Ephydridae (as retained by *Diastata*), and may possibly be due to synapomorphy between the Ephydridae and the Campichoetidae.

(4)   8th sternum ($\male$) completely lost.
(5)   7th sternum ($\male$) lost.

In the preparations which I studied the last ventral sclerite is the 5th sternum. But J. F. McAlpine's (1962a) figure of *Campichoeta obscuripennis* (Meigen) (fig. 5) seems to indicate the retention of the 6th sternum in that species.

In some respects the Campichoetidae seem little removed from the groundplan of the Drosophiloidea. They retain discrete 6th and 7th terga ($\male$) which are relatively large in some species (fig. 37), and the full complement of seven pairs of abdominal spiracles is present in both sexes. The modifications of the female reproductive system characteristic of the Ephydridae (including *Diastata*) are lacking, and two sclerotized spermathecae are retained. The proclinate *ors* in Campichoetidae (and *Euthychaeta*) lies between the reclinate *ors* and the eye-margin, as in the groundplan of the Ephydridae. It is not known whether this

110

condition should be considered apomorphous with respect to the groundplan of the Drosophiloidea.

C. 7. Family Ephydridae (including Diastata)

Sources of information

The available comparative information on the male postabdomen and genitalia of the Ephydridae as a whole is still inadequate, although a helpful study on part of the family has been prepared by CLAUSEN (1965). BOLWIG (1940) published a detailed study of *Scatophila unicornis* CZERNY. DAHL (1959) discussed the variation among Scandinavian Ephydridae *s.s.*, but gave no figures.

I studied preparations of *Diastata vagans* LOEW, *Hydrellia* sp. and *Psilopa* sp.

Description of male postabdomen and genitalia (figs. 41-42)

6th abdominal spiracles symmetrically situated, usually in membrane but within 5th tergum in some Parydrinae. 7th spiracles absent.

Postabdomen fully symmetrical. The last large tergum is the 5th tergum: 6th and 7th terga developed as narrow discrete sclerites in *Diastata*, but absent in all other genera. 8th sternum absent. 5th sternum the last ventral slcerite before the hypopygium.

Periandrium large in *Diastata*, but relatively small and strongly deflexed below 5th tergum in most other genera, usually bearing discrete telomeres. Interparameral sclerites consisting of pair of broad plates fused posteriorly *(Diastata)*: whether homologous sclerites are present in other genera is not clear. Cerci usually discrete. Hypandrium sometimes quadrate anteriorly and bearing posterior processes (pregonites), as in many other Drosophiloidea; but variously reduced or modified in some groups of Ephydridae *s.s.* Aedeagal apodeme variably developed, sometimes rod-like (e.g. *Diastata*), small and wedge-shaped in many Parydrinae; either free from hypandrium or linked to it anteriorly (e.g. *Ephydra*). Aedeagus short, variable in form, often bearing lobes or processes and with lateral sclerites more or less differentiated in its walls; of variable orientation, sometimes posteriorly or posteroventrally directed (e.g. *Diastata, Psilopa*), sometimes anteriorly directed (e.g. *Scatophila* and *Hydrellia*, in which it forms an acute angle with the aedeagal apodeme); not able to be swung through wide arc against aedeagal apodeme. Postgonites well developed in many groups, but sometimes apparently absent. Slender ejaculatory apodeme present in *Diastata*, but none found in other genera.

Characterization and discussion

I propose to widen the customary limits of the Ephydridae to include *Diastata*,

111

which has been classified by most recent authors in a family 'Diastatidae' together with *Campichoeta* and *Euthychaeta*. However STURTEVANT (1925-26) discovered that *Diastata* shares certain major structural modifications of the female reproductive system with the Ephydridae *sensu stricto*. Since these modifications are not shown by *Campichoeta*, I can only conclude that the 'Diastatidae' was not a monophyletic group because *Diastata* is more closely related to the Ephydridae *s.s.* than to *Campichoeta*. HENNIG (1958) unfortunately did not consider STURTEVANT's paper in his analysis.

With the inclusion of *Diastata*, which shows more plesiomorphous conditions than other Ephydridae in several respects, the following characterization of the family is offerred in terms of groundplan conditions which are apomorphous with respect to the groundplan of the Drosophiloidea.

(1) Spermathecae (♀) rudimentary; ventral receptacle heavily sclerotized (STURTEVANT 1926).

The ventral receptacle is usually called a 'spermatheca' in the literature on Ephydridae, which is clearly objectionable as the structure is not homologous with the spermathecae of other Diptera. STURTEVANT (1926) reported finding sperm only in this ventral receptacle. The true spermathecae thus appear to have lost their former function of storing sperm, though it is possible that they retain a glandular function. STURTEVANT found a closely similar condition in the Ephydridae *s.s.* and in *Diastata*: even the curious curvature of the ventral receptable of *Diastata* is not without parallel among some of the Ephydridae *s.s.*, as will be seen if STURTEVANT's figure of *Diastata* is compared with that given by CLAUSEN (1965) for *Pelina truncatula* LOEW. *Diastata* and the Ephydridae *s.s.* thus share a complex apomorphous condition of the female reproductive system which does not occur in any other group.

(2) 7th abdominal spiracles absent in both sexes (HENNIG 1958).

This is probably an autapomorphous condition of the Ephydridae, since the 7th pair of spiracles is retained in the Curtonotidae, Campichoetidae and most Drosophilidae.

(3) 6th and 7th terga (♂) reduced, much shorter than preceding terga (fig. 41).

The 6th and 7th terga are both retained as distinct sclerites only in *Diastata*, in which they are very short in comparison with the preceding terga. A similar condition is found in the groundplan of the Curtonotidae, but it is doubtful whether this is due to synapomorphy with the Ephydridae. CLAUSEN (1965) has suggested that the 6th tergum may have become fused with the 5th tergum in some Parydrinae which show two spiracles within this sclerite. But since this spiracular arrangement clearly does not belong to the groundplan of the Ephydridae *s.s.*, I think his alternative explanation (that the 6th spiracles have migrated forwards onto the 6th tergum) much more probable.

(4) 6th and 7th sterna (♂) lost.

The last pregenital sternum in all male Ephydridae, as far as I am aware, is the 5th sternum. What CLAUSEN (1965) has called 'sternite 6' is clearly the hypandrium.

(5) 8th sternum (♂) completely lost.
(6) Anal cross-vein (cu_{1b}) angulate (see HENNIG 1958).

The complete anal cell shown by *Diastata* I consider to indicate the groundplan condition for the Ephydridae. In this genus the anal cross-vein is distinctly angulate, a condition which is possibly due to synapomorphy with the Campichoetidae. In other Ephydridae a more strongly apomorphous condition is found, the anal cell being open distally due to reduction of the anal cross-vein.

(7) Subcosta becoming fused with r_1 distally (HENNIG 1958).

In most Ephydridae the base of m_4 ('tb') has been lost, as in many other Drosophiloidea. But this is clearly not a groundplan condition of the family, as *Diastata* retains this vein section.

Since *Diastata* shows more plesiomorphous conditions than those shown by other Ephydridae in respect of several character sequences, I think it most probable that this genus is the sister-group of all other Ephydridae. HENNIG's (1958) conclusion that the Ephydridae *s.s.* is a monophyletic group seems to me well justified, and my proposal to widen the limits of the Ephydridae to include *Diastata* does not conflict with that conclusion. From the standpoint of phenetic classification *Diastata* would probably be considered intermediate between the Ephydridae *s.s.* and less modified drosophiloids, and hence pose a problem to the classifier. In phylogenetic systematics, where relationship is inferred only from resemblances in respect of apomorphous conditions, *Diastata* must clearly be grouped with the Ephydridae.

A. 4. Superfamily Nothyboidea

HENNIG (1958) tentatively proposed a group Nothyboidea, in which he included the Megamerinidae, Nothybidae, Diopsidae, Psilidae, Tanypezidae and Strongylophthalmyiidae. It is apparent from my studies that this group is heterogenous. In this paper I refer the Megamerinidae, Diopsidae and Tanypezidae (including Strongylophthalmyiidae) to different prefamilies of Muscoidea (the Sciomyzoinea, Diopsioinea and Tanypezoinea respectively). In my revised sense of Nothyboidea I include the Nothybidae, Psilidae, Teratomyzidae and Periscelididae (including *Somatia*).

My conclusion that the Nothyboidea in my revised sense is probably monophyletic, is based on the following groundplan conditions which are apomorphous with respect to the groundplan of the Schizophora.
(1) Aedeagus ($\male$) articulated with aedeagal apodeme, swung through wide arc from anteriorly directed rest position to copulatory position, with basal phallophore and variable differentiation of distal sclerotization in its walls.

I ascribe to the groundplan of the Nothyboidea and Muscoidea the presence of a well developed mechanism for swinging the aedeagus through a wide arc against the aedeagal apodeme (see further below in my discussion of Muscoidea). The common presence of this mechanism can probably be ascribed to synapomorphy. Such a mechanism is always lacking in the Lonchaeoidea, Lauxanioidea and Drosophiloidea. This swinging mechanism is well developed in three families of Nothyboidea, the Nothybidae, Teratomyzidae and Psilidae. The other family, Periscelididae, shows a type of aedeagus which I interpret as highly apomorphous and not indicative of the groundplan of the superfamily. I am not able to suggest the precise condition of the aedeagus in the groundplan of the Nothyboidea beyond the presence of the swinging mechanism. The types of aedeagus shown by the Nothybidae, Terato-

113

myzidae and Psilidae are divergent and possibly all apomorphous to some degree. In Nothybidae part of the aedeagus is flexible and ribbon-like, a condition comparable with that shown by the Periscelididae (in which the aedeagus is flexible and ribbon-like along most of its length). But the presence of a flexible area of the aedeagus is probably not a groundplan condition of the Nothyboidea, as the aedeagus of Teratomyzidae is rigid throughout (as I postulate also for the groundplan of the Muscoidea).

(2) Only a single dorsal sclerite present between 6th segment and hypopygium (♂); no ventral sclerites present between 6th sternum and hypopygium (figs. 43, 46 and 51).

The homology of this dorsal pregenital sclerite requires clarification. Two interpretations seem possible. If it is postulated that the postabdominal structure of the Nothyboidea is derived from the condition ascribed to the groundplan of the Muscoidea (see below), then the dorsal sclerite preceding the hypopygium may be interpreted as the fused 7th and 8th sterna. However this interpretation is problematical because in most Nothyboidea this sclerite is symmetrical, and I have not found any trace of a suture line indicating that fusion of separate areas of sclerotization has occurred. It is true that in *Somatia* (Periscelididae) this sclerite is asymmetrical, more strongly developed on the left side; but this asymmetry is probably secondary, not ascribable to the groundplan of the Nothyboidea or Periscelididae, since the structure of the abdomen in *Somatia* is highly modified in many respects. In all Nothyboidea except *Somatia* the dorsal pregenital sclerite is symmetrical. An alternative interpretation must therefore be considered, namely that the dorsal pregenital sclerite represents an unreduced 7th tergum. The position of the 7th spiracles in Nothyboidea conforms well with this interpretation, since these are always situated (when present) below or within the dorsal pregenital sclerite. If this latter interpretation is correct, then the postabdominal structure of the Nothyboidea cannot be derived from the condition shown in the groundplan of the Muscoidea, in which the sclerites of the 7th segment have been asymmetrically reduced. The homology of the dorsal pregenital sclerite is thus of crucial importance for the classification here proposed. My recognition of the Nothyboidea as a full superfamily rests on the postulate that the second explanation (that the Nothyboidea show an unreduced 7th tergum) is more probably correct. If, to the contrary, it can be shown that the postabdominal structure of the Nothyboidea is in fact derived from the condition shown in the groundplan of the Muscoidea, then the Nothyboidea should be reclassified as a prefamily of Muscoidea.

In all species which I have studied the last ventral sclerite before the hypopygium is the 6th sternum, which is symmetrical and, except in Nothybidae, of similar size to the preceding sterna. VERBEKE (1952) reports that in some *Loxocera* species (Psilidae) an additional narrow band of ventral sclerotization is present, which he interprets as the 7th sternum. If the presence of a symmetrical 7th sternum can be ascribed to the groundplan of the Nothyboidea, then the view that the condition of the postabdomen shown by the Nothyboidea has not been derived from the condition shown in the groundplan of the Muscoidea will be strengthened, because the 7th sternum is asymmetrically developed in the groundplan of the Muscoidea (becoming fused with the inverted 8th sternum on the left side). However, since this apparent 7th sternum has so far only been reported in a few species of a single genus, I think it more probable that it represents a secondary sclerotization. The characterization given above is therefore based on the postulate that the last ventral sclerite present in the groundplan of the Nothyboidea was an unmodified 6th sternum.

Clearly the considerations here presented have not definitely settled the question of what modifications of the 7th and 8th abdominal segments occurred in the groundplan of the Nothyboidea. Relevant information from ontogenetic studies or studies on intersexes would be of great interest. In view of the doubt about the homology of the last dorsal sclerite before the hypopygium, I refer to this as the 'pregenital sclerite' in this work.

(3) Aedeagal apodeme (♂) fused with the body wall posteriorly.

In Nothybidae, Teratomyzidae and some Psilidae the aedeagal apodeme is fused with the body wall posteriorly (between the hypandrial arms), as far as its ventral spur (figs. 44 and 50); only the anterior end of the apodeme remains free in the body cavity. This condition seems more plesiomorphous than the complete fusion of the apodeme with the body wall shown by most Periscelididae and by some groups of Psilidae. Only in *Somatia* (Periscelididae) is the aedeagal apodeme free from the body wall along its entire length; this condition is probably secondary, associated with reduction of the hypandrium.

(4) Only one postalar bristle *(pa)* present.

HENNIG (1965b, 1969a, in press) reports this as an apomorphous condition of the Nothybidae, Teratomyzidae, Periscelididae and Psilidae.

The Nothyboidea retain (at least in their groundplan) many conditions which are regarded as plesiomorphous for the Schizophora. These include an unmodified 6th segment in the male, a complete subcosta, the absence of vibrissae, and an apparently unmodified female postabdomen (as shown by *Nothybus;* see HENNIG 1958, fig. 78).

Whether the costa was broken or unbroken in the groundplan of the Nothyboidea is not clear. The costa in unbroken in Nothybidae and Periscelidinae, but broken in Psilidae, Teratomyzidae and *Somatia* (Periscelididae).

I do not propose a division of the Nothyboidea into prefamilies at the present time, since I am in doubt as to which families are most closely related to one another. The Periscelididae and Teratomyzidae share certain apomorphous conditions of the head. However, certain modifications in the structure of the wing are shown by the Teratomyzidae and Nothybidae, but not by the Periscelididae. The Psilidae are highly modified in some respects, and it is difficult to judge whether they are the sister-group of all other Nothyboidea or most closely related to one of the other families. Some of the characters on which the family classification is based are indicated on figure 16.

HENNIG (in press) classifies the Periscelididae and Teratomyzidae together with the Aulacigastridae and Asteiidae in a subgroup (Periscelidea) of the Anthomyzoinea. In my opinion the 'Periscelidea' are polyphyletic, or at least diphyletic. The Aulacigastridae and Asteiidae show the typical muscoid morphology of the male postabdomen, and I accept their classification in the Anthomyzoinea (Anthomyzoidea of HENNIG). But the structure of the male postabdomen in Periscelididae and Teratomyzidae is fundamentally different, and seems to me incompatible with their classification in the Anthomyzoinea. Since the apomorphous conditions on which HENNIG's proposal is based (large peristomal opening; 2 *ors* in groundplan; only 1 *pa*) are all of rather widespread distribution, I think that his proposal should be rejected on grounds of incompatibility with the distribution of the additional characters which I treat in this work.

C. 8. Family Nothybidae

Sources of information

ACZÉL's (1955) paper is the main source of information on this group. I studied preparations of *Nothybus longithorax* RONDANI.

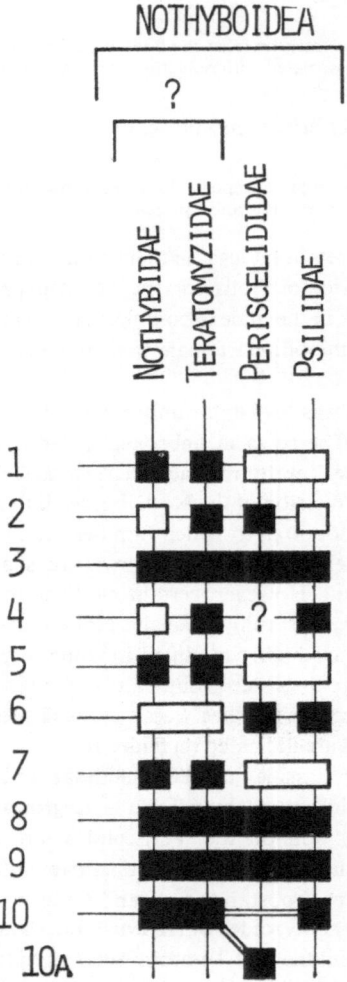

Fig. 16. Distribution of some character states among the families of Nothyboidea. The black rectangles indicate groundplan conditions interpreted as apomorphous with respect to the groundplan of the Schizophora. Where common possession of such apomorphous conditions is interpreted as due to synapomorphy, a continuous black rectangle is shown against the families concerned. Corresponding plesiomorphous conditions are shown by the white rectangles. The condition indicated at 10a is considered to be a further modification of that at 10. Some autapomorphous conditions of single families have been omitted. The list of characters is as follows.

Apomorphous conditions (black rectangles)	Plesiomorphous conditions (white rectangles)
Head	
1. Postverticals reduced	Divergent postverticals present
2. Only 1 *ors* present	2 *ors* present
Thorax	
3. Only 1 *pa* present	

116

Wing
  4. Subcosta not reaching wing margin   Subcosta distinct to wing margin
     as distinct vein
  5. Axillary lobe and alula reduced   Axillary lobe and alula normal
  6. Anal vein cut off apically   Anal vein long, almost or fully reaching
                                           wing margin
  7. Lower basal cell open anterodistally Lower basal cell fully closed

Male postabdomen
  8. Only a single dorsal sclerite between
     6th segment and hypopygium; no
     ventral sclerites between 6th sternum
     and hypopygium
  9. Aedeagal apodeme fused with body
     wall posteriorly
 10. Aedeagus swung through wide arc
     against aedeagal apodeme
10a. Aedeagus ribbon-like,
     without swinging mechanism

Description of male postabdomen and genitalia (figs. 43-45)

6th and 7th abdominal spiracles symmetrically situated, 6th pair at edge of 6th tergum, 7th pair in membrane below pregenital sclerite.

Postabdomen fully symmetrical. 6th tergum much shorter than preceding terga. Between the 6th tergum and the hypopygium lies one further dorsal sclerite (here called 'pregenital sclerite'). The last ventral sclerite before the hypopygium is the 6th sternum, which is much shorter than the preceding sterna.

Periandrium bearing discrete telomeres, which are joined by a crescentic interparameral sclerite. Cerci discrete, conspicuously projecting. Hypandrium with setose lateral lobes (pregonites): above the aedeagus lies a pair of sclerites ('x' on fig. 45) which are contiguous with the hypandrium anteriorly. Aedeagal apodeme rodlike, fused with body wall posteriorly (after the ventral spur). Aedeagus with large phallophore at base; beyond the phallophore flattened (ribbon-like) and flexible, coiled when at rest, strengthened by paired strips of sclerotization; apically forked and ornamented with pair of large lobes; the phallophore seems able to be swung through a wide arc against the aedeagal apodeme. Small postgonites present. Large ejaculatory apodeme present.

Characterization and discussion

This family contains only the Oriental genus *Nothybus*, characterized by the following apomorphous conditions with respect to the groundplan of the Nothyboidea.
  (1) Axillary lobe and alula of wing reduced; anal vein close to wing margin (Aczél 1955, Hennig 1958).
  (2) Base of media (m) interrupted, so that the lower basal cell is open anterodistally (Aczél 1955).

117

Conditions of the wing similar to those stated above (characters 1 and 2) are also shown by the Teratomyzidae. It is not clear whether these resemblances can be ascribed to synapomorphy, since there are other characters which suggest that the Teratomyzidae are more closely related to the Periscelididae than to the Nothybidae.

(3) Ocellar and postvertical bristles absent (ACZÉL 1955, HENNIG 1958).
(4) Thorax elongated anteriorly, with front coxae situated far posterior to head (ACZÉL 1955, HENNIG 1958; see also the figure of the prosternum given by SPEIGHT 1969, fig. 15).
(5) Postscutellum conical and very large, longer than scutellum (ACZÉL 1955).
(6) Sterna 2-5 long and narrow (ACZÉL 1955).
(7) 6th tergum and pregenital sclerite (♂) much shorter than preceding terga (fig. 43); 6th sternum much shorter than 5th sternum.

The groundplan condition of the postabdomen for the Nothyboidea is probably indicated by the condition shown by the Periscelididae, in which the sclerites of the 6th segment are unreduced (almost as long as those of the 5th segment) and the pregenital sclerite also relatively large.

(8) Membrane above base of aedeagus (♂) strengthened by pair of sclerites which are contiguous with the hypandrium proximally ('x' on fig. 45).
(9) Aedeagus (♂) with flexible ribbon-like basal section, forked apically and ornamented with pair of large lobes (fig. 44).

HENNIG (1958) has given as an additional apomorphous character '*vte* oder *vti* fehlt'. I prefer to omit this as the homology of the bristles concerned is in doubt. Possibly ACZÉL (1955) was right in considering that both vertical bristles are retained (so that only two bristles are considered to represent fronto-orbitals).

When ACZÉL (1955) first proposed the superfamily Nothyboidea, with Nothybidae as the sole included family, he assumed that this was the only family of Schizophora in which the male postabdomen is completely symmetrical. This is incorrect, since a completely symmetrical male postabdomen is also shown by most Drosophiloidea and Lauxanioidea, as well as by members of the other families here referred to the Nothyboidea. ACZÉL's conclusion that *Nothybus* does not possess a *hypopygium circumversum* was also unwarranted, as it was based on the false premise that hypopygial rotation necessarily produces external asymmetry (see section 3.2). On the basis of these misconceptions ACZÉL argued that the Nothyboidea (Nothybidae only) should be recognized as of equal rank (superfamily) to the Calyptratae and 'Acalyptratae'. I am in full agreement with HENNIG (1958) that these classificatory proposals should be rejected.

C. 9. Family Teratomyzidae

Sources of information

The only written source of information on the male postabdomen and genitalia

118

of the Teratomyzidae which I have traced is VOCKEROTH's (in press) work entitled 'Diptera of Nepal. Anthomyzidae'. I examined preparations of an undescribed Australian species of *Teratomyza sensu lato*.

Description of male postabdomen and genitalia (figs. 50-51)

6th abdominal spiracles symmetrically situated within margin of 6th tergum. 7th spiracles absent.

Postabdomen fully symmetrical. 6th segment unmodified, with tergum and sternum of similar size and length to those of preceding segments. Between the 6th segment and the hypopygium lies only a single sclerite, the 'pregenital sclerite': this is broad dorsally, becoming linear ventrally, either forming a complete ring around the postabdomen or narrowly divided at the centre-line of the venter.

Periandrium bearing large articulated telomeres, whose inner sides are confluent with a broad interparameral sclerite. Cerci discrete. Aedeagal apodeme rod-like, fused with body wall posteriorly (after the anteroventral process or pair of processes). Aedeagus with complex sclerotization, swung against its apodeme through wide arc from anteriorly directed rest position to copulatory position: epiphallus variably developed. Two articulated processes present near base of aedeagus (probably representing pregonites and postgonites). Slender ejaculatory apodeme present.

Characterization and discussion

The name Teratomyzidae has been proposed by COLLESS & D. K. MCALPINE (1970). Formerly members of this group were referred either to the Anthomyzidae (following MALLOCH 1933) or to the Opomyzidae (HARRISON 1959). Both the latter families are classified in the Anthomyzoinea (Muscoidea) in this work, and I do not think that the Teratomyzidae can be included in that group for the reasons give above in my opening discussion of the Nothyboidea.

My present treatment of the Teratomyzidae does not take account of the morphology of *Neogeomyza*, which HENNIG (in press) refers to this family. HENNIG mentions only the structure of the head and wing in his discussion of this genus. I think that additional information is needed before this genus can be classified with confidence.

Many of the species of Teratomyzidae in collections remain undescribed. The limited information currently available suggest that the family may be characterized by the following apomorphous conditions with respect to the groundplan of the Nothyboidea.

(1) Axillary lobe and alula of wing reduced: anal vein (cu_{1b} + 1a) close to wing margin.

HENNIG (in press) argues that the long anal vein shown by many Teratomyzidae is probably secondary. However the similarities in the morphology of the wings of Nothybidae and Teratomyzidae suggest to me that the long anal vein is probably a groundplan condition of the latter. A firm judgment on this question will probably not be possible until some comprehensive treatment of the Teratomyzidae is available.

(2) Base of media (m) weak or interrupted on anterodistal boundary of lower basal cell (see the illustrations given by VOCKEROTH).

(3) Costa ending at or just beyond end of vein r_{4+5} (VOCKEROTH, in press).

(4) Subcosta fading distally or becoming fused with r_1 (see the illustrations give by VOCKEROTH).

(5) Posterior cross-vein (m-m) displaced towards wing base, level with or only shortly distal to the anterior cross-vein (r-m).

(6) Postvertical bristles *(pvt)* reduced, at most represented by minute setulae (variably directed).

(7) Only one well-developed fronto-orbital bristle *(ors)* present.

HENNIG (in press) refers to the presence in some Australian species of a pair of fine setulae which possibly represent a reduced anterior pair of *ors*.

(8) Peristomal opening large (HENNIG, in press).

(9) Vibrissae present.

(10) Two spermathecae (♀) (VOCKEROTH, in press).

(11) 7th abdominal spiracles lost in both sexes.

(12) 5th and 6th abdominal spiracles (♂) lying within margins of respective terga.

(13) Pregenital sclerite (♂) extending ventrally, forming complete ring around postabdomen or only narrowly divided at centre-line of venter (fig. 51).

The pregenital sclerite also extends ventrally in Periscelididae. Whether the condition shown by the Teratomyzidae can be considered synapomorphous with that shown by the Periscelididae is not clear. It is perhaps possible that some degree of ventral extension of the pregenital sclerite was shown in the groundplan of the Nothyboidea, but has been secondarily lost in Nothybidae and Psilidae. But in the absence of clear evidence for such a hypothesis, I prefer to interpret the condition shown by the Teratomyzidae as apomorphous with respect to the groundplan of the Nothyboidea.

The aedeagi of the species from Nepal figured by VOCKEROTH (in press) show an asymmetrical distal area, and are consequently rather different in appearance from the symmetrical flattened type of aedeagus shown by the Australian species which I studied (fig. 50). Without information on additional groups of species, I am not able to suggest what type of aedeagus should be ascribed to the groundplan of the family.

C. 10. Family Periscelididae (including Somatiidae)

The correct form of the family name based on *Periscelis* (the Greek word περισκελίς, genitive περισκελίδος, meaning 'leg-band') is Periscelididae, not 'Periscelidae'.

Sources of information

There is little information on the male postabdomen and genitalia of Periscelididae available in the literature, but some relevant discussion and figures were given by HENNIG (1952b) for *Scutops*, and by STEYSKAL (1958a) and HENNIG (1958, fig. 235) for *Somatia*. I studied preparations of the following: – *Periscelis annulata* (FALLÉN); *Scutops maculipennis* MALLOCH; *Somatia* sp.

Description of male postabdomen and genitalia (figs. 46-47)

Postabdominal spiracles symmetrically situated, 6th pair in membrane below 6th tergum, 7th pair within pregenital sclerite.

Postabdomen fully symmetrical (Periscelidinae) or with pregenital sclerite asymmetrical *(Somatia)*. 6th tergum large, of similar length to 5th tergum. 6th sternum large, not much shorter than 5th sternum (Periscelidinae), but very small (like preceding sterna) in *Somatia*. The 6th sternum is the last ventral sclerite before the hypopygium. Between the 6th tergum and the hypopygium lies the large 'pregenital sclerite' which extends towards the venter on either side (where it includes the 7th spiracles); this sclerite is symmetrical in Periscelidinae, but asymmetrical (more strongly developed on the left side) in *Somatia*.

Periandrium with slender, more or less discrete telomeres (Periscelidinae): in *Somatia* periandrium forming complete ring around genital segment, with neither telomeres nor hypandrium clearly differentated from this. Interparameral sclerites absent. Cerci discrete, rather long and pointed apically (Periscelidinae), fused ventrally in *Somatia*. Hypandrium and aedeagal apodeme more or less fused, forming cavity into which the aedeagus is folded when at rest (Periscelidinae): in *Somatia* no discrete hypandrium is present, the aedeagal apodeme is free and rod-like, and the aedeagus lies exposed below the abdomen when at rest. Aedeagus slender and ribbon-like, supported by broad strip of flexible sclerotization in *Somatia* and *Scutops*, but membranous along most of its length in *Periscelis*; in *Somatia* and *Scutops* bearing a minute apical spine; phallophore weakly differentiated, probably not able to be swung against aedeagal apodeme through wide arc. Small postgonites present in Periscelidinae, absent in *Somatia*. Large ejaculatory apodeme present.

Characterization and discussion

The definition of this family has been discussed by STURTEVANT (1954) and HENNIG (1969a). The latter accepts the inclusion of four genera; *Periscelis* (including *Marbenia*), *Scutops*, *Neoscutops* and *Diopsosoma*. Only *Periscelis* has

121

a wide distribution, the other three genera being exclusively neotropical. The Periscelididae in Hennig's sense doubtless constitute a monophyletic group (here called Periscelidinae), but I think that the limits of the family can be widened to include the neotropical genus *Somatia*.

The relationships of *Somatia* have long remained in doubt. Hennig (1958) suggested possible relationships with the Richardiidae or Heleomyzidae, and Steyskal (1958a) also thought that *Somatia* is related to the Richardiidae. Rohdendorf (1964) regarded *Somatia* as constituting a monobasic superfamily ('Somatiidea'), sharply distinguished from other Schizophora by certain features which he interpreted as 'archaic', namely the large size of the anal cell and lower basal cell, the weak development of bristles on the head, and the structure of the abdominal terga which 'do not extend to the ventral side'. However all these suggestions should in my opinion be rejected, because there are clear synapomorphies between *Somatia* and the Periscelidinae. The aedeagus of *Somatia* is in all respects very similar to that of *Scutops*, even down to the presence of a minute apical spine in both genera; the 7th abdominal spiracles ($\male$) in *Somatia* are situated within the pregenital sclerite, as in Periscelidinae; and the structure of the female postabdomen in *Somatia* also agrees fully with that of the Periscelidinae, with the 7th tergum and sternum fused to form a complete ring which includes the 7th spiracles. The absence of a retractile ovipositor ($\female$) and the retention of an unreduced 6th tergum ($\male$) clearly exclude the possibility of *Somatia* being closely related to the Richardiidae (Tephritoinea). Rohdendorf's evaluation of certain conditions shown by *Somatia* as 'archaic' I regard as mistaken. Since the conditions of the wing venation, frontal chaetotaxy and abdominal structure shown by the Periscelidinae are more similar to those shown by most other groups of Schizophora, the unusual conditions of *Somatia* which Rohdendorf calls 'archaic' should in my opinion be interpreted as apomorphous.

Hennig (in press) now recognizes that *Somatia* belongs to the Nothyboidea. But he does not associate it with the Periscelidinae, which he classifies in the Anthomyzoinea (Muscoidea).

The Periscelididae, in the widened sense here proposed, are characterized as a monophyletic group by the following conditions in their groundplan which are apomorphous with respect to the groundplan of the Nothyboidea.

(1) Anal vein (cu_{1b} + 1a) abruptly cut off apically, not reaching wing margin.

The anal vein is also cut off apically in the Psilidae.

(2) Only one fronto-orbital bristle *(ors)* present: ocellar bristles standing near ocellar prominence, not between ocelli (Hennig 1958, 1969a).

Further reduction of the frontal chaetotaxy is shown by *Somatia*.

(3) 7th abdominal tergum and sternum ($\female$) fused, forming ring which includes the 7th pair of spiracles.

HENNIG (1958) has figured this condition for *Periscelis* and *Somatia*. I have also confirmed that the same condition is shown by *Scutops*. Some groups of Teratomyzidae show a similar condition, but probably through homoplasy, since the 7th tergum and sternum are discrete in *Teratoptera* according to VOCKEROTH (in press).

(4) Pregenital sclerite ($\male$) extending ventrally on either side; 7th abdominal spiracles lying within this sclerite (fig. 46).

See the comment under Teratomyzidae, which show a similar structure of the pregenital sclerite (but lack the 7th spiracles).

(5) Aedeagus ($\male$) slender and ribbon-like, supported by broad strip of flexible sclerotization (fig. 47).

Extrusion of the aedeagus in Periscelididae is presumably effected by pressure of body fluid, since the swinging action between the phallophore and aedeagal apodeme (as retained by the other families of Nothyboidea) seems poorly developed. I regard the presence of a strip of flexible sclerotization in the aedeagus, as shown by *Scutops* and *Somatia*, as probably indicative of the groundplan condition of the family. The aedeagus of *Periscelis* differs in being membranous beyond the basal phallophore.

The Periscelididae in the sense of HENNIG (1969a) is a clearly defined monophyletic group, which I propose to call subfamily Periscelidinae. *Somatia* may be recognized as constituting a second subfamily, the Somatiinae. These two groups show an alternance of apomorphous and plesiomorphous conditions, indicating a probable sister-group relationship, as follows:

Periscelidinae	Somatiinae
Apomorphous	*Plesiomorphous*
Costa to r_{4+5}	Costa to m_{1+2}
Anal cross-vein faded	Anal cross-vein normal
Plesiomorphous	*Apomorphous*
Alula well developed	Alula reduced
2nd basal and anal cells normal	2nd basal and anal cells elongate
Costal cell broad	Costal cell narrow
Abdomen normal: sterna large	Abdomen broad: sterna small, separated by wide membrane from terga
Frontal setae normal	Frontal setae reduced
Pregenital sclerite symmetrical ($\male$)	Pregenital sclerite asymmetrical ($\male$)
Telomeres present ($\male$)	Telomeres lost ($\male$)
Hypandrium well developed ($\male$)	Hypandrium reduced ($\male$)
3 spermathecae ($\female$)	2 spermathecae ($\female$)

C. 11. Family Psilidae (= Loxoceridae)

Although the name Loxoceridae has strict priority, SABROSKY'S (1946) recommendation that the name Psilidae should continue to be used for this family has been followed by all subsequent authors.

Sources of information

The male postabdomen and genitalia of the Psilidae are still poorly known, although some information has been give by HENNIG (1941b) and VERBEKE (1952). I studied preparations of the following: – *Psila rosae* (F.); *Psila fimetaria* (L.); *Loxocera cylindrica* SAY; *Chyliza erudita* MELANDER.

Description of male postabdomen and genitalia (figs. 48-49)

6th and 7th abdominal spiracles symmetrically situated in membrane.

Postabdomen fully symmetrical. 6th tergum large, of similar length to 5th tergum. 6th sternum large, not much shorter than 5th sternum. The 6th sternum is the last ventral sclerite before the hypopygium in the species studied by me, but VERBEKE (1952) reports the presence of a narrow 7th sternum in some *Loxocera* species. A narrow pregenital sclerite lies between the 6th tergum and the hypopygium in some *Loxocera* and *Psila* spp., but this is absent in many species.

Periandrium with *(Chyliza)* or without *(Psila, Loxocera)* discrete telomeres. Interparameral sclerites absent. Cerci more or less discrete. Hypandrium forming floor of cavity whose roof is formed by the aedeagal apodeme; the latter broad and ill-defined *(Psila)* or more or less rod-like *(Chyliza and Loxocera)*, fused with body wall throughout its length or with only its anterior end free from body wall *(Loxocera)*. In *Loxocera* the hypandrium and aedeagal apodeme are turned so that the anterior end of the latter is directed posteriorly. Aedeagus usually short and sclerotized only at its base, but long and 'pubescent' in *Psilosoma:* base of aedeagus able to be swung through wide arc against aedeagal apodeme. Postgonites present (short and broad). Ejaculatory apodeme minute or absent.

Characterization and discussion

The Psilidae are well characterized as a monophyletic group by the following groundplan conditions which are apomorphous with respect to the groundplan of the Nothyboidea.

(1) Subcosta cut off distally, connected with wing margin by hyaline stria.

The above wording is based on VERBEKE (1952:5).

(2) Anal vein (cu_{1b} + la) abruptly cut off apically, not reaching wing margin (see HENNIG 1958).

(3) Spermathecae ($\mathcal{Q}$) not sclerotized, represented by branched tubes.

This highly apomorphous condition described by STURTEVANT (1926) is not known out-side the Psilidae. Probably it represents an autapomorphous groundplan condition of the fam ly, since HENNIG (1958: 543) failed to find sclerotized spermathecae in representatives of any of six genera studied. DUFOUR's (1851) report of the presence of two sclerotized spermathecae in *Loxocera* was doubtless in error. I have checked a specimen of *Loxocera ichneumonea* (L.) and can find no sclerotized spermathecae, and HENNIG (1958) has reported the same for *L. elongata* MEIGEN.

(4) Female postabdomen elongate, forming retractile ovipositor (HENNIG 1958, figs. 80, 81 and 83).

(5) Ejaculatory apodeme ($\mathcal{S}$) reduced.

In *Psila rosae* (F.) and *Chyliza erudita* MELANDER the ejaculatory apodeme is minute. No sclerotized structure was seen in my preparations of *P. fimetaria* (L.) and *Loxocera cylindrica* SAY. Reduction of the ejaculatory apodeme probably constitutes an autapomorphous ground-plan condition of the Psilidae, since the ejaculatory apodeme is retained in the other families of Nothyboidea.

(6) Pregenital sclerite ($\mathcal{S}$) reduced, much shorter than 6th tergum (fig. 48).

The reduced pregenital sclerite is retained as a distinct sclerite in some *Loxocera* and *Psila* species. However in many members of the family this sclerite has either been entirely lost or further reduced to an inconspicuous trace of sclerotization.

The condition of the aedeagus in the groundplan of the Psilidae requires clarification. In the species which I studied the aedeagus is somewhat reduced, sclerotized only at its base. However HENNIG (1941b, fig. 8) indicates that *Psilosoma* possesses a long aedeagus, bearing fine cuticular processes ('pube-scence'). Unfortunately I have not been able to examine material of this genus, and do not know whether the aedeagus is rigid or flexible. The evolution of aedeagal pubescence in this case must have been independent of the Tephrito-inea (Muscoidea), as the structure of the male postabdomen in the Psilidae seems incompatible with their classification in that prefamily.

The South American genus *Schizostomyia*, referred to the Psilidae by MAL-LOCH (1934), does not belong to this family (see below under Anthomyzoinea).

A. 5. Superfamily Muscoidea

The name Muscoidea has been used in a variety of senses. It has often been used for the whole of the Schizophora, corresponding with COQUILLETT's (1901) original proposal except that he excluded the Pupipara (Hippoboscidae *s.l.*). However the use of the name in this wide sense is superfluous, since the well-known name Schizophora has priority. Recent authors have generally applied the superfamily category to subordinate groups of the Schizophora, and I think it preferable to follow this tradition, at least until sufficient information is available for applying the time criterion of rank. Some recent authors have restricted the name Muscoidea to part of the Calyptratae, following ROBACK (1951). However ROBACK's proposed group is probably not monophyletic (see my comments below on the Calyptratae), and I am in any case reluctant to use the superfamily category at such a low level in the sequence of subordination because of the proliferation of superfamily names in other parts of the classification which this would entail. Accordingly I apply the superfamily category in this work at a much higher level than ROBACK has proposed, but at a lower level than COQUILLETT. My concept of Muscoidea is substantially new, based on characters of the male postabdomen and genitalia which have not previously been analysed in sufficient detail to be used for classification. Although no previous authors have proposed a group corresponding to the whole of the Muscoidea in my sense, the relationships between many of the included families have been recognized. For instance, HENDEL's (1936) groups Trypetides and Sciomyzides consisted largely of series of muscoid families, although a few heterogenous families were also included. HENNIG (1958) made considerable progress towards defining some of the major subordinate groups of Muscoidea, but the possibility of delimiting a wider group of the kind now proposed was not indicated by the characters considered in his analysis.

I consider the Muscoidea in the sense here proposed to be monophyletic on the basis of the following conditions of their groundplan which are apomorphous with respect to the groundplan of the Schizophora.

(1)   7th abdominal sternum (♂) strongly asymmetrical, extending dorsally on left side until becoming fused or contiguous at its dorsal extremity with the inverted 8th sternum.

This characteristic condition of the 7th sternum can be readily recognized in most Muscoidea, except for a few groups in which the sclerotization of the postabdomen has been reduced or simplified. Because the 7th sternum extends dorsally on the left side, some authors have maintained that this sclerite represents the 7th tergum. But this view is hardly reconcilable with its position between the spiracles of the segment; and the sternal nature of the sclerite is further shown by its frequent retention of the *sensilla trichodea* (which form a homonomous series as far as the 7th sternum in the groundplan of the Cyclorrhapha). I have noted such sensilla on the 7th sternum of Muscoidea in members of the Tanypezidae and Heteromyzidae (Tanypezoinea), Dryomyzidae (fig. 89) (Sciomyzoinea), Heleomyzidae (fig. 101) (Anthomyzoinea), Clusiidae (Agromyzoinea), and Tephritidae *s.l.* and Odiniidae (Tephritoinea). This list is probably not exhaustive, as the sensilla may be present in members of other families but be too small to be resolved with a binocular microscope. From the above considerations I think that the homology of the 7th sternum in Muscoidea is clear beyond reasonable doubt.

(2)   7th tergum (♂) asymmetrically reduced to small vestige situated lateroventrally on right side: 7th right spiracle displaced towards centre-line.

The vestige of the 7th tergum has sometimes been overlooked, as it is usually infolded between membrane and often cannot be detected unless the abdomen has been macerated and extended. According to my observations this tergum vestige is present as a discrete sclerite in members of the following families: Micropezidae (fig. 71), Neriidae and Cypselosomatidae (Micropezoinea), Coelopidae (fig. 81), Phaeomyiidae (fig. 85), Dryomyzidae, Megamerinidae (fig. 77) and Cremifaniidae (fig. 99) (Sciomyzoinea), Heleomyzidae (fig. 101) and Borboropsidae (fig. 111) (Anthomyzoinea), Clusiidae (fig. 119) (Agromyzoinea), Chiropteromyzidae (fig. 122), Carnidae (fig. 134), Piophilidae *s.l.*, Tephritidae *s.l.* and Richardiidae (fig. 142) (Tephritoinea). This wide distribution convinces me that such a vestigial sclerite should be ascribed to the groundplan of the Muscoidea. I interpret the vestige as

126

representing the right lateral extremity of the 7th tergum, which has been asymmetrically reduced. It cannot be derived from the sternum, since it always lies to the right of the 7th right spiracle.

In those Muscoidea in which the 7th tergum vestige is retained, the adjacent 7th right spiracle is usually displaced towards the centre-line in relation to the preceding spiracles.

(3) 6th sternum ($\male$) asymmetrical, more strongly developed on left side where it extends towards the 7th sternum.

The 6th sternum is asymmetrical in nearly all Muscoidea, except those in which it has been modified through fusion with the 7th and 8th sterna. Only in the Heteromyzidae (Tanypezoinea) does a large symmetrical discrete 6th sternum occur, similar to the sterna of the preceding segments. However, since members of the Tanypezidae, which I include with the Heteromyzidae in the Tanypezoinea, show an asymmetrical 6th sternum, I think it possible that the apparently unmodified condition of the 6th sternum shown by the Heteromyzidae is secondary (apomorphous). Accordingly I postulate that asymmetrical development of the 6th sternum can be ascribed to the groundplan of the Muscoidea. I think that a symmetrical 6th sternum should be ascribed to the groundplan of all other superfamilies of Schizophora.

(4) Hypandrium ($\male$) linked to anterior margin of inverted 8th sternum by unpaired muscle on left side (*musculus hypandriotergalis* of SALZER 1968).

This muscle has been called the *musculus hypandriotergalis* (M 33) by SALZER (1968). I use SALZER's term in this work, although it is not wholly appropriate because it implies that the inverted 8th sternum is a tergum. Other names which have been applied to this muscle are *M. introtractor longus anterior* (SCHRÄDER 1927), M 8 (HENNIG 1936a), and muscle 'j' (GOODING & WEINTRAUB 1960). SALZER states that it is the only unpaired muscle in the postabdomen of *Calliphora* which does not show indications of paired origin. Since I can find no such muscle in the Syrphidae or Platypezidae (the most relevant non-schizophorous families of Cyclorrhapha because of their retention of a large 8th sternum), I am convinced that its presence in many Muscoidea is an apomorphous condition. No homologous muscle has been reported in members of any other superfamily of Schizophora. The list of the groups of Muscoidea for which the presence of this muscle has been reported in the literature is as follows: Micropezidae (*Calycopteryx*, HENNIG 1936a), Tachinidae *s.l.* (*Ernestia*, PETZOLD 1927; *Calliphora*, SCHRÄDER 1927 and SALZER 1968; *Hypoderma*, GOODING & WEINTRAUB 1960). On the basis of serial sections which I have made, I can add *Pherbellia* (Sciomyzidae), *Nemopoda* (Sepsidae), *Clusia* (Clusiidae), *Anthomyza* (Anthomyzidae) and *Leptocera* (Sphaeroceridae). Only for the Tephritoinea is there definite information that this muscle is absent (see my comments on that prefamily below). Since the presence of the *musculus hypandriotergalis* has now been confirmed for five prefamilies of Muscoidea, I postulate that this muscle was present in the groundplan of the Muscoidea and has been secondarily lost in the Tephritoinea. Of course, information on members of additional groups is needed to test the validity of this hypothesis.

(5) Aedeagus ($\male$) articulated with aedeagal apodeme, swung through wide arc from anteriorly directed rest position (fig. 52) to copulatory position (fig. 53), with basal phallophore and at least a pair of lateral sclerites (paraphalli) differentiated in its walls.

It is clear that the Muscoidea are characterized in their groundplan by a large aedeagus which is directed anteriorly (folded into a genital pouch beneath the hypandrium and aedeagal apodeme) in the rest position, but swung out posteriorly through a wide arc for copulation. The copulatory position corresponds with the position of the aedeagus in Lonchaeoidea, Drosophiloidea, Lauxanioidea and non-schizophorous Diptera, in which no swinging mechanism has been developed. SALZER (1968: 204-206) has recently described in detail the mechanisms involved in extruding and retracting the aedeagus of *Calliphora erythrocephala* MEIGEN, a species in which a strong swinging action is shown. His figures (reproduced as my figs. 52 and 53) indicate movement of the aedeagus through an angle of about 160° about its articulation with the aedeagal apodeme. In the groundplan of the Muscoidea the aedeagus was probably a rigid structure whose extrusion was achieved primarily by muscular action; probably the only well-defined sclerites were a basal ring or partial ring on which muscles insert (the phallophore), with which were joined a pair of elongate sclerites (paraphalli) strengthening the walls of the aedeagus. The varied types of aedeagus found in different muscoid families can all be interpreted as further developments from such a basic type. Additional distal sclerites of the aedeagus are differentiated in many groups of Muscoidea, but the development of such sclerites is very variable, even between closely related groups. Whether the presence of such distal sclerites can be ascribed to the groundplan of the Muscoidea seems to me doubtful, since they are not differentiated, for instance, in the Tanypezoinea. The development of this characteristic mechanism for swinging the base of the aedeagus through a wide arc against the aedeagal apodeme by means of muscles inserted on the phallophore, probably constitutes a synapomorphy between the Muscoidea and Nothyboidea. This mechanism is lacking in the Lonchaeoidea, Lauxanioidea and Drosophiloidea, in which the aedeagus remains in the same direction relative to the aedeagal apodeme when at rest.

My use of the term paraphalli for the pair of lateral sclerites in the walls of the aedeagus follows the practice of authors on Calyptratae. The term was originally proposed in the singular ('paraphallus') by LOWNE (1893), since in *Calliphora* (Tachinidae *s.l.*) these sclerites are fused basally where they join the phallophore. WESCHÉ (1906) amended this usage to the plural paraphalli.

The Muscoidea exhibit in their groundplan some significant plesiomorphous conditions in comparison with other superfamilies of Schizophora. Most noteworthy is the condition of the inverted 8th abdominal segment ($\circlearrowleft$), which is clearly defined and retains a large dorsal sternum and a small ventral vestige of the 8th tergum (as also in Platypezidae). The 8th tergum vestige is frequently lost, but its retention in the groundplan of the Muscoidea is indicated by its presence as a distinct sclerite in members of the Scatophagidae and Anthomyiidae (fig. 61) (Calyptratae), Cypselosomatidae (fig. 71), Neriidae and Micropezidae (Micropezoinea), Coelopidae (fig. 81) and Phaeomyiidae (fig. 85) (Sciomyzoinea), Heleomyzidae (fig. 101) (Anthomyzoinea) and Richardiidae (fig. 142) (Tephritoinea). The only group of Schizophora apart from the Muscoidea in which the 8th tergum vestige is retained is the Lonchaeidae (Lonchaeoidea). Another plesiomorphous condition of the male abdomen in the groundplan of the Muscoidea is the retention of an unmodified 6th tergum, fully as long as the preceding 5th tergum. This condition is well shown, for instance, by the Micropezoinea (fig. 72), Australimyzoinea (fig. 74) and many Agromyzoinea (fig. 120). In respect of the characters of the chaetotaxy, wing venation and female postabdomen analysed by HENNIG (1958), the Muscoidea include groups which show both plesiomorphous and apomorphous conditions. These characters do not aid in the definition of the group, at least on the basis of existing analyses, but only in its further subdivision.

Within some subordinate groups of Muscoidea there is a strong tendency for the structure of the postabdomen to become simplified through loss of the 7th and 8th tergum vestiges and fusion of the 7th and 8th sterna. Such changes have occurred independently in many families. Nevertheless, I wish to emphasize that all the highly modified conditions of the male postabdomen shown by particular groups here included in the Muscoidea are in my opinion secondarily derived from the condition ascribed to the groundplan of the Muscoidea. I have not included in the Muscoidea any group with symmetrical male postabdomen for

128

Fig. 17. Distribution of some character states among the families of Muscoidea. This figure is intended to summarize the distribution of conditions shown by a range of families. All entries refer to the postulated groundplan condition for each family. Various autapomorphous conditions of single families or family-groups have been omitted; many characters of the female postabdomen have also been omitted because comprehensive information on them is not available. The black rectangles indicate conditions interpreted as apomorphous with respect to the groundplan of the Schizophora. Where common possession of such apomorphous conditions is interpreted as due to synapomorphy, a continuous black rectangle is shown against the families concerned. Corresponding plesiomorphous conditions are shown by the white rectangles. Further modifications of a primary apomorphous condition are shown in separate rows, with synapomorphy at the primary level indicated by linking the rectangles. The list of characters is as follows.

Apomorphous conditions (black rectangles)	Plesiomorphous conditions (white rectangles)
Head	
1. 2nd antennal article with cleft	2nd antennal article without cleft
2. Ocellar bristles reduced	Ocellar bristles well developed
3. *Ori* present	*Ori* absent
4. Only 1 *ors* present	At least 2 *ors* present
4a. *Ors* absent	
5. 1 or no *vt* present	2 *vt* present
6. *Pvt* parallel or convergent	Divergent *pvt* present
7. *Pvt* reduced or absent	(as 6)
8. External mouth opening with prestomal teeth	Prestomal teeth absent
9. Hyoid present	Hyoid absent
10. Pseudotracheae opening into one or two main channels	Pseudotracheae opening directly into external mouth opening
Wing	
11. Costa broken at end of sc or r_1	Costa unbroken
11a. Costa broken twice	
11b. Costa broken near humeral crossvein	
11c. Costa secondarily unbroken	
12. Subcosta faded or fused with r_1 distally, not reaching wing margin as distinct vein	Subcosta distinct to wing margin
13. Anal vein not reaching wing margin as distinct vein	Anal vein distinct to wing margin
13a. Anal vein and anal cell lost	
14. Lower cross-vein lost	Lower cross-vein present
Preabdomen	
15. 2nd to 5th spiracles within terga	2nd to 5th spiracles in membrane
15a. 3rd to 5th spiracles secondarily in membrane	
Male postabdomen	
16. 6th tergum shorter than 5th tergum	6th tergum as long as 5th tergum
16a. 6th tergum asymmetrically reduced	
16b. 6th tergum reduced to two fragments	
16c. 6th tergum lost	
16d. 6th tergum fused with 8th sternum	

129

17. 6th sternum more strongly
    developed on left side
17a. 6th sternum secondarily
    symmetrical
17b. 6th sternum triangular
17c. 6th sternum linked to pregenital
    sclerite on both sides, or forming
    part of this sclerite
17d. 6th sternum reduced or lost
18. 7th sternum asymmetrical, ex-
    tending dorsally on left side
18a. 7th sternum forming complete
    ventral band
18b. 7th sternum not delimited, absent
    or fully fused with 8th sternum to
    form a composite pregenital sclerite
18c. Pregenital sclerite divided into la-
    teral plates
19. 8th sternum or pregenital sclerite   8th sternum (or pregenital sclerite) large
    reduced to narrow band
20. 7th tergum reduced to lateroventral
    vestige
20a. 7th tergum vestige lost
21. 8th tergum vestige lost   8th tergum vestige present
22. 7th spiracles lost (both sides)   6th and 7th spiracles present in mem-
                                               brane on both sides
23. 6th and 7th spiracles lost (both sides)   (as 22)
24. 7th left spiracle within 7th sternum   (as 22)
25. 7th right spiracle within 8th sternum   (as 22)
26. 6th and 7th spiracles on both sides   (as 22)
    within pregenital sclerite
27. various other modifications in posi-   (as 22)
    tion and/or numbers of postabdom-
    inal spiracles
28. Telomeres lost or fused with perian-   Telomeres discrete, not double or bifid
    drium
29. Telomeres double or bifid   (as 28)
30. Processus longi well differentiated   Processus longi poorly differentiated or
                                               absent
30a. Processus longi secondarily lost
31. Cerci linked to telomeres   Cerci not linked to telomeres
31a. Cerci reduced (not linked to telo-
    meres)
32. Aedeagal apodeme linked or fused   Aedeagal apodeme free from body wall
    with the hypandrium or the area be-
    tween the hypandrial arms
33. Hypandrial arms fused above base   Hypandrial arms not fused, nor forming
    of aedeagus   a trough
34. Hypandrial arms forming a trough   (as 33)
    or sheath around aedeagal apodeme
    and base of aedeagus
35. Articulated pregonites present   Pregonites absent or not articulated
35a. Pregonites secondarily reduced
36. Postgonites setose   Bare postgonites present
37. Postgonites reduced   (as 36)

130

38. Base of aedeagus swung through
    wide arc against aedeagal apodeme
38a. Aedeagus directed posteroventrally
    at rest (swinging mechanism sec-
    ondarily lost)
39. Aedeagus long and flexible, coiled   Aedeagus not of this type
    when at rest
39a. Aedeagus reduced
40. Epiphallus present   Epiphallus absent
40a. Epiphallus secondarily lost
41. Ejaculatory apodeme reduced   Ejaculatory apodeme well developed
41a. Ejaculatory apodeme lost
Female postabdomen
42. 7th spiracles displaced anteriorly   7th spiracles on 7th segment
43. 7th spiracles lost   (as 42)
44. 2 spermathecae   3 spermathecae
44a. Spermathecae reduced

which there is no clear evidence that this condition is secondary because other
characters indicate a close relationship to groups showing the characteristic
muscoid structure. My conclusion that the Muscoidea represent a monophyletic
group is based not on any vague distinction between asymmetrical and symmetrical types of postabdomen, but on a particular and precisely definable
asymmetrical condition of their groundplan, as indicated in my characterization
above. Asymmetrical conditions of some of the postabdominal sclerites are also
shown by a few members of other superfamilies, notably *Eurychoromyia*
(Lauxanioidea) and *Somatia* (Periscelididae, Nothyboidea), but these conditions
appear to be derived from symmetrical conditions in the groundplan of those
superfamilies. It is possible that the male postabdomen was also asymmetrical
(but probably only weakly so) in the groundplan of the Lonchaeoidea, as
previously discussed. The male postabdomen was probably fully symmetrical in
the groundplan of the remaining superfamilies of Schizophora (Lauxanioidea,
Drosophiloidea and Nothyboidea).

I have found the subdivision of the Muscoidea into subordinate groups
problematical, because of the basic similarity in the structure of the male postabdomen and genitalia in the groundplan of many families. Further analyses of
other organ systems are desirable in order to improve the classification. The
groups Sciomyzoinea and Anthomyzoinea are particularly in need of critical
study, since they are based on weak characterization. The possibility that these
groups may be paraphyletic because they include families more closely related
to one of the other prefamilies, should be borne in mind in future studies. The
other groups here recognized as prefamilies (Tanypezoinea, Calyptratae,
Micropezoinea, Australimyzoinea, Diopsioinea, Agromyzoinea and Tephritoinea) seem to me to represent probable monophyletic groups on the basis of the
characterization given below. Some of the characters on which the family classification is based are indicated on figure 17.

B. 5. Prefamily Tanypezoinea

In this group I include two families, the Tanypezidae (including *Strongyloph-thalmyia*) and the Heteromyzidae. This grouping is new, although perhaps anticipated by BRAUER's (1880) reference of both families to the 'Schizometopa'. The agreement in apomorphous conditions between these two families is substantial, and I see no good reason to ascribe it to homoplasy. I characterize the Tanypezoinea by the following groundplan conditions which are apomorphous with respect to the groundplan of the Muscoidea.

(1) 7th abdominal spiracles lost in both sexes.

Loss of spiracles occurs in only a few groups of Muscoidea, notably in certain Anthomyzoinea and Tephritoinea, and such reduction has not always proceeded in parallel in both sexes. Other families in which parallel reduction of the 7th spiracles in both sexes has occurred are the Trixoscelididae, Opomyzidae and Asteiidae (Anthomyzoinea) and the Eurygnathomyiidae, Mormotomyiidae and Cnemospathidae (Tephritoinea). Doubtless these other cases are due to homoplasy with the condition shown by the Tanypezoinea, as the other characters of the groups concerned do not suggest that they are most closely related to the Tanypezoinea. Probably the loss of the 7th spiracles in Tanypezoinea is an autapomorphous condition.

(2) 7th and 8th abdominal segments (♀) elongate, forming slender ovipositor which can be retracted within 5th and 6th segments (see HENNIG 1958, figs. 79, 84 and 242).

(3) 7th and 8th tergum vestiges lost (♂).

(4) Sclerotized fold (epiphallus) projecting posteriorly from base of aedeagus (♂) (figs. 54 and 58).

(5) Costa broken at end of subcosta (see HENNIG 1958).

The above apomorphous conditions are combined with certain noteworthy plesiomorphous conditions. The 6th segment of the male abdomen is little modified, bearing a large tergum and sternum (figs. 55 and 56). In Tanypezidae this 6th sternum is distinctly asymmetrical, more strongly developed on the left side, while in Heteromyzidae it appears almost symmetrical. It is not clear which condition should be ascribed to the groundplan of the Tanypezoinea. If the large, almost symmetrical 6th sternum shown by the Heteromyzidae corresponds to the groundplan condition of the Tanypezoinea, then this is the only prefamily of Muscoidea showing an almost symmetrical 6th sternum in its groundplan (apart from groups showing conditions which are obviously secondary). However, if the 6th sternum was large and almost symmetrical in the groundplan of the Muscoidea, it is puzzling that such a condition has not been retained by any group other than the Heteromyzidae. I think it more parsimonious, at least on present information, to postulate that the distinctly asymmetrical condition of the 6th sternum shown by the Tanypezidae corresponds to the groundplan condition for the Tanypezoinea and for the Muscoidea as a whole (as indicated in my characterization of the Muscoidea above).

Another noteworthy condition of the Tanypezoinea which is probably plesiomorphous for the Muscoidea, is sexual dimorphism in the size of the eyes (resulting in a narrower frons in the male). Such a condition I consider to be plesiomorphous for the Schizophora, as indicated above in my discussion of the

132

Lonchaeoidea. To my knowledge the only groups of Muscoidea which show a pronounced sexual dimorphism of this kind are the Calyptratae and Tanypezoinea, apart from the almost certainly secondary case of *Neomaorina* (Tephritidae family-group). The dimorphism is shown by many members of both families of Tanypezoinea, although lost in a few groups (such as *Strongylophthalymia*).

The aedeagi of the northern hemisphere genera of Tanypezidae may at first sight seem very different from those of the Heteromyzidae. But the morphological gap is bridged by the neotropical *Neotanypeza* (fig. 54), which has a rather short aedeagus whose sclerotization includes a pair of lateral paraphalli and a weakly sclerotized semicylindrical area (probably part of the wall of the ejaculatory duct) lying between the paraphalli. This condition of the aedeagus distinctly resembles the condition shown by the Heteromyzidae (fig. 58), in which the ejaculatory duct is conspicuously sclerotized. However it is not clear whether the Tanypezoinea can be characterized as apomorphous in their groundplan in respect of sclerotization of the ejaculatory duct, as such sclerotization is scarcely developed in *Tanypeza* and *Strongylophthalmyia*.

The Tanypezidae have divergent postvertical bristles, while in the Heteromyzidae these are convergent. Which condition should be ascribed to the groundplan of the Tanypezoinea is not clear. The same doubt applies to the development of oral vibrissae, which are more or less differentiated in the Heteromyzidae but not in the Tanypezidae.

Aczél (1951) placed the Tanypezidae with the Micropezoinea in a group 'Tanypezidiformes', but I am unable to find any firm grounds for supposing that the Tanypezidae (or the Tanypezoinea as a whole) are more closely related to the Micropezoinea than to other Muscoidea. Aczél did not give any convincing definition of the 'Tanypezidiformes' in terms of synapomorphous conditions, and I therefore support Hennig (1958) in rejecting his proposal.

C. 12. Family Tanypezidae (including Strongylophthalmyiidae)

Sources of information

The male postabdomen and genitalia of the Tanypezidae were treated by Hennig (1936b). I studied preparations of the following: – *Tanypeza luteipennis* Knab & Shannon; *Strongylophthalmyia angustipennis* Melander; *Neotanypeza elegans* (Wiedemann).

Description of male postabdomen and genitalia (figs. 54-55)

6th abdominal spiracles symmetrically situated in membrane below 6th tergum: 7th spiracles absent.

Postabdomen asymmetrical. 6th tergum large, of similar length to 5th tergum. 6th sternum large, slightly asymmetrical (extending towards 7th sternum on left side). 7th and 8th terga absent. 7th sternum asymmetrically developed on left side, more or less fused with large 8th sternum.

Telomeres not discrete, represented by setose lobes of periandrium. Interparameral sclerotization consisting of a pair of bacilliform sclerites (processus longi). Cerci discrete. Aedeagal apodeme long and rod-like, linked to hypandrium by ventral process or pair of ventral processes arising from about its middle. Conspicuous downwardly projecting lobes (? pregonites) arising from posterior part of hypandrium in *Strongylophthalmyia*, but absent in *Tanypeza* and *Neotanypeza*. Aedeagus variable in length, able to be swung through wide arc against aedeagal apodeme to anteriorly directed rest position: in *Neotanypeza* the aedeagus is short, with a pair of lateral paraphalli and a weakly sclerotized semicylindrical internal area (probably part of the wall of the ejaculatory duct) lying between the paraphalli: the aedeagus of *Tanypeza* is longer, with the paraphalli the only well differentiated sclerites: in *Strongylophthalmyia* the aedeagus is extremely long, with additional distal sclerites as well as the paraphalli: a sclerotized epiphallus is well developed in all genera. Postgonites absent. Slender ejaculatory apodeme present.

Characterization and discussion

The Tanypezidae are well characterized as a monophyletic group by the following apomorphous conditions with respect to the groundplan of the Tanypezoinea.

(1)   Telomeres (♂) fused with periandrium.

   In the Heteromyzidae the large telomeres remain discrete.

(2)   Epiphallus (♂) much enlarged (fig. 54).
(3)   Aedeagal apodeme (♂) linked to hypandrium by ventral process, or pair of ventral processes, arising from about its middle (fig. 54).

   In the Heteromyzidae the aedeagal apodeme is free from the hypandrium. The apodeme is long and rod-like in both families.

(4)   Postgonites (♂) lost.
(5)   Anal vein not reaching wing margin (HENNIG 1958).
(6)   Legs very long and slender.

The genus *Strongylophthalmyia*, formerly misplaced in the Psilidae, was treated as a distinct family by HENNIG (1958), who suggested that its affinities probably lay with the Tanypezidae. The additional characters considered in my present analysis confirm HENNIG's judgement; as far as I can see, *Strongylophthalmyia* is simply a genus of Tanypezidae in which sexual dimorphism of the eyes has been lost. The male postabdomen and genitalia of *Strongylophthalmyia* are in most respects very similar to those of *Tanypeza* and *Neotanypeza*, from which they differ obviously only in respect of the extreme elongation of the aedeagus. In none of the character sequences which I have considered does

134

Strcngylophthalmyia show relatively plesiomorphous conditions in comparison with other Tanypezidae, and therefore the recognition of the group as a separate family does not seem justified.

HENNIG (1958) placed the Tanypezidae (including *Strongylophthalmyia*) in his tentative group Nothyboidea, together with the Megamerinidae, Nothybidae, Diopsidae and Psilidae. However he was only able to suggest possible synapomorphies between the Tanypezidae and the Psilidae. The conditions in question were: (1) the presence of a costal break, and (2) shortening of the anal vein. These conditions are unconvincing in isolation, as they are of widespread occurrence among the Schizophora. The additional characters considered in my present analysis do not support HENNIG's delimitation of the Nothyboidea, and I have reappraised the affinities of the included families in this work.

C. 13. Family Heteromyzidae

Sources of information

No information on the male postabdomen and genitalia of this group was found in the literature. My treatment is based solely on study of preparations of *Tephrochlamys rufiventris* (MEIGEN) and *Heteromyza atricornis* MEIGEN.

Description of male postabdomen and genitalia (figs. 56-58)

6th abdominal spiracles symmetrically situated in membrane below 6th tergum: 7th spiracles absent.

Postabdomen asymmetrical. 6th tergum well developed, although shorter than 5th tergum. 6th sternum symmetrical, rather smaller than preceding sterna. 7th and 8th terga absent. 7th sternum asymmetrically developed on left side, more or less fused with large 8th sternum.

Periandrium bearing large discrete telomeres. Interparameral sclerite consisting of angulate band of sclerotization. Cerci discrete. Aedeagal apodeme long and rod-like, free from hypandrium. Aedeagus short, able to be swung through wide arc against aedeagal apodeme to anteriorly directed rest position; sclerotization of aedeagus including well-developed phallophore, from which an epiphallus arises, conspicuous internal sclerotization of ejaculatory duct, and ill-defined paraphalli which have a rather striate appearance. Small postgonites present. Slender ejaculatory apodeme present.

Characterization and discussion

The sense of Heteromyzidae here proposed is new, although the name has been used before in another sense (for instance by FREY 1921). Recent authors have included *Tephrochlamys* and *Heteromyza* in the Heleomyzidae, a group much in need of critical study. Since the available monographs on Heleomyzidae consider only a limited range of external characters and make no attempt to give details of the postabdominal morphology of either sex, it is scarcely possible to judge from them whether there are further genera which may be referable to the Heteromyzidae. Conventional keys to 'Heleomyzidae' are misleading in that they imply that *Heteromyza* and *Tephrochlamys* have a row of outstanding costal spines like true heleomyzids, when in fact these are only weakly or not at all differentiated. As far as I am aware, true Heleomyzidae never retain sexual dimorphism in the size of the eyes. Such dimorphism is well marked in *Heteromyza*, as also in Tanypezidae, though apparently not in *Tephrochlamys*.

Apomorphous conditions of the Heteromyzidae with respect to the ground-plan of the Tanypezoinea are as follows.

(1) 6th segment ($\male$) shortened, with its sclerites at most half as long as those of 5th segment; 6th sternum almost symmetrical (fig. 56).

In some Tanypezidae the sclerites of the 6th segment are fully equal in length to those of the 5th segment, while in those Heteromyzidae studied they are somewhat shorter. As previously discussed, I postulate that some degree of asymmetry of the 6th sternum can be ascribed to the groundplan of the Tanypezoinea and of the Muscoidea as a whole; hence I evaluate the absence of distinct asymmetry of this sternum in the Heteromyzidae as probably secondary (apomorphous).

(2) Aedeagus ($\male$) with conspicuously sclerotized ejaculatory duct (fig. 58).
(3) Spermathecal ducts ($\female$) with swollen sclerotized area adjacent to spermathecae.

Both the Tanypezidae and Heteromyzidae retain three spermathecae. In *Tephrochlamys* two of these spermathecae are served by a common duct. Each spermathecal duct is conspicuously swollen and sclerotized adjacent to the spermatheca or twin spermathecae respectively (see HENNIG 1958, fig. 27). This condition is unique, as far as known. Whether it characterizes all Heteromyzidae has not been confirmed.

B. 6. Prefamily Calyptratae (= Thecostomata FREY)

The name Schizometopa BRAUER is often listed as an additional synonym of the Calyptratae, but these groups only partly coincide since BRAUER (1883) included the Tanypezoinea in the Schizometopa but excluded the Scatophagidae.

The Calyptratae are one of the most surely grounded monophyletic groups within the Schizophora. The definition of the group has been studied by many workers, with the result that a long list of conditions can now be put forward as probable groundplan conditions of the Calyptratae which are apomorphous with respect to the groundplan of the Muscoidea.

(1) 2nd antennal article with longitudinal cleft or suture externally.

This condition is by no means confined to the Calyptratae (as implied in some identification keys), but also occurs widely among the Lonchaeidae and Drosophiloidea, and in some genera of Tephritoinea, Megamerinidae and Psilidae. However there seem to be no sufficient

136

grounds for considering any of these groups most closely related to the Calyptratae, and I therefore think that the presence of this suture is probably an autapomorphous condition of the Calyptratae, in agreement with HENNIG (1958). Important new information and discussion on the structure of the antennae in Schizophora is given by HENNIG (in press).

(2) Lower fronto-orbital bristles *(ori)* present (HENNIG 1958).

(3) Anterior upper fronto-orbital *(ors)* proclinate (HENNIG 1958).

(4) Vibrissae present (HENNIG 1958).

(5) Posthumeral or intra-alar bristles (or both) present (GIRSCHNER 1893, 1896).

It is probable that the presence of one or other of these bristles may be ascribed to the groundplan of the Calyptratae, but the sequence of conditions involved has not been clarified.

(6) Costa broken at end of subcosta.

HENNIG (1965a: 5) now accepts this as a groundplan condition of the Calyptratae, contrary to his earlier view (HENNIG 1958: 678).

(7) Abdominal spiracles 2-5 ($\male$) or 2-6 ($\female$) lying within side-margins of respective terga.

It is arguable that this condition does not belong to the groundplan of the Calyptratae, if the Hippoboscidae family-group is the sister-group of all other Calyptratae (see below). However, even if this hypothesis is correct, I am doubtful whether the characterization given above should be revised. I think that the situation of (at least) the 3rd to 5th spiracles in membrane in members of the Hippoboscidae family-group may well be secondary, resulting from expansion of the area of abdominal membrane to allow greater distension when a blood meal is obtained. I draw a distinction between the sexes in respect of this character, since the 6th spiracles lie in membrane in most male Calyptratae and there is no reason to doubt that this was also so in the groundplan of the group.

(8) 7th left spiracle ($\male$) lying within 7th sternum, displaced dorsally in relation to spiracles of preceding segments (figs. 59, 61 and 67).

In the groundplan of the Muscoidea the 7th left spiracle lay in membrane before the 7th sternum. Its inclusion within the 7th sternum in Calyptratae is apomorphous. A similar modification has occurred sporadically among other groups of Muscoidea, but is much less widespread than the situation of this spiracle in membrane.

(9) 7th tergum vestige lost ($\male$).

I found no trace of the 7th tergum in any Calyptratae. But I noted the presence of the 8th tergum vestige in *Cordilura* (Scatophagidae) and *Pegomya* (fig. 61) (Anthomyiidae).

(10) Large articulated pregonites present ($\male$) (figs. 60 and 70).

The pregonites of Calyptratae represent a second pair of paraphyses according to my interpretation of BLACK's (1966) observations (see section 3.1 above). Although pregonites (or analogous structures) are present in many groups of Schizophora, I do not know of any group other than the Calyptratae in which they are fully articulated. According to SALZER (1968) muscles originating from the aedeagal apodeme (M 38, *musculi phallapodemoparamerales anteriores*) are inserted on the base of the pregonites ('Proparameren').

137

(11) Epiphallus (♂) well developed (figs. 60, 62, 65 and 70).

A long epiphallus (posterior process of the phallophore) should probably be ascribed to the groundplan of the Calyptratae. Such a structure is widely distributed among all families except the Fanniidae and the Hippoboscidae family-group.

(12) Processus longi present (fig. 70).

The processus longi are paired sclerites articulated at either end with the vertical sections of the hypandrial arms and the outer basal processes of the telomeres. They are absent in the Hippoboscidae family-group, Muscidae and Fanniidae, but present in Tachinidae (*s.l.*), Anthomyiidae and Scatophagidae. This distribution suggests that such sclerites were probably present in the groundplan of the Calyptratae.

FREY (1921: 208) interpreted three conditions of the mouthparts as 'specializations' of the Calyptratae, as follows:

(13) External mouth-opening with prestomal teeth;

(14) Inner mouth-opening with hyoid (= theca, Gelenkkapsel); and

(15) Pseudotracheae with both dorsal and ventral main channel ('Sammlungs-rohr').

Since Frey studied members of seven genera of Scatophagidae, I regard the absence of character 15 in the scatophagid *Pogonota* reported by van EMDEN (1950) as secondary. The hyoid is secondarily lost in Stomoxyinae (Muscidae) and some Hippoboscidae.

The Calyptratae as here defined include the Pupipara (Hippoboscidae *s.l.*), which I classify with the Glossinidae in the Hippoboscidae family-group. The other families of Calyptratae here recognized are the Scatophagidae, Anthomyiidae, Fanniidae, Muscidae and Tachinidae *s.l.* I do not include the Mormotomyiidae, which HENNIG (1958) thought were aberrant Calyptratae. That family is discussed below under the Tephritoinea.

GIRSCHNER (1893) was the first to conclude that the Scatophagidae should be included in the Calyptratae, basing his conclusion on characters of the chaetotaxy. But many taxonomists resisted this change and continued to classify the family among the 'Acalyptratae'. GIRSCHNER'S views were vindicated by FREY (1921), who reached the same conclusion from analysis of an entirely different set of characters (the mouthparts). The structure of the male postabdomen and genitalia provides further strong evidence in support of the classification of the Scatophagidae in the Calyptratae, as indicated in my characterization above.

The name 'Calyptratae' refers to the large squamae (= calyptrae) shown by many members of this group. However much variation is shown in the size of the squamae, and they are relatively small in the Scatophagidae and some members of other groups. HENNIG (1965a) considers that the phylogenetic significance of size differences in the squamae is still obscure, and I am consequently not using this character to support the classification here presented. Another character not used in my characterization is the presence or absence of sexual dimorphism in the size of the eyes and frons width. A greater or lesser degree of dimorphism in this respect is shown by most species of all families except the Scatophagidae and the Hippoboscidae family-group. I think that such sexual dimorphism was

138

present in the groundplan of the Schizophora (see above under Lonchaeoidea), but it is arguable that its presence in some families of Calyptratae is secondary (see HENNIG, in press), since no dimorphism is apparent in the two families just mentioned. I myself doubt the need to postulate secondary evolution of such sexual dimorphism in the Calyptratae, since such dimorphism is also well developed in a second group of Muscoidea, the Tanypezoinea. But since the evaluation of this character is disputed, I do not base any argumentation upon it.

The question of how the Calyptratae should be subdivided has still not been settled. The primary subdivision should probably be either between the Scatophagidae and all other Calyptratae (as in the classification outlined by HENNIG 1955: 2), or between the Hippoboscidae family-group and all other Calyptratae. The latter hypothesis was implied in the comments of HERTING (1957) and is now favoured by HENNIG (in press). These different interpretations of the relationships between the families of Calyptratae are implicated with possible alternative evaluations of certain character sequences. The following conditions shown by the Glossinidae can be interpreted as groundplan conditions of the Calyptratae, if the Hippoboscidae family-group is postulated to be the sister-group of all other Calyptratae: (1) 6th tergum (♂) unreduced, as long as the 5th tergum (fig. 67); (2) 6th and 7th segments (♀) not retractile, with the 7th pair of spiracles lying on the 7th segment; (3) cerci (♂) not linked to the telomeres; and (4) 7th right spiracle (♂) lying within the 8th sternum (figs. 61 and 67). On the other hand there are grounds for suspecting that some of these conditions may be secondary, in particular because the development of pseudoplesiomorphous (secondary) conditions of the female postabdomen is well shown by certain larviparous groups of Tachinidae s.l. Since all members of the Hippoboscidae family-group are macrolarviparous or pupiparous, the external structure of the female postabdomen in this group can be explained as the result of a similar series of modifications leading to a pseudoplesiomorphous condition; reduction of the male cerci may have been correlated with such changes in the structure of the female postabdomen. Only the presence of an unreduced 6th tergum in the male is difficult to interpret as pseudoplesiomorphous; but the evidence of this condition seems inconclusive, since the difference in the relative size of the 6th tergum in the Glossinidae and in some Scatophagidae is not great.

The presence of the 7th right spiracle (♂) on the inverted 8th sternum is an undoubtedly apomorphous condition shown by members of all groups of Calyptratae except the Fanniidae (whose postabdominal morphology is evidently highly modified) and some Scatophaginae. In the latter group the position of the right postabdominal spiracles is variable. Both may be in the membrane of the genital pouch *(Cordilura latifrons* LOEW); the 6th at the edge of the 6th tergum, the 7th in membrane *(Cordilura confusa* LOEW); both within the 6th tergum *(Orthacheta* and *Spaziphora)*; or the 6th within the 6th tergum, the 7th within the 8th sternum *(Pogonota* and *Scatophaga)*. None of the species examined showed the arrangement characteristic of most other Calyptratae, in which the 6th right spiracle remains in membrane but the 7th right lies within the (inverted) 8th sternum. The question of whether the latter condition can be ascribed to the groundplan of the Calypratae is thus implicated with whatever view is taken of the relationship between the Scatophaginae and other Calyptratae. Only if the former are postulated to be the sister-group of all other Calyptratae (including Delininae), is it reasonable to maintain that both the right postabdominal spiracles lay in membrane in the groundplan of the Calyptratae. Movement of the 7th right spiracle onto the inverted 8th sternum can then be postulated as an autapomorphous groundplan condition of the second sister-group (other Calyptratae, including Delininae). Additional support for such an interpretation is provided by the relatively large size of the 6th tergum (♂) in Scatophaginae, to which CRAMPTON (1944a) drew attention. As a further consequence of the interpretation that the Scatophaginae are the sister-group of all other Calyptratae, the following conditions may be attributed to the ground-plan of the Calyptratae: (1) female postabdomen forming an ovipositor which can be retracted within the 5th segment (see HENNIG 1965a: 6); (2) 7th abdominal spiracles (♀) displaced anteriorly, lying on or near the posterior margin of the 6th tergum (see the information presented by HERTING 1957); and (3) cerci (♂) linked to the telomeres (see HENNIG, in press).

SALZER (1968) has described a link (articulation) between the telomeres and cerci in males of *Calliphora* (Tachinidae *s.l.*). I agree with HENNIG's (in press) view that this condition can be ascribed to the groundplan of the Calyptratae as a whole, or possibly with the exclusion of the Hippoboscidae family-group (if the latter is the sister-group of all other Calytratae). Such a link is absent in most other Schizophora, but I found a similar condition, probably through homoplasy, in the Diopsidae and a few genera of Heleomyzidae.

The classification of the Calyptratae proposed by ROBACK (1951) was incomplete, since he did not consider the Hippoboscidae *s.l.* ROBACK's paper provides much useful information, but I do not think that he gave sufficient justification for two of his main classificatory proposals, namely that the Scatophagidae ('Scopeumatinae') and Anthomyiidae be included in the 'Anthomyiidae' in a widened sense, and that the 'Anthomyiidae', Fanniidae and Muscidae be included in a superfamily 'Muscoidea'. While the possibility that the 'Muscoidea' and 'Anthomyiidae' in ROBACK's sense are monophyletic groups cannot be excluded on present information, no positive evidence for this in terms of synapomorphous conditions has yet been offered. The characterization of these groups given by ROBACK refers to symplesiomorphous conditions or conditions whose phylogenetic significance has not been clarified. Therefore I accept HENNIG's (1958) opinion that these proposals were not an improvement in the classification from the standpoint of phylogenetic systematics. The question of which of the groups here recognized as families are most closely related to one another is unlikely to be clarified until much more comprehensive information than hitherto is available on the Scatophagidae and Anthomyiidae.

The presence of the 7th right spiracle on the inverted 8th sternum in most male Calyptratae has caused some authors difficulty in interpreting the homologies of the sclerites. To clarify this question I therefore present this detailed note. At first sight my interpretation may seem self-contradictory, since although I claim to have shown that the 8th segment is inverted (see section 3.2), it is established that the 7th spiracles of Calyptratae are connected with the preceding spiracles on the same side in the normal manner (fig. 12). Crossing of the lateral tracheal trunks as a result of the rotation within the puparium occurs only after the last (7th) pair of spiracles. To understand this morphological paradox, it is necessary to consider the time sequence of the changes which occur within the puparium. The development of the imaginal tracheal system of *Calliphora erythrocephala* MEIGEN within the puparium has been treated by PIHAN (1969). The author has unfortunately not stated the relationship of the sequence of tracheal development to the rotation of the terminal segments. But if we assume that the speed of development was similar to that of SCHRÄDER's (1927) material of the same species, the synchronization appears to be roughly as follows. The terminal segments rotate on about the 5th day after pupation. At this time the larval tracheal epithelium has largely degenerated, but the imaginal tracheoblasts of the 4th to 7th abdominal segments are still small; the lateral commissures between them have not yet been regenerated, and (to judge from PIHAN's figure) the adult spiracles have not yet been formed. The larval transverse spiracular commissures of the 8th abdominal segment have degenerated by this time and are not regenerated in the adult. The time of formation of the adult spiracles and their linking with the rest of the system was unfortunately not established precisely, but it seems unlikely that this occurred before the eighth day. At any rate, though the possibility of some degree of error in correlating two authors' experimental data must be admitted, it seems clear that the 7th abdominal spiracles are not formed until after the completion of rotation. SCHRÄDER (1927) makes no mention of spiracles, which presumably means that no trace of these was detected in his serial sections of specimens fixed during the course of rotation. Thus it appears that the completion of rotation before the formation of the terminal abdominal spiracles has allowed one of these (the 7th right) to develop on a rotated segment without any change in its relation to the preceding spiracles. The ends of the lateral trunks are rotated (after the 7th spiracles), because they develop from the epithelium of the larval tracheal trunks which are carried across one another by the rotation while in the process of regeneration.

C. 14. Family Scatophagidae (= Cordiluridae)

I follow the advice of C. W. SABROSKY in accepting the family name Scatophagidae, in preference to Cordiluridae, on grounds of priority.

Sources of information

I found no detailed account of the male postabdomen and genitalia of Scatophagidae in the literature. My treatment is based on study of preparations of the following.

Scatophaginae: – *Cordilura (Cordilura) confusa* LOEW; *Cordilura (Cordilura) latifrons* LOEW; *Cordilura (Achaetella) varipes* (WALKER); *Spaziphora cincta* (LOEW); *Orthacheta cornuta* (LOEW); *Pogonota gilvipes* (LOEW); *Scatophaga aldrichi* (MALLOCH); *Scatophaga stercoraria* (L.).

Delininae: – *Chylizosoma vittatum* (MEIGEN); *Hexamitocera vittata* COQUILLETT; *Neochirosia* Sp.

Description of male postabdomen and genitalia

Scatophaginae (figs. 59-60)

6th left abdominal spiracle usually in membrane, but within 6th tergum in *Spazipora:* 7th left spiracle within synsternum (7 + 8) (displaced dorsally in relation to preceding spiracles). 6th and 7th right spiracles variable in position, sometimes both in membrane *(Cordilura)*, or both within 6th tergum *(Ortha-cheta, Spaziphora)*, or within 6th tergum and 8th sternum respectively *(Scato-phaga, Pogonota)*.

Postabdomen asymmetrical. 6th tergum relatively large in most species, half to two-thirds as long as 5th tergum: if 6th tergum less than half as long as 5th tergum (e.g. *Pogonota, Spaziphora*), synsternum (7 + 8) similarly narrow. 6th sternum usually represented by strongly sclerotized narrow band, except for expanded area on left side, asymmetrically linked to 7th sternum on left side, ventrally partly concealed by lobes of 5th sternum: however in *Spaziphora* the 6th sternum is represented only by a short strip of sclerotization on the left side, and it is apparently absent in *Pogonota*. 7th sternum fused on left side with 8th sternum, at most with weak indication of suture line between the two sclerites. 7th tergum absent. 8th tergum represented by large infolded sclerite on centre-line of venter in *Cordilura*, but absent in members of other genera examined.

Periandrium bearing discrete telomeres which are variable in shape. Cerci fused or more or less discrete, linked ventrally to telomeres. Telomeres connected with hypandrium by pair of processus longi. Aedeagal apodeme more or less rod-like, free from hypandrium. Hypandrium bearing articulated pregonites posteriorly. Aedeagus able to be swung through wide arc against aedeagal apodeme to anteriorly directed rest position, its sclerotization consisting of the basal phallophore, from which arises a well-developed epiphallus, and variable more distal sclerotization. Postgonites usually well developed (but small in *Spaziphora*). Ejaculatory apodeme present, often fan-shaped.

Delininae

6th abdominal spiracles both in membrane: 7th left spiracle within 7th stern-um (displaced dorsally in relation to preceding spiracles): 7th right spiracle within anteroventral corner of 8th sternum.

Postabdomen asymmetrical. 6th tergum represented by narrow band of sclero-tization fused to 8th sternum (with traces of a suture visible). 6th sternum re-presented by narrow band with expanded area on left side. 7th sternum fully fused with 8th sternum on left side, at most with weak indication of suture line between the two sclerites. 7th and 8th terga absent.

Hypopygium as described for Scatophaginae, except that the processus longi are absent (telomeres and hypandrial arms linked directly).

Characterization and discussion

This family has never been satisfactorily defined from the standpoint of phylogenetic systematics. The diagnostic conditions given in conventional keys to identification are either more plesiomorphous than the conditions shown by other Calyptratae, or of uncertain phylogenetic significance. My impression is that two distinct groups can be recognized within the Scatophagidae as current-ly delimited. These are distinguished as the Scatophaginae and Delininae in the description above. This distinction was first made by VOCKEROTH (in STONE *et*

142

al. 1965). The Delininae in this sense (Delinini of VOCKEROTH) include only those species whose larvae are leaf-miners on Orchidaceae and Liliaceae.

I am not satisfied that the Delininae are monophyletic with the Scatophaginae, but do not offer any firm conclusion on this point because of unresolved difficulties in the evaluation of some of the relevant character sequences. As indicated in previous discussion, it is possible to interpret the Scatophaginae as the sister-group of all other Calyptratae through the retention in their groundplan of a relatively large 6th tergum ($\circ$) and the situation of the 7th right spiracle ($\circ$) in membrane (as in *Cordilura*). If further reduction of the 6th tergum and movement of the 7th right spiracle onto the 8th sternum are autapomorphous conditions of one of the postulated sister-groups (all other Calyptratae), then the Delininae are probably referable to the latter and should be removed from the Scatophagidae. They could be reclassified as a separate family or perhaps included in the Anthomyiidae. if grounds for the latter course could be found. However the position of the right postabdominal spiracles in male Scatophaginae is variable, and the condition shown by *Cordilura* (in which the postabdominal spiracles are unusually large) may be secondary.

I am not able to suggest any common apomorphous conditions (with respect to the groundplan of the Calyptratae) which suggest that the Scatophaginae and Delininae may be correctly grouped together, except perhaps loss of the visible suture between the 7th and 8th sterna ($\circ$) on the left side (fig. 59). However loss of the visible suture between these sclerites seems to have occurred independently in many groups of Calyptratae, and I do not think that much weight can be placed on the evidence of this condition in isolation.

CRAMPTON (1944a) suggested, on the basis of examination of the male postabdomen of *Scatophaga stercoraria* (L.), that this genus should be placed in the 'Scatophagidae' in a restricted sense because its terminalia more closely resemble those of the Anthomyiidae than those of the 'Cordiluridae'. However the reduced 6th tergum shown by that species has probably evolved independently, because I find a large 6th tergum (fully two-thirds of the length of the 5th tergum) in *Scatophaga aldrichi* (MALLOCH). CRAMPTON's proposal thus seems unwarranted. The name Cordiluridae in the sense of authors other than CRAMPTON is a nomenclatural synonym of Scatophagidae, not referring to a different group.

C. 15. *Family Anthomyiidae*

Sources of information

I found no detailed account of the male postabdomen and genitalia of Anthomyiidae in the literature. My treatment is based mainly on study of preparations of the following: – *Pegomya versicolor* (MEIGEN); *Hylemya (Delia) pilitarsis* STEIN; *Fucellia ariciiformis* (HOLMGREN).

Description of male postabdomen and genitalia (figs. 61-62)

6th abdominal spiracles both in membrane: 7th left spiracle within 7th sternum (displaced dorsally in relation to preceding spiracles): 7th right spiracle within anteroventral corner of 8th sternum.

Postabdomen asymmetrical. 6th tergum less than half as long as 5th tergum, and often obviously shorter than 8th sternum (but scarcely so in *Fucellia*). 6th sternum represented by strongly sclerotized narrow band (in *Fucellia* with expanded area on left side), asymmetrically linked to 7th sternum on left side, ventrally partly concealed by lobes of 5th sternum. 7th sternum asymmetrically developed on left side, where it is fused with the large 8th sternum along a conspicuous suture. 7th tergum absent. 8th tergum represented by small sclerite on centre-line of venter *(Pegomya)* or absent *(Hylemya, Fucellia)*.

Periandrium bearing long discrete telomeres. Cerci partly or completely fused, linked ventrally to telomeres. Telomeres connected with vertical sections of hypandrial arms by pair of short processus longi *(Pegomya)*, or linked directly (without discrete processus longi) *(Hylemya, Fucellia)*. Aedeagal apodeme more or less rod-like, free from hypandrium. Hypandrium bearing articulated pregonites. Aedeagus usually able to be swung through wide arc against aedeagal apodeme to anteriorly directed rest position, its sclerotization consisting of basal phallophore, from which arises an epiphallus, and variable more distal sclerotization: but in *Fucellia* the epiphallus is absent and the aedeagus remains posteriorly directed when at rest. Postgonites well developed. Ejaculatory apodeme present (variable in shape).

Characterization and discussion

The limits of the Anthomyiidae require clarification since no autapomorphous conditions can be put forward to demonstrate that the family, as presently delimited, is a probable monophyletic group. All Anthomyiidae studied by me, as well as those figured by CRAMPTON (1944a), may be characterized as apomorphous with respect to the groundplan of the Calyptratae in respect of shortening of the 6th tergum ($\male$), which is less than half as long as the 5th tergum. However such reduction is also shown by the Delininae (Scatophagidae), Fanniidae, Muscidae and Tachinidae *s.l.*, and therefore seems of little value for clarifying the limits of the Anthomyiidae. The extension of the anal vein to the wing margin used as a diagnostic character of the Anthomyiidae in keys to identification is doubtless a plesiomorphous condition for the Calyptratae as a whole. It is HENNIG's stated intention to complete his current revision of the Anthomyiidae in 'Die Fliegen der paläarktischen Region' with an 'allgemeine Teil', where questions of the relationships of the included genera will be considered. Until this work has been completed it will not be possible to judge whether the existing limits of the Anthomyiidae are justifiable.

144

C. 16. Family Fanniidae

Sources of information

The male postabdomen and genitalia of the Fanniidae were treated by CHILL-COTT (1960). I studied a preparation of *Fannia canicularis* (L.).

Description of male postabdomen and genitalia (fig. 63)

6th and 7th abdominal spiracles symmetrically situated within composite pregenital sclerite.

Postabdomen fully symmetrical. 6th tergum absent. 6th to 8th sterna fused to form symmetrical pregenital sclerite with broad lateral areas representing 7th sternum (delimited from the 8th sternum by conspicuous sutures) equally developed on both sides. 7th and 8th terga absent.

Periandrium bearing discrete telomeres of variable shape. Cerci fused, without ventral link with telomeres. In some species a variously modified process, called 'processus bacilliformis' by HENNIG (1965a), arises from the base of the cercal plate. Processus longi absent. Hypandrium without articulated pregonites. Aedeagal apodeme short, sometimes angulate, free from hypandrium or linked to it posteriorly. Aedeagus small, often largely or completely membranous but distinctly sclerotized in *Euryomma*, *Platycoenosia*, *Piezura* and species of *Fannia glaucescens* subgroup (see CHILLCOTT 1960): epiphallus absent: as far as I am aware, the aedeagus remains posteriorly directed when at rest and cannot be swung through a wide arc against the aedeagal apodeme in any species of Fanniidae. Postgonites either developed as small sensory lobes, or forming strongly sclerotized processes which shield base of aedeagus (see CHILLCOTT 1960). Ejaculatory apodeme absent.

Characterization and discussion

The Fanniidae have usually been considered a subfamily of Muscidae. HENNIG (1965a) accepted this classification, and concluded that the 'Fanniinae' represent the sister-group of all other Muscidae. On this interpretation the question of whether the group should be accorded subfamily or family rank has only formal significance. However I think it fair to point out that HENNIG in that paper did not attempt to demonstrate the monophyly of the Muscidae in the wide sense. From his analysis it seems to me that, while the monophyly of each of his groups 'Fanniinae' and 'Muscidae *s. str.*' is supported by strong morphological evidence, the association of these two groups in the 'Muscidae *s.l.*' rests on weaker foundations. The only possible synapomorphies (with respect to the groundplan of the Calyptratae) between these groups, as far as I am aware, are: (1) shortening of the anal vein; (2) loss of the processus longi; and (3) reduction of the 6th tergum (♂). Since these conditions also occur among other

145

groups of Calyptratae, I do not think that they provide conclusive evidence for the monophyly of the Muscidae *s.l.* in HENNIG's sense. I think it more satisfactory, at least on present information, to follow ROBACK (1951) in recognizing the Fanniidae as a full family, thereby restricting the name Muscidae to a group which is more firmly based from the standpoint of phylogenetic systematics.

The characterization of the Fanniidae (as Fanniinae) has been well discussed by HENNIG (1965a), who lists ten apomorphous conditions. I do not repeat these here. Additional apomorphous conditions which I wish to emphasize are as follows.

(11) Ejaculatory apodeme (♂) lost.

(12) Sterna 6-8 (♂) fused to form symmetrical pregenital sclerite ('composite pregenital segment', CHILLCOTT 1960), containing both 6th and 7th spiracles (fig. 63).

Symmetry of the male postabdomen in Fanniidae has been restored in a very characteristic manner, not known among the other families of Calyptratae. The normal structure of the right side of the postabdomen has been suppressed and replaced by a mirror image of the left side. The broad lateral area of the pregenital sclerite containing the spiracles clearly represents the 7th sternum, and is delimited from the 8th sternum posteriorly by a suture. What is remarkable is that the sternum and suture are developed on both sides of the body, not on the left alone as normally among the Muscoidea. The 6th sternum also appears to have become fused into the pregenital sclerite, being represented by the narrow ventral connection. CHILLCOTT's (1960) figures indicate that this complex apomorphous condition is probably universal among the Fanniidae.

C. 17. Family Muscidae

Sources of information

The most comprehensive treatment of the male postabdomen and genitalia of any muscid is RIVOSECCHI's (1958) work on *Musca domestica* L. Some comparative information was given by HENNIG (1965a). I studied preparations of the following: – *Muscina assimilis* (FALLÉN); *Myospila meditabunda* (F.); *Haematobia irritans* (L.).

Description of male postabdomen and genitalia (figs. 64-65)

6th abdominal spiracles both in membrane: 7th left spiracle usually within 7th sternum, but in membrane between 6th tergum and synsternum (7 + 8) in *Haematobia:* 7th right spiracle within anteroventral corner of 8th sternum *(Muscina)*, or in membrane near that corner *(Musca, Myospila)*, or in membrane between 6th tergum and synsternum (7 + 8) *(Haematobia)*.

Postabdomen asymmetrical. 6th tergum less than half as long as 5th tergum, obviously shorter than 8th sternum. 6th sternum represented by strongly sclerotized narrow band with expanded area on left side, asymmetrically linked to 7th sternum on left side. 7th sternum asymmetrically developed on left side, where it is fused with the 8th sternum: the suture between the 7th and 8th sterna is sometimes conspicuous (e.g. *Muscina*), but may not be apparent (e.g. *Myospila, Musca, Haematobia*). 7th and 8th terga absent.

Periandrium usually bearing large discrete telomeres (rarely the telomeres are absent, as in some species of the *Lispe*-group). Cerci large, discrete or partly fused, linked ventrally to telomeres. Telomeres linked directly to vertical sections of hypandrial arms (without discrete processus longi). Aedeagal apodeme rod-like, free from hypandrium. Hypandrium usually bearing articulated pregonites, but these are absent in the *Lispe*-group and *Lispocephala*-group. Aedeagus able to be swung through wide arc against aedeagal apodeme to anteriorly directed rest position, its sclerotization consisting of basal phallophore, from which usually arises an epiphallus (but this is absent in a few groups), and variable more distal sclerotization: HENNIG (1965a) reports that the distal part of the aedeagus is membranous in the Phaoniinae, Mydaeinae, Limnophorinae and Coenosiinae, but this is not the case in the *Myospila* species studied by me. Postgonites well developed. Ejaculatory apodeme usually present, but not traced in *Haematobia*.

Characterization and discussion

The characterization of this family has been discussed by HENNIG (1965a) (as 'Muscidae *s. str.*'), and is not repeated here. In the groundplan of the family the structure of the male postabdomen and external genitalia appears to have been little removed from that of the groundplan of the Calyptratae, except that the processus longi were lost (as also in Fanniidae), the 8th tergum vestige lost, and the 6th tergum shortened. The position of the postabdominal spiracles in *Muscina* (see description above) is the same as in most other Calyptratae (except Scatophaginae), and probably indicative of the groundplan condition for the Muscidae.

HENNIG (1965a) suggested that the Hippoboscidae family-group (*Glossina* + Pupipara) was the sister-group of the Stomoxyinae (Muscidae) and hence a subordinate group of the Muscidae in a phylogenetic system. But he has since abandoned this view (HENNIG, in press). I agree that the hypothesis of a sister-group relationship between the Stomoxyinae and the Hippoboscidae family-group should be rejected, in particular because of the presence of male accessory glands in the latter group. These glands are absent in Stomoxyinae and, as far as known, reduced or absent in all other Muscidae. There is no reason for postulating secondary redevelopment of the male accessory glands after loss, since they are not homologous with any structure present in the female. As far as known the law of irreversibility of evolution applies without restriction in cases of the loss of complex organs, since no case of the reappearance of complex homologous organs has ever been demonstrated (REMANE 1952).

The genus *Eginia* has been considered to represent a distinct family (Eginiidae) by HENNIG (in press), though this classification seems to reflect uncertainty about the relationships of the group rather than any conviction that it merits family rank. It seems to me that the structure of the male postabdomen and genitalia of *Eginia* is compatible with the view that the genus belongs to the Muscidae, but I do not offer a firm opinion as the comparative information available to me on the genitalia of Muscidae is inadequate. Points in favour of such a classification are as follows: (1) the 5th sternum (♂) in *Eginia* is a large plate which is only weakly lobed posteriorly and does not conceal the 6th sternum (as in Fanniidae and Muscidae); (2) the fused cerci (♂) are large, and the telomeres are more or less ventrally directed (as in most Muscidae); (3) the pregonites (♂) are reduced (not articulated) and the epiphallus scarcely developed (as in certain subordinate groups of Muscidae); (4) the telomeres (♂) are directly linked to the hypandrial arms, without intervening processus longi (as in Fanniidae and Muscidae); and (5) the sclerotization of the short aedeagus (♂) is rather weakly differentiated, comparable with the conditions shown by some groups of Muscidae. HENNIG (in press) states that he has found no abdominal spiracles in *Eginia*. Possibly this character is subject to individual variation, since in a preparation of *Eginia ocypterata* (MEIGEN) I found that the 2nd pair of abdominal spiracles was well developed; the atria and apodemes of the 3rd to 5th pairs of spiracles could also be seen, but these did not appear to open to the exterior. The postabdominal spiracles were completely absent, in agreement with HENNIG's observation.

C. 18. Family Tachinidae (sensu lato) (= Calliphoroidea HENNIG 1958)

(including Calliphoridae, Sarcophagidae, Oestridae, Hypodermatidae, Cuterebridae and Gasterophilidae)

The name Tachinidae was used in the present wide sense in the works of GIRSCHNER (1893, 1896), and therefore has precedence over GRUNIN's (1964, footnote on page 30) suggestion that the name Calliphoridae may be so used.

Sources of information

The outstanding work of SALZER (1968) deals with the musculature, innervation and function of the male postabdomen and genitalia of *Calliphora*, as well as with the skeletal parts. Among other recent works containing detailed treatments of the male postabdomen of some of the subordinate groups of Tachinidae *s.l.* may be mentioned those of ROHDENDORF (1967), VERBEKE (1963) and GRUNIN (1964, 1966, 1969). The work of MORRISON (1941) contains much information on the origins of terminology, but some of his proposals for standardizing the terminology are no longer acceptable (for instance he calls the inverted 8th sternum the '7th tergite'). I studied preparations of the following: – *Eucalliphora lilaea* (WALKER); *Pollenia rudis* (F.); *Hypoderma lineatum* (VILLERS).

148

Description of male postabdomen and genitalia (figs. 52, 53 and 70)

6th abdominal spiracles both in membrane: 7th left spiracle, when present, within 7th sternum; 7th right spiracle, when present, within anteroventral corner of 8th sternum. In *Hypoderma* 7th spiracles absent.

Postabdomen asymmetrical. 6th tergum usually present, less than half as long as 5th tergum and usually shorter than 8th sternum (more or less fused with 8th sternum in *Hypoderma*): in a few genera the 6th tergum is absent (e.g. *Gonia*, *Thelaira*, *Boettcheria*). 6th sternum represented by strongly sclerotized narrow band (slightly expanded on left side), which is asymmetrically linked to the 7th sternum on the left side. 7th sternum asymmetrically developed on left side, where it is fused with the 8th sternum: the suture between the 7th and 8th sterna is sometimes conspicuous (e.g. *Calliphora*), but may not be apparent (e.g. *Eucalliphora*, *Pollenia*, *Hypoderma*): 8th sternum or synsternum (7 + 8) usually well developed, but rather weakly sclerotized in *Hypoderma*. 7th and 8th terga absent.

Periandrium bearing discrete, usually elongate telomeres. Cerci fused, long and pointed in most groups, linked ventrally to telomeres, Telomeres connected with hypandrial arms by pair of processus longi. Aedeagal apodeme more or less rod-like, free from hypandrium. Hypandrium bearing articulated pregonites posteriorly. Aedeagus able to be swung through wide arc against aedeagal apodeme to anteriorly directed rest position, its sclerotization consisting of basal phallophore, which usually bears a distinct epiphallus, and variable (often complex) more distal sclerotization (see VERBEKE 1963 for an account of the variation of the aedeagus in Tachinidae *s.str.*). Postgonites usually well developed. Ejaculatory apodeme usually present (but reported to be absent in *Gasterophilus*).

Characterization and discussion

The available information indicates that the Tachinidae *s.l.* are characterized by the following groundplan conditions which are apomorphous with respect to the groundplan of the Calyptratae.

(1) Hypopleuron with strong bristles below metathoracic spiracle.

This condition was already used to define the Tachinidae in the wide sense by GIRSCHNER (1893, 1896). The only calyptrate with hypopleural pubescence which I exclude from the Tachinidae is *Eginia* (discussed above under Muscidae). In some 'Gasterophilidae' the hypopleural bristles have been secondarily lost, as is indicated by their presence (although weakly developed) in a few species (HENNIG 1952: 444).

(2) 8th sternum (♀) entire (HERTING 1957).

In the great majority of other Calyptratae the 8th sternum is divided into a pair of sclerites. It seems likely that an undivided condition has been secondarily restored in the Tachinidae.

(3) Vein m_{1+2} sharply bent towards r_{4+5} apically.

149

This groundplan condition is further modified in many groups by the development of an appendix ('Aderanhang') from the bend of m_{1+2}. In a very few genera (notably *Gasterophilus*) a straight vein m_{1+2} has been secondarily restored (HENNIG 1958: 680). Some Muscidae also show a similar bending of vein m_{1+2}, but this condition was not present in the groundplan of the family (HENNIG 1965a).

(4)   Anal vein not reaching wing margin.

HENNIG (1958: 680) has concluded that the anal vein was probably shortened in the ground-plan of this group (as also in Muscidae and Fanniidae). However, in a few genera the vein has been secondarily extended to the wing margin.

(5)   6th tergum ($\male$) shortened, less than half as long as 5th tergum.
(6)   8th tergum vestige ($\male$) lost.

The limits of the Tachinidae in the wide sense here adopted are the same as those of LINDNER's (1949) family 'Larvaevoridae', except that I exclude *Eginia* (see under Muscidae). The practice of most recent authors has been to split this group into a varying number of families. This question of formal rank cannot be settled at the present time (see section 6.1), and I thus have no grounds for protesting at least at the division of this group into a small number of families. However the recent tendency to split the group of mammal parasites, Oestridae in the wide sense, into a series of 'families' (Hypodermatidae, Cuterebridae, Gasterophilidae and Oestridae in a narrow sense) seems to me unwarranted. The granting of family rank to these groups is almost certainly incongruent with the general use of the family category in the classification of Schizophora, because of their low position in any phylogenetic sequence of subordination. The use of the name Oestridae in the old wider sense is surely preferable, and there are strong grounds for believing that this is a monophyletic group (see HERTING 1957). The insect parasitoids (Tachinidae in the sense of most recent authors) have also been split into a series of families by some authors, but fortunately such treatment has not gained wide acceptance.

ROBACK (1951) included the more generalized groups of Tachinidae *s.l.* (those with mostly free-living larvae) in the 'Sarcophagoidea', and ROHDENDORF (1967) uses the name 'Sarcophagidea' in the same sense. HENNIG (1958: 680) criticized the proposal of this group as based on plesiomorphous characterization, so that its validity in the phylogenetic system is open to doubt. The possibility that the more specialized groups whose larvae are parasites or parasitoids have a closer relationship with some genera of 'Sarcophagidea' than with others cannot be excluded on present information. Despite the many detailed works now available on particular groups of the Tachinidae *s.l.*, the phylogenetic relationships within this group as a whole have still not been clarified.

C. 19. Hippoboscidae family-group (= Glossinoidea HENNIG, *in press)*

The monophyly of the Glossinidae and Hippoboscidae *s.l.* (or Pupipara) now seems established beyond all reasonable doubt. The most recent review of this subject is that given by HENNIG (1965a), who states that most of the conspicuous apomorphous conditions of the Pupipara are already present 'als Vorstufen' (at a lower level of apomorphy) in *Glossina*. The most difficult morphological gap to bridge in analyses of character sequences now lies not between *Glossina* and the Hippoboscidae (Pupipara), but between *Glossina* and other Calyptratae. In the present work it seems unnecessary to present another extensive review of the many profound modifications shown by these insects, associated with the development of adenotrophic macrolarvipary or pupipary. I restrict my remarks mainly to interpretation of the structure of the male abdomen and external genitalia.

Most recent authors have treated the groups here included in the Hippoboscidae family-group as constituting four families (Glossinidae, Hippoboscidae, Nycteribiidae and Streblidae). In the present treatment I recognize only two families, Glossinidae and Hippoboscidae *s.l.* (= Pupipara). In view of the uncertainty presently attending questions of formal rank I have no grounds for objecting to a further subdivision of the latter into Hippoboscidae *(sensu stricto)* and Nycteribiidae (including Streblidae), if this is desired. However the further subdivision of the bat parasites into 'Nycteribiidae' and 'Streblidae' seems unwarranted, since the differentiation of these groups can hardly have preceded the early Tertiary radiation of the bats (Chiroptera); a lower rank than family seems appropriate in accordance with the time criterion of ranking (see section 2.1).

In respect of the structure of the male postabdomen and genitalia the Glossinidae show more plesiomorphous conditions than those shown by the Hippoboscidae *s.l.* In attempting to understand the structure of members of the Hippoboscidae family-group, it thus seems reasonable to proceed first by comparing the conditions shown by *Glossina* with those shown by other Calyptratae. The position of the 7th pair of abdominal spiracles on the male pregenital sclerite of *Glossina* (fig. 67) is particularly interesting, since it suggests the character sequence which has led to the formation of this composite sclerite. The 7th right spiracle lies within the right anteroventral corner of the pregenital sclerite, corresponding with the position of this spiracle within the anteroventral corner of the inverted 8th sternum in most other Calyptratae. The 7th left spiracle also lies within the pregenital sclerite, but is situated more dorsally than the 7th right spiracle (as already noted by PATTON 1934). A similar asymmetrical position of the 7th pair of spiracles within a pregenital sclerite of composite origin is shown by many Muscidae and Tachinidae *s.l.*, in which the suture between the 7th and 8th sterna on the left side is no longer visible. The condition shown by *Glossina* is readily interpretable as a further modification of such a condition through the additional involvement in the pregenital sclerite of the 6th sternum (which is discrete in all Muscidae and Tachinidae which I examined).

The identity of the male cerci and telomeres in members of the Hippoboscidae family-group is a source of difficulty, and the terminology here proposed differs in some respects from that of previous authors. To avoid possible confusion I now set out the differences between my terminology and that used for the same structures by other recent authors, including HENNIG, who gives a review of this question in his paper in press.

Glossinidae (figs. 68 and 69)

Present work	Previous works
Cerci	Adanalia (HENNIG, in press)
Telomeres	Cerci (PATTON 1934; ZUMPT 1936; HENNIG 1937c & in press)
Edita	Ninth coxite (PATTON 1934), Edita, (ZUMPT 1936), Dististyli (HENNIG 1937c), Edita or Surstyli (HENNIG, in press)

Hippoboscinae (Hippoboscidae)

Present work	Previous works
Cerci	Analrahmen (THEODOR 1963; THEODOR & OLDROYD 1964), Adanalia (HENNIG, in press)
Telomeres	Seitenfortsätze (THEODOR 1963; THEODOR & OLDROYD 1964), Surstyli (HENNIG, in press)

Nycteribiinae (Hippoboscidae)

Present work	Previous works
Cerci	Surstyli (JOBLING 1951), Adanalia (HENNIG, in press)
Telomeres	Inferior claspers (JOBLING 1951), Claspers (THEODOR & MOSCONA 1954a; THEODOR 1954b), fingerförmige Vorsprünge (THEODOR 1954c), Surstyli (HENNIG, in press)

I regard the interpretation of the identity of the telomeres and cerci given by HENNIG (in press) as improbable. In all members of the Hippoboscidae family-group there is one well-developed pair of articulated clasping lobes. HENNIG postulates that in 'Streblidae' and Hippoboscinae these claspers are the telomeres ('surstyli'), but that the latter have been functionally replaced by the cerci in the Glossinidae and 'Nycteribiidae' *(sensu stricto)*. This interpretation of the homologies involves postulating that functional replacement of the telomeres as claspers by the cerci has occurred twice (since the 'Nycteribiidae' are not most closely related to the Glossinidae). Such a hypothesis is hardly parsimonious, particularly when it leads to the further conclusion that the sclerites immediately adjacent to the anus are neomorphous ('Adanalia'). The need to postulate the evolution of neomorphous sclerites ('Adanalia') and functional replacement of one pair of clasping lobes by another can be removed simply by homologizing the main pair of claspers with the telomeres in all members of the Hippoboscidae family-group, and accepting the 'Adanalia' as reduced cerci. If this interpretation is correct, then the small setose lobes of *Glossina* called 'edita' may be interpreted as secondary structures differentiated from the periandrial fold. A final clarification of these homologies would probably emerge from detailed comparison of the hypopygial musculature of *Glossina* with that of other Calyptratae, and until such information is available the interpretation which I offer

152

should be regarded as only provisional. I regard my interpretation as more probable than HENNIG's on the basis of the available information, because it involves postulating fewer major morphological changes in the course of evolution.

The evolutionary evaluation of some of the conditions shown by members of the Hippoboscidae family-group is perplexed by the unresolved question of whether this group or the Scatophaginae is the sister-group of all other Calyptratae (see my opening discussion of the Calyptratae). For instance, it is not known whether the position of the right postabdominal spiracles ($\male$) shown in the groundplan of the Hippoboscidae family-group (see fig. 67) corresponds to the groundplan condition for the Calyptratae as a whole or is due to synapomorphy with other groups of Calyptratae (except Scatophaginae). However I think that the following groundplan conditions of the Hippoboscidae family-group can probably be considered apomorphous with respect to the groundplan of the Calyptratae. I discuss only external characters of the male abdomen. For other characters see the review by HENNIG (1965a).

(1) 3rd and 4th sterna reduced (scarcely differentiated from membrane); 3rd to 5th spiracles lying in membrane.

Both these conditions reflect an increase in the membranous area of the abdomen, thus allowing its greater distension when a blood meal is obtained. In some Hippoboscidae s.l. further increases in the area of membrane have occurred, resulting in reduction or loss of some of the terga. In Glossina the 2nd pair of abdominal spiracles also lie in membrane, but it is doubtful whether the situation of this pair of spiracles in membrane can be ascribed to the groundplan of the Hippoboscidae family-group, as these spiracles lie within the margin of the 2nd tergum, or syntergum ($1 + 2$), in many Hippoboscidae. These conditions of the preabdomen are of course also shown by females.

(2) 6th to 8th sterna ($\male$) fused, forming composite pregenital sclerite (fig. 67).

See previous discussion. The conditions of the postabdomen shown by members of the Hippoboscidae s.l. can be interpreted as further modifications of the condition shown by Glossina.

(3) 8th tergum vestige ($\male$) lost.
(4) Vertical sections of hypandrial arms ($\male$) expanded, forming trough (open dorsally) or complete sheath around aedeagal apodeme and base of aedeagus (fig. 66).

In Glossina the vertical sections of the hypandrial arms (called 'tergosternum' by PATTON 1934) broadly fuse above the aedeagal apodeme. In the Hippoboscidae s.l. the hypandrium is usually of much more elongate form than in Glossina, but the same expansion of the area equivalent to the vertical sections of the hypandrial arms is evident. The resulting sheath or trough around the aedeagus base has been called the 'phallobase' in THEODOR's works. This term is open to misunderstanding, since it has been used to denote the base of the aedeagus or phallophore (= basiphallus) in other groups of Diptera. Whether complete fusion of the hypandrial arms or the formation of a trough should be ascribed to the groundplan of the Hippoboscidae family-group is not clear. In Glossina and some Hippoboscidae complete fusion occurs, but only a trough (open dorsally) is formed in many genera of Hippoboscidae.

(5) Pregonites ($\male$) not articulated.

Setose lobes which possibly represent the pregonites are present in Glossina (fig. 66), but these are not articulated.

153

(6) Epiphallus (♂) lost.

(7) Processus longi (♂) not discrete; at most ill-defined interparameral sclerites present, whose sclerotization is continuous with the inner sides of the telomeres.

(8) Ejaculatory apodeme (♂) lost.

(9) Ccerci reduced (♂), represented by narrow setose strips of sclerotization (see previous discussion).

D. 1. Family Glossinidae

Sources of information

The male postabdomen and genitalia of Glossinidae were treated by PATTON (1934), ZUMPT (1936) and HENNIG (1937c). I studied a preparation of *Glossina morsitans* WESTWOOD.

Description of male postabdomen and genitalia (figs. 66-69)

6th abdominal spiracles symmetrically situated in membrane below 6th tergum: 7th spiracles asymmetrically situated within pregenital sclerite (fused 6th to 8th sterna), with left spiracle lying further dorsally than right.

Postabdominal sclerites nearly symmetrical, with pregenital sclerite showing only slight asymmetry. 6th tergum large, as long as 5th tergum. 5th sternum well developed. Postabdominal sterna (6-8) fused to form pregenital sclerite which completely encircles postabdomen. There are no traces of the 7th and 8th terga.

Periandrium rather elongate, bearing discrete telomeres: sclerotization of inner side of telomeres continuous with pair of ill-defined interparameral sclerites, which also bear setose lobes known as 'edita'. Cerci represented only by narrow setose strips of sclerotization. Hypandrium large, with its arms broadly fused above aedeagal apodeme: pregonites not discrete, possibly represented by setose lobes of hypandrium. Aedeagal apodeme long, linked to hypandrium by ventral process from about its middle. Aedeagus short, with complex sclerotization, able to be swung against its apodeme at least to limited degree. Postgonites variably developed. Ejaculatory apodeme absent.

Chararcterization and discussion

All Recent Glossinidae are classified in the single genus *Glossina*, confined to the Ethiopian region. The group is stenomorphous, and there is no reason to doubt that it is monophyletic. In respect of most character sequences analysed the Glossinidae show less modified conditions than the Hippoboscidae *s.l.*, and

154

it is difficult to judge whether some of these conditions are part of character sequences leading to conditions shown by the Hippoboscidae or autapomorphous conditions of the Glossinidae. However, on the basis of my studies of the male postabdomen and genitalia, I can suggest two conditions as probably autapomorphous for the Glossinidae.

(1) Aedeagal apodeme ($\male$) linked to hypandrium by ventral process from about its middle (fig. 66).

The aedeagal apodeme is rod-like and free from the hypandrium in Nycteribiinae (Hippoboscidae). Probably this condition should be ascribed to the groundplan of the Hippoboscidae family-group. The modified condition of the aedeagal apodeme shown by Hippoboscinae does not seem derived from the condition shown by Glossinidae.

(2) Edita ($\male$) present (see preceding discussion of the Hippoboscidae family-group).

D. 2. Family Hippoboscidae (sensu lato) (= Pupipara) (including Nycteribiidae and Streblidae)

Sources of information

Information on the male postabdomen and genitalia of the Hippoboscinae was given by THEODOR (1936) and THEODOR & OLDROYD (1964), and of the Nycteribiinae (including 'Streblidae') by JOBLING (1951), THEODOR & MOSCONA (1954a), THEODOR (1954b, 1954c) and ZEVE & HOWELL (1963).

Description of male postabdomen and genitalia

6th abdominal spiracles usually in membrane (except in some Hippoboscinae in which they lie within the paired lateral plates which represent the 6th tergum). 7th spiracles more or less symmetrically situated within pregenital sclerite or corresponding area, or in membrane when the latter sclerite is absent (some Hippoboscinae).

Postabdominal sclerites symmetrical. 6th tergum developed as complete dorsal plate (Nycteribiini, *Lipoptena*), represented by paired lateral plates (many Hippoboscinae) or completely absent (Streblini). Pregenital sclerite (fused protandrial sterna) represented by paired lateral plates (some Streblini, some Hippoboscinae), fused with periandrium (Nycteribiini, some Streblini) or absent (some Hippoboscinae). There are no traces of the 7th and 8th terga.

Periandrium elongate in Nycteribiinae: in Hippoboscinae most of the dorsal surface of the genital segment is membranous, and the periandrium is either absent or developed only as a small anteriorly situated U-shaped sclerite (well separated from the telomeres). Telomeres discrete, relatively large in most genera but sometimes represented only by finger-like processes (most Streblini, a few Hippoboscinae). Cerci developed as setose strips of sclerotization (Hippoboscinae, Nycteribiini) or as small setose plates (Streblini). Hypandrium usually elongate, with vertical sections of its arms forming subconical trough (open dorsally) or complete sheath ('phallobase') around aedeagal apodeme and base of aedeagus; pregonites not differentiated. Aedeagal apodeme more or less rod-like (expanded anteriorly only) without link to hypandrium (Nycteribiinae), or represented by short, more or less triangular plate which is articulated anteriorly with the hypandrium (Hippoboscinae). Aedeagus elongate, usually of simple tubular or conical shape, supported by longitudinal band of sclerotization or paired longitudinal sclerites; when at rest retracted within abdomen and remaining posteriorly or posteroventrally directed. Postgonites usually well developed. Ejaculatory apodeme absent.

Characterization and discussion

The above description does not take account of the highly aberrant genera *Cyclopodia* and *Eucampsipoda* (see THEODOR 1954b).

On the basis of my studies of the male postabdomen and genitalia I can add the following autapomorphous conditions to the evidence already available (see HENNIG 1965a) that the Hippoboscidae *(sensu lato)* are a monophyletic group.

(1) Pregenital sclerite (♂) reduced, at most represented by pair of lateral plates.

(2) Hypandrium (♂) elongate.

(3) Aedeagus (♂) elongate, retracted within abdomen and remaining posteriorly or posteroventrally directed when at rest.

The articulation (swinging mechanism) between the aedeagus and aedeagal apodeme is well developed in Hippoboscinae, but it is the orientation of the apodeme (not of the aedeagus) which changes when the aedeagus is extruded or retracted. The mechanism involved is described by THEODOR (1963). Whether any articulation between the aedeagus and the aedeagal apodeme is retained in Nycteribiinae is not known to me.

B. 7. Prefamily Micropezoinea

This group is equivalent to the Micropezoidea *s.l.* of HENNIG (1958), with the inclusion of the additional genera discussed by D. K. MCALPINE (1966) and HENNIG (1969a). I divide the group into three families: Cypselosomatidae, Neriidae and Micropezidae. I do not accept MCALPINE's suggestion that the Megamerinidae may also belong to this prefamily. I refer that family to the Sciomyzoinea. The Micropezoinea are well characterized as a monophyletic group by the following groundplan conditions which are apomorphous with respect to the groundplan of the Muscoidea.

(1) Periandrium (♂) elongate (figs. 71 and 72): aedeagus borne on long basal cone ('Basalkegel des Aedeagus', HENNIG 1958), which is strengthened by a pair of sclerites below ('x' on figs. 71-73) (the 'st 2' of HENNIG 1934 and 1936a) and an unpaired sclerite above ('y' on fig. 73) ('st 1' of HENNIG 1936a).

The supporting sclerites in the basal cone of the aedeagus are usually absent in Taeniap-terinae (HENNIG 1934), but I think that they were secondarily lost here. They are usually present in all other families and subfamilies, and are particularly well developed in those genera in which the structure of the postabdomen seems to be the most plesiomorphous in other respects (for instance *Micropeza* and *Heloclusia*). I therefore think it highly probable that these sclerites belong to the groundplan of the Micropezoinea. The presence of these sclerites is probably an autapomorphous condition of the Micropezoinea, since they have no demonstrated homologues in any other group.

(2) Aedeagus ($\male$) elongate, largely supported by very long paraphalli (figs. 72 and 73).

This condition is correlated with the modifications of the genital segment stated above. In contrast with the Tephritoinea, with which the Micropezoinea were classified by HENDEL (1936), the aedeagus of Micropezoinea is largely supported by long rigid paraphalli and cannot be coiled when at rest (except for the membranous terminal processes present in a few species). Protection of the aedeagus when at rest has been achieved by elongation of the genital segment. The sclerotization of the apical area of the aedeagus ('glans') is variable.

(3) 6th sternum ($\male$) triangular, more or less symmetrical (fig. 71).

The condition of the 6th sternum shown by the Cypselosomatidae and Taeniapterinae (Micropezidae) is probably indicative of the groundplan condition for the Micropezoinea. In these groups the 6th sternum is triangular (with the apex of the triangle directed anteriorly) and more or less contiguous posteriorly with the 7th sternum. The latter is asymmetrically developed (extending dorsally on the left side, where it becomes fused with the inverted 8th sternum), as normally in Muscoidea. In Micropezinae and Calobatinae (Micropezidae) the 6th sternum is enlarged, but retains its triangular shape. This condition is probably a further modification of the postulated groundplan condition. Only in Neriidae does the derivation of the condition of the 6th sternum seem problematical. In the single species studied by me the 6th sternum is rather small (like the preceding sterna) and widely separated from the 7th sternum by membrane. If the Neriidae and Micropezidae are monophyletic, I think that the reduced condition of the 6th sternum in Neriidae is probably apomorphous.

(4) 7th tergum and sternum ($\female$) lengthened to form more or less closed ovipos-itor sheath (HENNIG 1958).

This condition has been confirmed in the genera subsequently added to the Micropezoinea (under the family name Pseudopomyzidae) by D. K. MCALPINE (1966) and HENNIG (1969a).

The characteristic apomorphous conditions of the Micropezoinea are combin-ed with certain noteworthy plesiomorphous conditions; in particular the reten-tion of a large unmodified 6th tergum ($\male$) (all species) and the retention of the 7th and 8th tergum vestiges ($\male$) in at least some genera of each family. HENNIG's and MCALPINE's studies indicate that the Recent members of the Micropezoinea are probably divisible into two sister-groups, the Cypselosomatidae (in which I include the Pseudopomyzidae) and the Micropezidae family-group (called 'Micropezoidea s. str.' by HENNIG 1958).

C. 20. *Family Cypselosomatidae (including Pseudopomyzidae)*

Sources of information

D. K. MCALPINE (1966) and HENNIG (1969a) have given information on the

157

male postabdomen and genitalia of the Cypselosomatidae. I studied preparations of the following: – *Heloclusia imperfecta* MALLOCH; *Pseudopomyza atrimana* (MEIGEN); *Formicosepsis tinctipennis* DE MEIJERE.

Description of male postabdomen and genitalia (fig. 71)

6th abdominal spiracles symmetrically situated in membrane or within 6th tergum (e.g. *Heloclusia*): 7th left spiracle in membrane near margin of 7th sternum *(Heloclusia)* or within composite pregenital sclerite *(Pseudopomyza, Formicosepsis)*: 7th right spiracle in membrane near 7th tergum vestige *(Heloclusia)* or within pregenital sclerite *(Pseudopomyza, Formicosepsis)*.

Postabdomen obviously asymmetrical in *Heloclusia*, but almost symmetrical in other genera. 6th tergum large, of similar length to 5th tergum. 6th sternum triangular *(Heloclusia, Formicosepsis)* or H-shaped *(Pseudopomyza)*, more or less symmetrical. 7th tergum represented by small lateroventral sclerite on right side in *Heloclusia* only. 7th sternum asymmetrically developed on left side, where it is fused dorsally with the large 8th sternum *(Heloclusia)*: in *Formicosepsis, Cypselosoma* and *Pseudopomyza* 8th sternum and sclerites of 7th segment fully fused to form more or less symmetrical pregenital sclerite, which extends around the venter to form a complete ventral ring in *Pseudopomyza* and *Cypselosoma australis* MCALPINE. 8th tergum well developed in *Heloclusia*, absent in other genera.

Periandrium elongate, bearing discrete telomeres. Interparameral sclerites ill-defined, hardly differentiated from membrane. Cerci fused anteriorly or more or less discrete. Aedeagal apodeme long and rod-like, free from hypandrium. Hypandrium small, with its arms linked to longitudinal sclerite in dorsal wall of basal cone of aedeagus: ventral wall of basal cone supported by pair of longitudinal sclerites ('x' on fig. 71) extending posteriorly from hypandrium. Aedeagus borne at end of long 'basal cone', able to be swung through wide arc against aedeagal apodeme to anteriorly directed rest postion, usually slender, with its sclerotization consisting mainly of long rigid paraphalli, with flexible whip-like apical section; but in *Rhinopomyzella* and *Pseudopomyza* the aedeagus is rather short, without such an apical section. Postgonites situated at end of ventral longitudinal sclerites of basal cone, near base of aedeagus. Small ejaculatory apodeme present.

Characterization and discussion

The monophyly of the Cypselosomatidae (= Cypselosomatidae + Pseudopomyzidae of D. K. MCALPINE 1966 and HENNIG 1969a) seems established by certain apomorphous conditions of the wing venation, well illustrated in MCALPINE's and HENNIG's papers, as follows.
(1) Subcosta reduced, failing to reach wing margin as distinct vein.
(2) Base of m_{3+4} (= 'tb', lower cross-vein) lost, so that the discal and lower basal cells are confluent.

(3) Anal vein and second median (m_4) fading apically, not reaching wing margin.

I have not followed HENNIG in listing the presence of a costal break as an apomorphous condition of this family, but prefer to leave this question open. The interpretation of this character in this case is disputable, and I do not exclude the possibility that the presence of the costal break is a groundplan condition of the Micropezoinea, as MCALPINE maintained. Long vibrissae are present in the Cypselosomatidae, but absent in most species of the Micropezidae family-group. Again it is not clear which condition should be ascribed to the groundplan of the Micropezoinea. Further discussion of these characters and clarification of his views is given by HENNIG (in press).

My studies of the male postabdomen do not indicate any additional evidence which can be used to demonstrate the sister-group relationship between the Cypselosomatidae and the Micropezidae family-group. The structure of *Heloclusia* (fig. 71) is the most plesiomorphous shown by any of the Cypselosomatidae which I studied, differing little from that of some members of the Micropezidae family-group. No autapomorphous conditions of the Cypselosomatidae were detected. Within the family there is clearly a tendency to restore symmetry of the postabdomen. In both *Pseudopomyza* and *Formicosepsis* symmetry has been almost restored by fusion of the sclerites of the 7th segment with the inverted 8th sternum to form a composite pregenital sclerite. In *Pseudopomyza* this pregenital sclerite extends around the venter to form a complete ventral ring (as also described for *Cypselosoma australis* MCALPINE). These genera also lack the 8th tergum vestige retained by *Heloclusia*. The presence of apparently synapomorphous conditions in the postabdomen of *Pseudopomyza* and *Cypselosoma/Formicosepsis* clearly raises serious doubts about whether the 'Pseudopomyzidae' in the sense of HENNIG (1969a) and MCALPINE (1966) is monophyletic. Neither MCALPINE nor HENNIG have attempted to justify their retention of the subdivision of the Cypselosomatidae + Pseudopomyzidae into two families. Most probably the 'Cypselosomatidae' *(Cypselosoma + Formicosepsis)* have a closer relationship to some genera of 'Pseudopomyzidae' than to others. To my mind this conclusion is strengthened by HENNIG's (1969) description of the characteristic long bristles on the pregenital sclerite, as shown by 'Cypselosomatidae', also in the genus *Pseudopomyzella* which he classifies in the Pseudopomyzidae. HENNIG (in press) suggests that these bristles were possibly present in the groundplan of the Cypselosomatidae/Pseudopomyzidae and lost secondarily in some genera of 'Pseudopomyzidae'. However the need to postulate secondary loss only arises if there are grounds for believing that the Pseudopomyzidae, as currently delimited, are a monophyletic group. If *Pseudopomyza* is less closely related to *Heloclusia* than to the 'Cypselosomatidae' *(sensu stricto)*, then there are no grounds for postulating secondary loss of the bristles in question in *Heloclusia*. From the above considerations I conclude that a satisfactory subdivision of the Cypselosomatidae *sensu lato* is not yet available.

C. 21. Micropezidae family-group

This group was called the Micropezoidea *s.str.* by HENNIG (1958). All its members show reduction or loss of the ocellar bristles, which HENNIG (1958) interprets as an autapomorphous condition of the group. Since the apomorphous conditions of the wing venation characteristic of the Cypselosomatidae are absent in most species, a sister-group relationship between this group and the Cypselosomatidae seems indicated by an alternance of plesiomorphous and apomorphous conditions.

The further subdivision of this group has been a source of difficulty. HENNIG (1958) proposed a provisional classification, but this has been rejected by D. K. MCALPINE (1966), who reverts to the previous classification of ACZÉL (1951). The differences between these workers' classifications are summarized as follows.

HENNIG (1958)

A Taeniapteridae
B (unnamed)
   B1 Trepidariidae
   B2 (unnamed)
      B2a Micropezidae
      B2b Neriidae

ACZÉL (1951) and D. K. MCALPINE (1966)

A Neriidae
B Micropezidae
   B1 Taeniapterinae
   B2 Calobatinae (= Trepidariinae)
   B3 Micropezinae

A major difference between these two classifications lies in the assessment of the relationship of the Neriidae. MCALPINE considers the relatively unmodified 5th sternum ($\male$) in Neriidae as plesiomorphous, while HENNIG thinks that the 'Kopulationsgabel' (enlarged forked 5th sternum) has been secondarily reduced in this family. The arguments of neither author seem to me conclusive, because the evolutionary interpretation of some of the conditions concerned is problematical. The structure of the genitalia and postabdomen of this group is in many respects rather uniform (stenomorphous), and a study of many genera will be needed before firm conclusions can be drawn. The additional characters of the male postabdomen and genitalia considered in my present study do however suggest that HENNIG's grouping of the Neriidae with the Micropezidae *(sensu stricto)* was incorrect. It seems to me that the Calobatinae and Micropezinae (Micropezidae *sensu* HENNIG) show the following synapomorphous conditions: (1) the presence of an anterior ventral process (usually forked) of the aedeagal apodeme ($\male$) (fig. 73), and (2) enlargement of the 6th sternum ($\male$). If this interpretation is correct, then the Micropezinae cannot be monophyletic with the Neriidae. HENNIG's alternative suggestion (HENNIG 1958: 558), that the Neriidae may be most closely related to the 'Micropezidae + Trepidariidae' (= Micropezinae + Calobatinae) does not conflict with the evidence of these additional characters, but has not been supported by any convincing demonstration of synapomorphous conditions. Pending further information I propose to follow

160

provisionally the more widely accepted classification of Aczél and McAlpine, and to divide the Micropezidae family-group into two families, the Neriidae and Micropezidae *(sensu lato)*. This classification allows me to accept, at least provisionally, the absence of the 'Kopulationsgabel' in the Neriidae as a plesiomorphous condition. However I emphasize that the validity of classifying the Taeniapterinae in the Micropezidae has not been conclusively demonstrated, and needs further investigation.

D. 3. Family Neriidae

Sources of information

No detailed information on the male postabdomen and genitalia of the Neriidae was found in the literature. My treatment is based solely on study of preparations of *Gymnonerius fuscus ceylanicus* HENNIG.

Description of male postabdomen and genitalia (fig. 72)

6th abdominal spiracles symmetrically situated in membrane: 7th spiracles absent.

Postabdomen asymmetrical. 6th tergum large, of similar length to 5th tergum. 6th sternum small, subtriangular. 7th tergum represented by narrow lateroventral sclerite on right side. 7th sternum asymmetrically developed on left side, where it is fused dorsally with the large 8th sternum. 8th tergum well developed.

Periandrium elongate, bearing discrete telomeres. Interparameral sclerites hardly differentiated from membrane. Cerci large, more or less discrete. Aedeagal apodeme long and rod-like, free from hypandrium. Hypandrium divided ventrally into two halves, its arms linked to longitudinal sclerite in dorsal wall of basal cone of aedeagus: ventral wall of basal cone supported by pair of longitudinal sclerites ('x' on fig. 72) extending posteriorly from hypandrium halves. Aedeagus borne at end of long 'basal cone', able to be swung through wide arc against aedeagal apodeme to anteriorly directed rest position, very slender, slightly expanded distally, supported by long rigid strip of sclerotization (paraphallus). Postgonites situated at end of ventral longitudinal sclerites of basal cone, near base of aedeagus. Slender ejaculatory apodeme present.

Characterization and discussion

The Neriidae show certain characteristic autapomorphous conditions, particularly in the structure of the head, as indicated in the characterization given by HENNIG (1958). I do not repeat this here.

161

In *Gymnonerius* the structure of the male postabdomen and genitalia agrees well with that of *Heloclusia* (Cypselosomatidae). The vestiges of the 7th and 8th terga are both well developed; and the aedeagal apodeme is very long, without any anterior ventral process. In these respects *Gymonerius* seems little removed from the groundplan of the Micropezoinea. Apomorphous conditions of the male postabdomen shown by *Gymnonerius* are: (1) loss of both spiracles of the 7th segment, and (2) reduction of the 6th sternum (see the comments on the condition of the 6th sternum in my characterization of the Micropezoinea). Whether these apomorphous conditions can be ascribed to the groundplan of the Neriidae requires clarification, as relevant comparative information is not available. All other members of the Micropezoinea which I have studied retain the full complement of seven pairs of abdominal spiracles, and a better developed 6th sternum.

D. 4. Family Micropezidae (including Calobatidae and Taeniapteridae)

Sources of information

Information on the male postabdomen and genitalia of the Micropezidae was given by HENNIG (1934, 1936a) (as 'Tylidae'). I studied preparations of the following: – *Calobata petronella* (L.); *Compsobata pallipes* (SAY); *Compsobata univitta* (WALKER); *Micropeza lineata* VAN DUZEE; *Cnodacophora nasoni* (CRESSON); *Taeniaptera trivittata* MACQUART.

Description of male postabdomen and genitalia (fig. 73)

6th abdominal spiracles symmetrically situated in membrane: 7th left spiracle in membrane near margin of 7th sternum or within 7th sternum *(Micropeza)*: 7th right spiracle in membrane (near 7th tergum vestige when this is present).

Postabdomen asymmetrical. 6th tergum large, of similar length to 5th tergum. 6th sternum triangular, more or less symmetrical, very large in Micropezinae and Calobatinae. 7th tergum represented by lateroventral sclerite on right side *(Micropeza, Taeniaptera, Calobata)*, absent in *Compsobata* and *Cnodacophora*. 7th sternum asymmetrically developed on left side, where it is fused dorsally with the large 8th sternum. 8th tergum represented by small ventral plate in *Micropeza*, absent in members of other genera studied.

Periandrium elongate, bearing discrete telomeres except in Taeniapterinae. Interparameral sclerotization more or less H-shaped, linked to base of telomeres and hypandrial arms (Micropezinae, Calobatinae); ill-defined in *Taeniaptera*. Cerci fused or more or less discrete. Aedeagal apodeme long and rod-like, free from hypandrium in Taeniapterinae but linked to hypandrium by ventral process (usually forked) in Calobatinae and Micropezinae. Hypandrium small, in Calobatinae and Micropezinae with its arms linked to longitudinal sclerite ('y'

162

on fig. 73) in dorsal wall of basal cone of aedeagus: ventral wall of basal cone supported by pair of longitudinal sclerites ('x' on fig. 73) extending posteriorly from hypandrium in Calobatinae and Micropezinae: but such sclerites are not present in the basal cone in most Taeniapterinae. Aedeagus borne at end of long 'basal cone', able to be swung through wide arc against aedeagal apodeme to anteriorly directed rest position, slender, its sclerotization consisting mainly of long rigid paraphalli (with slender discrete apical section in *Taeniaptera*, but not in other genera examined by me). Postgonites situated at end of ventral longitudinal sclerites of basal cone, near base of aedeagus (Calobatinae, Micropezinae), absent in *Taeniaptera*. Ejaculatory apodeme present.

Characterization and discussion

The only apomorphous condition of the Micropezidae *(sensu lato)* which is indicated by HENNIG's (1958) discussion is: (1) the modication of the 5th sternum (♂) to form a copulatory fork ('Kopulationsgabel'). This is absent only in a few genera, where it has probably been secondarily lost. D. K. MCALPINE (1966) has drawn attention to another possibly apomorphous condition: (2) front legs short and weak. The other conditions listed for the Micropezidae in MCALPINE's table (pages 681-682) are either plesiomorphous or in need of clarification with respect to their evolutionary significance. HENNIG (in press) points out that sternopleural bristles are reported to be present in some Neriidae, as well as in Micropezidae, so that MCALPINE's statements about this character require correction.

The structure of the male postabdomen and genitalia does not indicate any autapomorphous conditions in the groundplan of Micropezidae. The full complement of spiracles is retained in all the species studied by me (contrast Neriidae). In *Micropeza* the full complement of abdominal sclerites is retained, including clearly defined vestiges of the 7th and 8th terga. One or both of these has been lost in the species of other genera studied.

B. 8. Prefamily Australimyzoinea, Australimyzidae familia nova (type-genus Australimyza HARRISON 1953)

Sources of information

My present treatment is based solely on study of a preparation of *Australimyza anisotomae* HARRISON. No detailed information on the male postabdomen and genitalia of this group was available in the literature.

Description of male postabdomen and genitalia (fig. 74-76)

6th abdominal spiracles within 6th tergum: 7th spiracles absent.

Postabdomen asymmetrical. 6th tergum large, slightly asymmetrical (larger on right side), fully as long as 5th tergum. 6th sternum asymmetrical, linked to 7th sternum on left side. 7th sternum extending onto dorsum on left side, fused both dorsally and ventrally with 7th tergum (which is represented by a narrow strip of sclerotization on the right side), so that a complete ring of sclerotization is formed about the 7th abdominal segment. 8th sternum large, narrowly fused with 7th sternum near centre-line of dorsum. 8th tergum absent.

Periandrium divided into two halves: telomeres not discrete, delimited from periandrium only by partial suture. Interparameral sclerites absent. Cerci more or less fused, forming projecting mesolobus. Aedeagal apodeme rod-like, free from hypandrium. Hypandrium extending anteriorly, of similar length to aedeagal apodeme. Aedeagus long and slender, supported by very long paraphalli, able to be swung through wide arc against aedeagal apodeme. Postgonites large and flattened. Ejaculatory apodeme slender, rather large.

Characterization and discussion

The genus *Australimyza* is distributed in New Zealand (HARRISON 1959), Australia (COLLESS & D. K. MCALPINE 1970) and neighbouring subantarctic islands (Campbell Island, Antipodes Island and Macquarie Island) (HARRISON 1959). HARRISON classified the genus in the Milichiidae, but this is clearly untenable because the structure of the male postabdomen and genitalia of *Australimyza* is fundamentally different. The male postabdomen of *Australimyza* (figs. 74-76) retains a highly plesiomorphous structure, little removed from the postulated groundplan condition for the Muscoidea. The 6th tergum is unreduced, fully as long as the 5th tergum (as for instance in the Micropezoinea), and the asymmetrical 6th and 7th sterna are large and clearly defined. The aedeagal apodeme is free and rod-like. These plesiomorphous conditions are combined with some characteristic apomorphous conditions, as stated below. As far as I can see, the combination of apomorphous and plesiomorphous conditions shown by *Australimyza* provides no basis for including the genus in any previously described family or in any of the other prefamilies of Muscoidea.

The Australimyzoinea (Australimyzidae) are characterized by the following apomorphous conditions with respect to the groundplan of the Muscoidea.

(1) Postverticals reduced.

According to HARRISON (1959) the small postverticals may be parallel or slightly convergent or divergent.

(2) Vibrissae present.
(3) Costa broken twice, shortly beyond humeral cross-vein and at end of subcosta.
(4) Subcosta becoming contiguous or fused with vein r_1 apically.
(5) Spiracles 2-6 ($\male$) lying within margins of terga; 1st and 7th spiracles lost.

This characterization requires confirmation from study of additional species. In the species studied by me the spiracles are extremely small and difficult to locate. Only spiracles 2-6 could be traced.

164

(6) 7th sternum (♂) and 7th tergum vestige fused both dorsally and ventrally, forming complete ring of sclerotization about 7th abdominal segment (figs. 74 and 75).

This condition is one of the most unusual modifications shown in the species studied by me. The sclerotization on the right side of the 7th segment consists of a narrow strip. In my opinion this represents the 7th tergum vestige, which is represented by a similar (but discrete) strip of sclerotization in certain other Muscoidea which retain highly plesiomorphous post-abdominal structure (for instance some Micropezoinea and Coelopidae). I thus regard the condition of the sclerites of the 7th segment of *Australimyza* as a further modification of the groundplan condition of the Muscoidea. This modification consists of dorsal extension of the 7th sternum on the left side, so that this is fused with the 7th tergum vestige on the right side of the dorsum, as well as ventral fusion of the same two sclerites. A comparable dorsal fusion of the 7th sternum and 7th tergum vestige has evolved independently in *Helcomyza* (Dryomyzidae).

(7) 8th tergum vestige lost (♂).
(8) Periandrium (♂) divided into two halves (in other words, basimeres secondarily failing to meet across dorsum); telomeres more or less fused with periandrium, at most delimited by partial suture (fig. 76).
(9) Aedeagus (♂) long and slender, supported by very long paraphalli (fig. 76).

In *Australimyza* the apex of the aedeagus is filamentous, without any indication of differentiated apical sclerites ('glans'). The base of the aedeagus can be swung against the aedeagal apodeme, as normally in Muscoidea.

B. 9. Prefamily Diopsioinea

The two families which I include in this prefamily were placed by HENNIG (1958) in his tentative group Nothyboidea. As previously discussed, I regard HENNIG's Nothyboidea as probably heterogenous, and I present in this work a different classification of the included families. One clearly defined segregate of HENNIG's 'Nothyboidea' consists of the families Diopsidae and Syringogastridae. These are rather characteristically modified muscoid flies occurring mainly in tropical regions. Their distribution is vicariant, for the Diopsidae have an Old World distribution centre (with a single species in North America), while the Syringogastridae are confined to the neotropical region.

The Diopsioinea are characterized as a monophyletic group by the following groundplan conditions which are apomorphous with respect to the groundplan of the Muscoidea.

(1) Postvertical bristles absent.
(2) Only one fronto-orbital *(ors)* and one vertical bristle *(vte)* present.

This groundplan condition is shown by the Diopsidae (see SHILLITO 1950 and HENNIG 1958). In the Syringogastridae the remaining fronto-orbital has been lost.

(3) Only one postalar bristle *(pa)* present (see HENNIG 1965b and DO PRADO 1969).
(4) 1st and 2nd abdominal terga fused, forming syntergum which is the largest sclerite of the abdomen: basal segments of abdomen (especially the 2nd

165

segment) elongate, but distal segments becoming short and wide, so that the abdomen has a petiolate appearance (fig. 77).

The characteristic petiolate appearance of the abdomen in Diopsioinea seems to constitute an autapomorphous condition. The other groups included in HENNIG's (1958) 'Nothyboidea' all show a more uniform elongation of the abdominal segments, not a differential elongation of the basal segments.

(5) 7th sternum (♂) forming complete ventral band of sclerotization, fused with inverted 8th sternum on right side (as well as on left side as normally in Muscoidea) (fig. 77).

The structure of the male postabdomen of *Centrioncus* (the sister-group of all other Recent Diopsidae) is readily interpretable as a modification of the arrangement of postabdominal sclerites in the groundplan of the Muscoidea. In *Centrioncus* the expanded area of the 7th sternum on the left side is clearly delimited (fig. 78), as in most Muscoidea. But the condition shown by *Centrioncus* is apomorphous in respect of extension of the 7th sternum towards the right side, so that it forms a complete ventral band of sclerotization (fig. 77). In the other Diopsidae studied by me the sclerotization of the 7th and 8th segments is much reduced, and the ventral band is incomplete. However the condition shown by *Centrioncus* accords well with that described for the Syringogastridae (DO PRADO 1969), in which a complete ventral band is also present. Therefore I think it probable that the presence of the ventral band can be ascribed to the groundplan of the Diopsioinea. DO PRADO does not refer specifically to an expanded area of the 7th sternum on the left side in Syringogastridae, but his figure of *Syringogaster lanei* DO PRADO (fig. 12), if I interpret it correctly, indicates the delimitation of such an area (as in *Centrioncus*). In most other Muscoidea in which a complete ventral band of postabdominal sclerotization is formed, this condition has been achieved through expansion and fusion of the 6th sternum with the 7th and 8th sterna, although a similar ventral band formed by sclerites of the 7th segment is shown by some Cypselosomatidae. It is noteworthy that in the Diopsioinea the 6th sternum is not involved in the formation of the ventral band. In *Centrioncus* this sternum remains large and is slightly more expanded on the left side (as normally in Muscoidea). DO PRADO (1969) indicates that in Syringogastridae the 6th sternum is represented by a pair of small sclerites which are discrete from the ventral band.

(6) 7th and 8th tergum vestiges (♂) lost.
(7) 7th left spiracle (♂) lying within 7th sternum.

This condition is shown by the Syringogastridae and by *Centrioncus* (Diopsidae). In *Diopsis* and *Sphyracephala* the 7th left spiracle lies in membrane, but this condition may be secondary, consequent upon reduction of the sclerotization of the 7th and 8th segments.

(8) Aedeagus (♂) rather short, with a complex distal section which bears lobes or processes (figs. 79 and 80).

I have no firm opinion on which other group of Muscoidea may be most closely related to the Diopsioinea. HENNIG (1958) suggested possible synapomorphies with the Megamerinidae (in which *Syringogaster* was formerly included) in respect of the loss of certain frontal bristles. However I doubt whether this interpretation is correct, as it is difficult to reconcile with the evidence of other characters. The structure of the male postabdomen in the Megamerinidae (fig. 97) is very different, and I provisionally classify that family in the Sciomyzoinea in this work. In the Diopsioinea the 6th abdominal segment in the male is scarcely reduced in comparison with the preceding segments, a plesiomorphous condition comparable with that shown, for instance, by the Micropezoinea and Tanypezoinea. But in Megamerinidae the sclerites of the 6th segment (♂) are much reduced. The female postabdomen of Diopsioinea does not show the elongation of the terminal segments characteristic of the Megamerinidae (see the figures given by HENNIG 1958).

166

C. 22. Family Diopsidae

Sources of information

I studied preparations of the following: – *Centrioncus prodiopsis* SPEISER; *Diopsis* sp.; *Sphyracephala brevicornis* (SAY). Only limited information on the male postabdomen and genitalia of Diopsidae is available in the literature. HENNIG (1941a, 1941d) briefly discussed the genitalia of a few species. The paper by NAYAR & TANDON (1963) is misleading, because the authors have misunderstood the morphology of the aedeagus: in their figure 1 they portray the aedeagus in its anteriorly directed rest position, but indicate the 'ejaculatory opening' at its base.

Description of male postabdomen and genitalia (figs. 77-79)

6th abdominal spiracles symmetrically situated in membrane: 7th left spiracle within 7th sternum *(Centrioncus)* or in membrane *(Diopsis, Sphyracephala)*: 7th right spiracle within ventral band of sclerotization *(Centrioncus)* or in membrane *(Diopsis, Sphyracephala)*.

Postabdomen weakly asymmetrical *(Centrioncus)* or fully symmetrical *(Diopsis, Sphyracephala)*. 6th tergum well developed, over half as long as 5th tergum. 6th sternum large, asymmetrically linked to 7th sternum on left side *(Centrioncus)*: or represented only by pair of fragments on either side of centreline *(Diopsis, Sphyracephala)*. 7th sternum forming complete ventral band of sclerotization, with expanded area on left side where its fusion with the large 8th sternum is indicated by a visible suture *(Centrioncus)*: in *Diopsis* and *Sphyracephala* the 7th sternum is not delimited from the 8th sternum, which is represented by a narrow ill-defined dorsal sclerite which extends ventrally on either side but does not form a complete ventral band. 7th and 8th terga absent.

Periandrium bearing discrete telomeres, which are large and complexly lobed in *Centrioncus*. In *Centrioncus* and *Sphyracephala* the telomeres are connected with the hypandrium by broad *(Sphyracephala)* or rod-like *(Centrioncus)* interparameral sclerites (processus longi), but such sclerites seem absent in *Diopsis*. Cerci discrete, linked ventrally to telomeres. Aedeagal apodeme free anteriorly only, becoming fused with body wall posteriorly where it bears paired ventral processes which are contiguous or fused laterally with the hypandrium. Aedeagus with variable complex sclerotization, bearing distal lobes or processes, able to be swung through wide arc against aedeagal apodeme to anteriorly directed rest position. Small postgonites present. Ejaculatory apodeme present.

Characterization and discussion

The classification of the Diopsidae was recently reviewed by HENNIG (1965b), who accepts the limits of the family proposed by SHILLITO (1950). The Diopsidae

167

are characterized as a monophyletic group by the following groundplan conditions which are apomorphous with respect to the groundplan of the Diopsioinea.

(1) Base of m₄ (= 'tb', lower cross-vein) absent (HENNIG 1958).
(2) Ocellar bristles lost.

Weak ocellar bristles are retained in some species of Syringogastridae, but apparently never in Diopsidae.

(3) Scutellum with pair of setigerous spines (SHILLITO 1950, HENNIG 1958).
(4) Front femora thickened and armed with double series of ventral tubercles (SHILLITO 1950, HENNIG 1958).

In Syringogastridae it is the hind femora which are thickened, while the front femora are slender.

(5) Aedeagal apodeme (♂) free anteriorly only, becoming fused with body wall posteriorly where it bears paired ventral processes which are contiguous or fused laterally with the hypandrium (fig. 79).

In the species of *Syringogaster* studied by me the aedeagal apodeme is free from the hypandrium throughout its length.

(6) Cerci (♂) linked ventrally to telomeres.

In the preparations studied by me the lower outer corner of each cercus articulates with a process of the telomere, a condition similar to that shown by the Calyptratae (presumably through homoplasy).

C. 23. Family Syringogastridae

Sources of information

Do PRADO's (1969) revision includes information on the male postabdomen and genitalia of the Syringogastridae. I studied preparations of *Syringogaster rufa* CRESSON.

Description of male postabdomen and genitalia (fig. 80)

6th abdominal spiracles symmetrically situated within 6th tergum: 7th spiracles lying within ventral band of sclerotization formed by 7th sternum.

Postabdomen almost symmetrical. 6th tergum well developed, almost as long as 5th tergum. 6th sternum represented only by pair of fragments on either side of centre-line. 7th sternum forming complete ventral band of sclerotization, fused with narrow 8th sternum (sometimes with fusion line indicated by visible suture on left side). 7th and 8th terga absent.

Periandrium bearing discrete telomeres. Interparameral sclerites absent. Cerci discrete, not linked to telomeres. Hypandrium bearing large setose posterior lobes (pregonites), its arms fused to form bridge above aedeagal apodeme. Aedeagal apodeme rod-like, free from hypandrium. Aedeagus with variable complex sclerotization, bearing distal lobes or processes, able to be swung through wide arc against aedeagal apodeme to anteriorly directed rest position. Postgonites well developed. Ejaculatory apodeme present.

Characterization and discussion

I accept DO PRADO's (1969) proposal of a new family based on *Syringogaster* CRESSON, which was previously classified in the Megamerinidae. DO PRADO raises the number of described species to eight, all of which he refers to the single genus *Syringogaster*. The Syringogastridae are characterized by the following conditions which are apomorphous with respect to the groundplan of the Diopsioinea.

(1) Orbital bristles *(ors)* completely lost.
(2) Hind femora thickened.
(3) Alula reduced.

The alula is also reduced in some groups of Diopsidae, but is retained in the groundplan of that family (see HENNIG 1965b: 58).

(4) 6th spiracles lying within 6th tergum in both sexes; 7th spiracles (on both sides) lying within 7th tergum (♀), or within sclerotized area formed by fusion of 7th and 8th sterna (♂).

HENNIG (1958, fig. 85) indicates that the 7th spiracles are situated within the 7th tergum in the female of *Megalabops* (Diopsidae), but whether this is a general condition of female Diopsidae is not known to me. At least the inclusion of the 6th spiracles within the 6th tergum (in both sexes) may be considered an autapomorphous condition of the Syringogastridae, as these lie in membrane in both sexes of Diopsidae.

B. 10. Prefamily Sciomyzoinea

I include in this group all the families referred to it by HENNIG (1958) (as Sciomyzoidea), and tentatively add two small families, the Cremifaniidae and Megamerinidae. The Cremifaniidae were not considered in HENNIG's (1958) analysis; the Megamerinidae were included in his group 'Nothyboidea', which was heterogenous in my opinion. The complete list of families now referred to the Sciomyzoinea is as follows: Coelopidae, Phaeomyiidae, Dryomyzidae, Sciomyzidae, Helosciomyzidae, Ropalomeridae, Sepsidae, Megamerinidae and Cremifaniidae.

The reservations expressed by HENNIG (1958, 1965b) on the validity of this group still remain. The families included here show many plesiomorphous conditions for the Muscoidea, and I am not able to give any convincing characterization in terms of autapomorphous conditions. The included families show, at least in their groundplan, the following conditions which according to HENNIG's (1958) analysis are probably plesiomorphous (groundplan conditions of the Muscoidea): (1) the costa is unbroken, (2) the subcosta is complete, (3) the anal vein reaches the wing margin (except Sepsidae and Cremifaniidae), (4) vibrissae are absent or only weakly differentiated (in Sepsidae and Ropalomeridae), and (5) divergent postverticals are present (except Helosciomyzidae and Megamerinidae). The only possibly synapomorphous condition with respect to the groundplan of the Muscoidea which I can put forward is:

(1) 6th tergum (♂) somewhat shortened, not more than half as long as 5th tergum.

The best developed 6th tergum among the spieces studied by me is that of *Heterocheila* (here provisionally referred to the Coelopidae), in which the 6th tergum is about half as long as the 5th tergum (fig. 82).

However reduction of the 6th tergum is so widespread among the Muscoidea that little reliance can be placed upon this condition in isolation. In view of this lack of satisfactory characterization, the Sciomyzoinea as presently understood can only be considered a group whose limits require further investigation. Pending clarification it is convenient to maintain this group in the classification, since there is no evidence that the included families are more closely related to groups outside the Sciomyzoinea than to each other.

J. F. MCALPINE (1963) treated the Sciomyzoinea and Lauxanioidea as monophyletic, and this is also implied in the classification of COLLESS & D. K. MCALPINE (1970). However I do not find in these works any demonstration of synapomorphous conditions to support this view. The placing of these two groups together is difficult to reconcile with the presence of the typical muscoid structure of the male postabdomen in Sciomyzoinea. In Lauxanioidea the structure of the male postabdomen is of a different type (see above), and the mechanism for swinging the aedeagus against its apodeme is not developed.

The division of the Sciomyzoinea into families is in some respects problematical. Until recently much reliance was placed on the presence or absence of a 'precoxal bridge' in defining the Dryomyzidae, Coelopidae and Helcomyzidae. HENNIG (1958) postulated that the presence of a precoxal bridge was a synapomorphous condition of the Helcomyzidae and Ropalomeridae. However J. F. MCALPINE's (1963) comments imply that the presence of such a bridge is a groundplan condition of the Sciomyzoinea, and HENNIG (1965b: 78) has modified his previous view to the extent of postulating that the bridge has been secondarily lost in the Sepsidae. The interpretation of the precoxal bridge character has recently been discussed in detail by SPEIGHT (1969), who reports that the same type of precoxal bridge as shown by the Helcomyzidae (in which he includes *Heterocheila*) is also found in some Sciomyzidae and Coelopidae. SPEIGHT has not been able to offer any firm conclusion on whether this type of precoxal bridge ('variant V') should be ascribed to the groundplan of the Scio-myzoinea, but his view that 'the way in which the prosternum has been used in the definition of families in the Sciomyzoidea is inadequate' seems justified from the data presented. Because of these doubts about which conditions should be considered plesiomorphous and apomorphous in particular cases, I have decid-ed to omit reference to the shape of the prosternum in my characterization of the families of Sciomyzoinea, and attempt to base my interpretation of the relation-ships on other characters. SPEIGHT's (1969: 339) comment that 'Coelopa and Orygma possess similar basisterna, yet very dissimilar musculature' indicates that independent evolution of reduced prosternal basisterna, as well as of precoxal bridges, may be suspected. Evidently any phylogenetic hypotheses suggested by similarities in the shape of the prosternum should be carefully checked for compatibility with other characters before they can be accepted.

C. 24. Family Coelopidae

Sources of information

The male postabdomen and genitalia of *Coelopa* were previously treated by HENNIG (1937a) and STEYSKAL (1957a). STEYSKAL (1958b) also described the male postabdomen and genitalia of *Heterocheila* (= *Oedoparea*). I studied preparations of the following: – *Coelopa (Fucomyia) frigida* (F.); *Icaridion nasutum* LAMB; *Malacomyia sciomyzina* (HALIDAY); *Heterocheila hannai* (COLE).

Description of male postabdomen and genitalia (figs. 81 and 83)

6th abdominal spiracles in membrane: 7th left spiracle in membrane at or near edge of 7th sternum, displaced dorsally in relation to preceding spiracles: 7th right spiracle lying in membrane near 7th tergum vestige, slightly displaced centrally in relation to preceding spiracles.

Postabdomen asymmetrical. 6th tergum represented by broad lateroventral area of sclerotization on right side which continues as a narrow band or strip across the dorsum. 6th sternum asymmetrical, linked to 7th sternum on left side. 7th tergum represented by narrow lateroventral sclerite on right side. 7th sternum asymmetrically developed on left side, where it is fused dorsally with the large 8th sternum. 8th tergum represented by narrow ventral sclerite.

Periandrium bearing discrete telomeres, which are connected with the hypan-drial arms by a broad area of ill-defined interparameral sclerotization. Cerci more or less discrete. Aedeagal apodeme rod-like, free from hypandrium. Hypandrium bearing pair of posteriorly directed setose processes: in *Malaco-myia* and *Coelopa* a small discrete sclerite (intermedium) lies in membrane be-tween these processes. Aedeagus able to be swung through wide arc against aedeagal apodeme to anteriorly directed rest position, with variable complex sclerotization, in *Coelopa* and *Icaridion* with fan-like basal lobes; distal part of aedeagus short with spiniform processes in *Malacomyia*, but represented by slender filament (without such processes) in *Coelopa* and *Icaridion*. Small postgonites present in *Malacomyia* and *Coelopa;* absent in *Icaridion*. Slender ejaculatory apodeme present.

Characterization and discussion

The above description does not take account of the structure of *Heterocheila* (figs. 82 and 84), whose classification in this family needs confirmation. If *Heterocheila* is excluded, the Coelopidae may be characterized by the following conditions which are apomorphous with respect to the groundplan of the Sciomyzoinea and Muscoidea.
(1) Postverticals convergent (HENNIG 1958).
(2) 6th tergum (♂) asymmetrically reduced, with broad lateroventral area on right side but becoming narrower and ill-defined dorsally (fig. 81).
(3) Hypandrium (♂) with pair of posteriorly directed setose processes (fig. 81).

The structure of the male postabdomen of *Coelopa*, *Icaridion* and *Malacomyia* (fig. 81) is noteworthy for the retention of the complete set of sclerites present in the groundplan of the Muscoidea. Both the 7th tergum vestige (situated lateroventrally on the right side) and the (inverted) 8th tergum vestige are clearly defined. Among other groups of Muscoidea such a complete set of sclerites is retained only by *Pelidnoptera* (Phaeomyiidae), by *Orbellia* (Heleomyzidae) and by various members of the Micropezoinea.

The genus *Heterocheila* was formerly classified in the Coelopidae, but most recent workers have followed MALLOCH (1933) in including it in the 'Helcomyzidae' on account of its possession of a precoxal bridge. However SPEIGHT's (1969) findings cast doubt on the interpretation of this character, and (as indicated in MALLOCH's key) *Heterocheila* is divergent in several external characters from the other genera ascribed to the 'Helcomyzidae'. The structure of the male postabdomen and the form of the aedeagus in *Heterocheila* (figs. 82 and 84) are also substantially different from the conditions shown by *Helcomyza* (which I include in the Dryomyzidae), and I can find no firm basis for including *Heterocheila* in the latter family. The condition of the aedeagus shown by *Heterocheila* seems to me to indicate that *Heterocheila* was perhaps correctly included in the Coelopidae. The aedeagus is characterized by the presence of fan-like basal lobes (similar to those shown by *Coelopa* and *Icaridion* although absent in *Malacomyia*) and spiniform distal processes (shown also by *Malacomyia*, but not by *Coelopa* and *Icaridion* in which the distal part of the aedeagus is modified to form a slender filament). Neither of these two conditions has been reported in any other group of Sciomyzoinea. The possibility thus seems indicated that the Coelopidae *sensu lato* (including *Heterocheila*) can be characterized in their groundplan by a particular apomorphous condition of the aedeagus. But further investigation is needed before this hypothesis can be adequately assessed. In the meantime I propose to include *Heterocheila* in the Coelopidae on a tentative basis.

172

I̊ *Heterocheila* is correctly included in the Coelopidae, then the three apomorphous conditions of *Coelopa*, *Icaridion* and *Malacomyia* stated above are not referable to the groundplan of the family, and the only strongly apomorphous condition of the Coelopidae *(sensu lato)* among the character sequences so far analysed is the form of the aedeagus. In *Heterocheila* the postverticals are parallel, the 6th tergum (♂) is large and symmetrical (about half as long as the 5th tergum), and the hypandrium (♂) does not bear posteriorly directed setose processes. In all these respects *Heterocheila* shows relatively plesiomorphous conditions. Apomorphous conditions of *Heterocheila* not shown by *Coelopa* and *Malacomyia* are: (1) loss of the 8th tergum vestige (♂), and (2) the presence of only two spermathecae (♀) (HENNIG 1958).

C 25. Family Phaeomyiidae status novus

Sources of information

My treatment is based solely on study of a preparation of *Pelidnoptera nigripennis* (F.). No detailed information on the male postabdomen and genitalia of this family was found in the literature.

Description of male postabdomen and genitalia (figs. 85-87)

6th and 7th left abdominal spiracles within sterna of these segments, displaced dorsally in relation to preceding spiracles. 6th and 7th right spiracles displaced centrally, in membrane near 6th and 7th tergum vestiges.

Postabdomen asymmetrical. 6th tergum represented by slender band of sclerotization on right side. 6th sternum asymmetrical, extending onto dorsum on left side. 7th tergum represented by narrow lateroventral sclerite on right side. 7th sternum asymmetrically developed on left side, where it is fused dorsally with the large 8th sternum. 8th tergum represented by slender strip of ventral sclerotization.

Periandrium bearing double telomeres, the posterior pair furcate. Interparameral sclerotization ill-defined. Cerci large, discrete apically. Aedeagal apodeme rod-like, free from hypandrium. Aedeagus able to be swung through wide arc against aedeagal apodeme to anteriorly directed rest position, with small pubescent basal lobes and rather complex distal lobes and sclerotization. Postgonites hardly differentiated from some ill-defined sclerotization above base of aedeagus. Slender ejaculatory apodeme present.

Characterization and discussion

The genera *Pelidnoptera* and *Phaeomyia* have previously been classified in the

Sciomyzidae. STEYSKAL (1965) has emphasized the differences between this group and typical Sciomyzidae. He proposes to recognize the Phaeomyiinae as a subfamily of Sciomyzidae, and to place *Phaeomyia* in synonymy with *Pelidnoptera*. Thus according to STEYSKAL's proposal all Recent species should be included in a single genus.

My studies on the male postabdomen and genitalia provide additional characters for separating *Pelidnoptera* from true Sciomyzidae. The full complement of postabdominal sclerites is retained in *Pelidnoptera* (figs. 85 and 86), as also in Coelopidae. In the preparation studied the 7th tergum vestige appears as a conspicuous sclerotized band on the right side of the venter, and the 7th right spiracle lies on about the centre-line of the venter at the edge of this sclerite. The 8th tergum vestige is less well developed than in some Coelopidae, but is clearly represented by a small band-like sclerite posterior to the 7th tergum vestige. Both these sclerites have been lost in true Sciomyzidae. According to STEYSKAL *Pelidnoptera* retains three spermathecae ($\female$), which is a more plesiomorphous condition than the reduction of the number of spermathecae to two shown by true Sciomyzidae.

From the character sequences so far analysed I can see no evidence that *Pelidnoptera* is more closely related to the Sciomyzidae than, for instance, to the Coelopidae or Dryomyzidae. Consequently I think it preferable to treat the group as a separate family. Only three recent species are known, but HENNIG (1965b, 1969b) has already reported at least one and possibly two species in Baltic amber. The life-histories of the Recent species are still unknown.

Apomorphous conditions of the Phaeomyiidae with respect to the groundplan of the Sciomyzoinea and Muscoidea are as follows.
(1) Middle and hind tibiae with several medial bristles, as well as preapical bristle (STEYSKAL 1965, HENNIG 1965b).

HENNIG (1965b) states that this condition is not shown by other Sciomyzoinea, and can therefore probably be considered apomorphous.

(2) 6th tergum ($\male$) asymmetrically reduced, represented only by slender band on right side (figs. 85 and 86).

The 6th and 7th sterna in *Pelidnoptera* are asymmetrically developed on the left side, as normally in Sciomyzoinea and other Muscoidea. Since they are relatively large and extend to some extent onto the dorsum, they have sometimes been confused with terga. For instance SACK's (1939: 11) description 'Beim $\male$ sind das 6. und 7. Tergit kugelig gewölbt' refers to the sterna, not to the reduced terga.

(3) 6th left and 7th left spiracles ($\male$) lying within 6th and 7th sterna; both 7th spiracles strongly displaced to about centre-line of dorsum and venter respectively (figs. 85 and 86).

Both the 7th spiracles are strongly displaced in *Pelidnoptera*, the 7th left lying near the centre of the dorsum (within the 7th sternum), and the 7th right lying near the centre of the venter (at the edge of the 7th tergum vestige). Some degree of displacement of the 7th spiracles is probably a plesiomorphous condition for the Muscoidea, but the displacement shown by *Pelidnoptera* is exceptionally strong. A similar strong displacement of the 7th spiracles is shown by some Dryomyzidae (such as *Helcomyza*), but the position of the spiracles in that family is somewhat variable.

(4)   Telomeres ($\male$) double, the posterior pair furcate (fig. 85) (STEYSKAL 1965).
(5)   Aedeagus ($\male$) short, with small pubescent basal lobes (fig. 87).

The presence of small pubescent basal lobes of the aedeagus in *Pelidnoptera* may possibly represent an earlier stage in the evolution of the condition shown by some Coelopidae (*Heterocheila* and *Coelopa*), in which conspicuous fan-like basal lobes are present. If confirmed, this hypothesis might provide a basis for including *Pelidnoptera* in the Coelopidae. But it seems premature to form a judgement on this question, until more comparative information on the male genitalia of Coelopidae and Phaeomyiidae is available.

C. 26.   Family Dryomyzidae (including Helcomyzidae)

Sources of information

HENNIG (1937a) and STEYSKAL (1957b, 1958b) gave information on the male postabdomen and genitalia of this family. I studied preparations of the following: – *Dryomyza anilis* FALLÉN; *Helcomyza mirabilis* MELANDER; *Maorimyia bipunctata* (HUTTON).

Description of male postabdomen and genitalia (figs. 88-89)

6th left abdominal spiracle usually in membrane, but within 6th tergum in some *Dryomyza* species: 7th left spiracle usually within 7th sternum and displaced dorsally in relation to preceding spiracles (but in membrane in *Dryomyza bergi* STEYSKAL and absent in a few other *Dryomyza* species). 6th right spiracle in membrane: 7th right spiracle usually in membrane and displaced centrally in relation to preceding spiracles (but absent in some *Dryomyza* species).

Postabdomen asymmetrical. 6th tergum represented by narrow band or strip of sclerotization (slightly asymmetrical in some species). 6th sternum asymmetrical, linked to 7th sternum on left side. 7th tergum represented by narrow lateroventral sclerite on right side or absent. 7th sternum asymmetrically developed on left side where it is fused dorsally with the large 8th sternum: in *Helcomyza* the 7th sternum extends across the dorsum and becomes fused with the 7th tergum. 8th tergum absent.

Periandrium bearing discrete or more or less fused telomeres, which in most species are furcate or double (e.g. *Dryomyza anilis* FALLÉN): pair of ill-defined interparameral sclerites extending from base of telomeres. Cerci discrete. Aedeagal apodeme rod-like, free from hypandrium. In *Dryomyza* and *Oedoparena* the hypandrial arms fuse above the base of the aedeagus and form a posteriorly directed projection, but this is not so in *Helcomyza* and *Maorimyia*. Aedeagus elongate, able to be swung through wide arc against aedeagal apodeme to anteriorly directed rest position, densely clothed distally with fine cuticular processes ('pubescence'), usually with well differentiated phallophore and paraphalli; small epiphallus present in *Maorimyia*, but not in other genera examined. Postgonites broad and rounded. Ejaculatory apodeme present.

Characterization and discussion

The separation of the 'Helcomyzidae' from the Dryomyzidae was first proposed by MALLOCH (1933), because of the presence of a precoxal bridge in the Helcomyzidae. On the basis of this same character HENNIG (1958) suggested that the Helcomyzidae are the sister-group of the Ropalomeridae. Now that it has become clear that these interpretations rested on oversimplified assumptions about prosternal variation (see my opening discussion of the Sciomyzoinea) the way is open for a reappraisal of the affinities of the so-called Helcomyzidae. I have concluded in this study that *Heterocheila* is not monophyletic with the other genera referred to the Helcomyzidae, and have provisionally classified that genus in the Coelopidae. It seems to me that there are clear synapomorphies between the remaining genera of 'Helcomyzidae' *(Helcomyza, Maorimyia* and *Paractora)* and the Dryomyzidae *(sensu stricto)*, and I therefore propose to classify these in the Dryomyzidae. In effect I revert to the definition of Dryomyzidae which prevailed before MALLOCH's (1933) proposal of the family Helcomyzidae.

The Dryomyzidae in the sense here proposed are characterized by the following conditions which are apomorphous with respect to the groundplan of the Sciomyzoinea and Muscoidea.

(1) '♂ with thumbnail-like apical ventral process on at least fore and hind metatarsi' (MALLOCH 1933: 326).

MALLOCH has given this characterization for *Helcomyza, Maorimyia* and *Paractora*. The same condition is shown by *Oedoparena* and some *Dryomyza* species (STEYSKAL 1958b). It is puzzling that STEYSKAL gives 'male fore and hind basitarsi simple' as a character of 'Helcomyzidae'. This must be a mistake, as pointed out by HENNIG (1965b: 76). I have confirmed that the apical processes are present in *Helcomyza*, as stated by MALLOCH.

(2) 6th tergum (♂) reduced, less than one-third of length of 5th tergum.

The figures given by STEYSKAL (1958b) indicate the range of variation of the 6th tergum.

(3) 8th tergum vestige lost (♂).

The 8th tergum vestige is the only sclerite of the postabdomen which has been lost in the groundplan of the Dryomyzidae. The 7th tergum vestige is retained in many species. In *Helcomyza* the 7th tergum vestige has become fused with the 7th sternum, so that a complete band of sclerotization extends across the dorsum.

(4) Aedeagus (♂) densely clothed with fine cuticular processes ('pubescence') distally (fig. 88).

The structure of the aedeagus provides evidence that the Dryomyzidae in the sense here proposed are monophyletic. In *Dryomyza* the sclerites and pubescence of the aedeagus are blackish and very conspicuous. In *Helcomyza* and *Maorimyia* the sclerites of the aedeagus are paler and the pubescence unpigmented. I do not think that the presence of aedeagal pubescence in Dryomyzidae can be due to synapomorphy with the Tephritoinea. The aedeagus of Dryomyzidae cannot be coiled like that of the Tephritoinea.

C. 27. Family Sciomyzidae

Sources of information

An introduction to the morphology of the male postabdomen and genitalia of Sciomyzidae was given by VERBEKE (1950). Good figures for particular genera and species are also given in numerous short papers by STEYSKAL. I studied preparations of the following: – *Pherbellia quadrata* STEYSKAL; *Pherbellia griseola* (FALLÉN); *Pherbellia albocostata* (FALLÉN); *Tetanura pallidiventris* (FALLÉN); *Sciomyza simplex* FALLÉN; *Pteromicra apicata* (LOEW); *Elgiva sundewalli* KLOET & HINCKS; *Tetanocera robusta* (LOEW).

Description of male postabdomen and genitalia (figs. 90-91)

6th left abdominal spiracle usually in membrane (but within 6th sternum in *Tetanocera*): 7th left spiracle displaced dorsally in relation to preceding spiracles, either in membrane near edge of 7th sternum *(Pteromicra, Pherbellia)*, within 7th sternum *(Elgiva, Tetanocera, Tetanura)* or within 6th sternum *(Sciomyza)*. 6th and 7th right spiracles in membrane (Sciomyzinae) or within 8th sternum (Tetanocerinae).

Postabdomen usually asymmetrical, but only weakly so in Tetanocerinae: fully symmetrical in 'Sepedoninae' according to VERBEKE (1950). 6th tergum represented by narrow strip of dorsal sclerotization *(Pteromicra, Pherbellia, Salticella)* or completely absent. 6th sternum asymmetrical, linked to or fused with 7th sternum on left side. 7th sternum asymmetrically developed on left side (relatively small in Tetanocerinae) where it is fused dorsally with the large 8th sternum. 7th and 8th terga absent (but a secondary sclerite on the ventral side of the 8th segment is found in *Tetanocera*).

Periandrium bearing discrete telomeres, which are double or furcate in many genera. Interparameral sclerotization absent or consisting of weakly sclerotized plates which extend from the base of the telomeres. Cerci variably developed, discrete or fused, very small in Tetanocerinae and *Tetanura* but large in other genera studied. Aedeagal apodeme rod-like, free from hypandrium. Aedeagus usually able to be swung through wide arc against aedeagal apodeme to anteriorly directed rest position (but not in *Pherbellia*), with variable sclerotization (very complex in many genera): in *Pherbellia* a complex epiphallus-like structure is present. Postgonites variably developed. Slender ejaculatory apodeme present.

Characterization and discussion

The Sciomyzidae in the sense here proposed exclude the Phaeomyiidae and Helosciomyzidae (including *Huttonina*), which were previously included in this family. The Sciomyzidae in this sense are characterized by the following ground-plan conditions which are apomorphous with respect to the groundplan of the Sciomyzoinea and Muscoidea.

(1) Clypeus separated from epistoma (= 'face') by large membranous area and not visible in profile when the proboscis is withdrawn.

This character requires further investigation, as stated by HENNIG (1965b).

(2) Two spermathecae (♀) (see STEYSKAL 1965 and HENNIG 1965b).

STEYSKAL (1965) has drawn attention to the presence of three spermathecae, a more plesio-morphous condition, in the groups which he has called the Huttonininae, Helosciomyzinae and Phaeomyiinae. None of these belong to the Sciomyzidae in the sense here proposed. As far as known all the malacophagous species which may be accepted as true Sciomyzidae possess two spermathecae with the exception of *Salticella*, for which STEYSKAL reports four. HENNIG's (1965b: 86) view that this condition has been derived from an earlier two-sperma-thecae condition seems to me the most probable interpretation.

(3) 6th tergum (♂) much reduced, represented at most by slender band (fig. 90).

A narrow remnant of the 6th tergum occurs in some *Pteromicra* spp. (STEYSKAL 1954), *Pherbellia albocostata* (FALLÉN) (STEYSKAL 1957a, fig. 2) and *Salticella* (STEYSKAL 1965). I have confirmed this condition from my own observations on the first two genera. It is not clear why STEYSKAL (1965) characterized his concept of Sciomyzinae (in which *Pteromicra* and *Pherbellia* are included) with the words '6th tergite fused with protandrium'. Since a free 6th tergum is retained by some members of the above genera, such characterization cannot refer to the groundplan of the Sciomyzidae. VERBEKE (1950) considered that in the Reno-cerinae, Tetanocerinae and Sepedoninae (all included in the Tetanocerini of STEYSKAL's classification) the 6th and 7th terga must have become fused with the (inverted) 8th sternum because of the situation of the 6th and 7th right spiracles within this sclerite. This inference seems to me unwarranted, because there are several examples among the Schizophora of the postabdominal spiracles having moved onto segments to which they did not originally belong. The simplest explanation of the observed structure is that the 6th tergum, which was already reduced to a narrow band in the groundplan of the family, has been lost completely in many groups (including all Tetanocerini *sensu* STEYSKAL).

(4) 7th and 8th tergum vestiges lost (♂).

In this respect the Sciomyzidae show a more apomorphous condition than do the Coelop-idae, Phaeomyiidae and Dryomyzidae, in which one or both of these sclerites are retained. I have noted a small sclerite in the centre of the genital pouch in *Tetanocera* which might be interpreted as the 8th tergum vestige. However, since the structure of the male postabdomen of *Tetanocera* is highly modified, I think it probable that this sclerotization is secondary. No possible trace of the 8th tergum was found in the other species examined.

(5) Larvae malacophagous, with serrate ventral arch (of labial origin) below mouth hooks.

The larvae of Sciomyzidae are now rather well known, due largely to the efforts of various workers at Cornell University. A detailed survey has been produced by KNUTSON (1963). The comparative table in that work (page 259) indicates that the characteristic serrate ventral arch is found in all the known larvae of true Sciomyzidae (but not of the misplaced 'Heloscio-myzinae').

HENNIG (1958) suggested as a further possible apomorphous condition of the Sciomyzidae that the 6th and 7th abdominal spiracles of the female always lie within the terga. However STEYSKAL (1965) has indicated that this is not the case in *Salticella*, and only '*usually*' the case in members of his concept of Sciomyzinae. Thus it seems doubtful whether such a condition can be ascribed to the groundplan of the Sciomyzidae.

The male postabdomen and genitalia of Sciomyzidae are diverse, but clearly derived from a groundplan condition with typical muscoid asymmetry of the 6th and 7th sterna and with the 6th and 7th right spiracles situated in the membrane of the genital pouch. The genera placed in Sciomyzini by STEYSKAL (1965) (= Sciomyzinae of VERBEKE 1950) all exhibit such a condition, as far as I am aware. In *Tetanocera* and related genera (Tetanocerini STEYSKAL 1965; Tetanocerinae VERBEKE 1950) the 7th sternum (on the left side) has been somewhat reduced, and the postabdomen is consequently more nearly symmetrical. Another note-worthy feature of the structure of the Tetanocerini, as previously noted, is that the 6th and 7th right spiracles lie within the inverted 8th sternum. They thus lie on a rotated segment. In my previous discussion of the similar displacement of the 7th right spiracle which has occurred in the Calyptratae, I suggest the follow-ing explanation for this paradoxical morphology. The posterior abdominal spiracles are not formed until after completion of the rotation of the terminal segments; consequently spiracles have been able to move onto the rotated 8th segment in the course of evolution, without the lateral tracheal trunks becoming crossed as a result of the rotation until *after* the last pair of spiracles. The ground-plan condition of the aedeagus in Sciomyzidae cannot be judged without more detailed analysis of the variation within the family. The species studied show a fairly short and broad aedeagus, with well-defined sclerites. A few species, such as *Pteromicra apicata* (LOEW), show distinct pubescence on the distal part of the aedeagus, a condition which, if it belongs to the groundplan of the Sciomyzidae, may possibly be due to synapomorphy with the Dryomyzidae. However this suggestion can only be tentative. I am not able to state any apomorphous con-ditions of the aedeagus of the Sciomyzidae on the basis of existing analyses. The aedeagal apodeme is rod-like and free from the hypandrium (plesiomorphous), as in most Sciomyzoinea.

STEYSKAL (1965) has presented a reclassification of the subfamilies of Scio-myzidae. His paper is most helpful, as he has clarified certain characters of some little-known groups which have been included in the Sciomyzidae. STEYSKAL treats these as new subfamilies of Sciomyzidae (the Huttonininae, Helosciomyz-inae and Phaeomyiinae). However the structure of these groups provides no firm basis for considering them more closely related to the Sciomyzidae proper (Sciomyzinae + Salticellinae in STEYSKAL's classification) than to other families of Sciomyzoinea. Therefore I treat them in this work as distinct families (Phae-omyiidae and Helosciomyzidae).

STEYSKAL (1965) has also recognized a subfamily Salticellinae, containing the single Recent genus *Salticella*. These insects must clearly be accepted as true Sciomyzidae, since they have malacophagous larvae with the typical sciomyzid structure (see KNUTSON 1963). It has not been established whether or not *Salticella* represents the sister-group of all other Sciomyzidae (Sciomyzinae *sensu* STEYSKAL). To infer the probable existence of a sister-group relationship from morphological evidence it is necessary to demonstrate that each of the groups under consideration shows some relatively plesiomorphous condition of a character sequence in respect of which the other shows a relatively apomorphous condition. The characterization of Salticellinae and Sciomyzinae given by STEYSKAL does not fully meet this criterion, because the relatively plesiomorphous conditions attributed to the Salticellinae (♂, '6th tergite a completely free slender band'; ♀, '6th and 7th spiracles well within membrane...') also occur in certain members of the 'Sciomyzinae'.

C. 28. Family Helosciomyzidae status novus

Sources of information

Some figures of the male genitalia of Helosciomyzidae were given by TONNOIR & MALLOCH (1928). I studied preparations of the following: – *Helosciomyza subspinicosta* TONNOIR & MALLOCH; *Huttonina abrupta* TONNOIR & MALLOCH.

Description of male postabdomen and genitalia (figs. 92-93)

6th abdominal spiracles in membrane: 7th left spiracle within 7th sternum: 7th right spiracle at margin of 8th sternum.

Postabdomen asymmetrical. 6th tergum represented by narrow band of dorsal sclerotization, discrete in *Helosciomyza*, more or less fused with 8th sternum in *Huttonina*. 6th sternum narrow, asymmetrical, linked to 7th sternum on left side. 7th sternum narrow, asymmetrically developed on left side where it is fused with the large 8th sternum. 7th and 8th terga absent.

Periandrium bearing discrete telomeres. Interparameral sclerotization consisting of pair of plates which extend anteriorly from the base of the telomeres (*Helosciomyza*), or of ill-defined transverse sclerite (*Huttonina*). Cerci discrete, well developed in *Helosciomyza* but scarcely differentiated in *Huttonina*. Aedeagal apodeme rod-like, free from hypandrium. Hypandrial sclerotization forming bridge above aedeagal apodeme in *Helosciomyza*, but not in *Huttonina*. Aedeagus bilobed distally, its sclerotization consisting mainly of basal phallophore and lateral paraphalli; able to be swung through wide arc against aedeagal apodeme to anteriorly directed rest position. Small postgonites present. Ejaculatory apodeme present.

180

Characterization and discussion

This family contains certain New Zealand and Australian genera which have usually been classified in the Sciomyzidae. KNUTSON (1963) reported that the larvae of *Helosciomyza* and *Xenosciomyza* do not show the ventral arch characteristic of sciomyzid larvae, and suggested that these genera should be classified in the Heleomyzidae (KNUTSON 1963: 261). However he gave no characterization in support of such a classification. I understand from correspondance with J. ABERCROMBIE that these larvae were obtained from fungi. STEYSKAL (1965) reported the retention of three spermathecae by members of these Australian and New Zealand genera, and proposed for them two new subfamilies of Sciomyzidae, the Helosciomyzinae and Huttonininae.

I have found no evidence that this group is more closely related to the Sciomyzidae than to the other families of Sciomyzoinea, and consequently think it preferable to treat it as a distinct family, rather than as a subordinate group of the Sciomyzidae. The genera included in the Helosciomyzidae as here proposed are *Huttonina*, *Helosciomyza*, *Xenosciomyza*, *Polytocus* and *Prosochaeta* (the last three not studied by me). The Helosciomyzidae are characterized in their groundplan by the following conditions which are apomorphous with respect to the groundplan of the Sciomyzoinea and Muscoidea.

(1) Postverticals parallel to slightly convergent (STEYSKAL 1965).

(2) 6th tergum (♂) reduced, less than one-third of length of 5th tergum (fig. 92).

The reduced 6th tergum is discrete in *Helosciomyza*, but virtually fused with the 8th sternum in *Huttonina*.

(3) 7th and 8th tergum vestiges lost (♂); 7th right spiracle lying at margin of 8th sternum.

In some Sciomyzidae the 7th right spiracle lies free in membrane, a condition which seems to belong to the groundplan of that family. Therefore the movement of the 7th right spiracle to the margin of the 8th sternum in Helosciomyzidae cannot be due to synapomorphy with the Tetanocerinae (Sciomyzidae), in which a similar movement has occurred.

(4) 7th left spiracle (♂) lying within 7th sternum (fig. 92).

(5) Aedeagus (♂) bilobed distally.

In *Prosochaeta* (HARRISON 1959, fig. 47) and *Huttonina* the distal section of the aedeagus is bilobed and bears fine scales or pubescence. In *Helosciomyza* (fig. 93) the aedeagus is also bilobed distally, but such pubescence is lacking.

C. 29. Sepsidae family-group

I refer to this group the Ropalomeridae and Sepsidae, which I treat as full families in accordance with current practice. The monophyly of these two families was first suggested by J. F. McALPINE (1963) and subsequently accepted by HENNIG (1965b). The table given on page 80 of HENNIG's work provides an up-to-date summary of the evidence for inferring that these two families are monophyletic, although I do not think that HENNIG was justified in placing the 'Helcomyzidae' as the sister-group of the Ropalomeridae + Sepsidae. I refer *Helcomyza* and related genera to the Dryomyzidae. HENNIG (in press) now acknowledges that his grounds for associating the 'Helcomyzidae' with the Sepsidae family-group were inconclusive.

On the basis of MCALPINE's and HENNIG's analyses, supplemented by my own studies on the male postabdomen and genitalia, I am able to characterize the Sepsidae family-group by the following groundplan conditions which are apomorphous with respect to the groundplan of the Sciomyzoinea and Muscoidea.

(1) Metastigmatal bristles present.

A group of conspicuous bristles posterior to each metathoracic spiracle is not found in any other group of Schizophora. An isolated setula in this position is reported in Eurychoromyidae (Lauxanioidea) by J. F. MCALPINE (1968b). I do not think that the presence of this setula can be due to synapomorphy with the Sepsidae family-group, because the structure of the male postabdomen and genitalia of *Eurychoromyia* seems to exclude the possibility that this genus belong to the Sepsidae family-group.

(2) Vibrissae present.

The vibrissae are only weakly differentiated from the peristomal bristles in members of the Sepsidae family-group. This character thus scarcely provides a clear distinction between the Sepsidae family-group and other Sciomyzoinea.

(3) Two spermathecae (♀) (STURTEVANT 1926, HENNIG 1958).
(4) 7th and 8th tergum vestiges lost (♂).

These sclerites have also been lost in the Sciomyzidae and Helosciomyzidae.

(5) Aedeagal apodeme (♂) expanded and fused with hypandrium anteriorly.

This characterization refers to the condition shown by the Ropalomeridae (fig. 94). In Sepsidae a more strongly modified condition is shown, in which the aedeagal apodeme has become completely fused with the hypandrium.

In the groundplan of the Sepsidae family-group the presence of a relatively large 6th tergum should be assumed, as shown by *Orygma* and the Ropalomeridae. In *Orygma* the 6th tergum is about half as long as the 5th tergum, as postulated for the groundplan of the Sciomyzoinea.

The condition of the aedeagus in the groundplan of the Sepsidae family-group was probably similar to the condition shown by *Orygma* (fig. 96), in which the basal area supported by the paraphalli is relatively long and the distal area contains differentiated sclerites. I postulate that the condition shown by the Ropalomeridae has been derived from this by reduction of the sclerotization of the distal area. In Sepsidae other than *Orygma* the basal area is shortened.

D. 5. Family Ropalomeridae

Sources of information

Information on the male postabdomen and genitalia of the Ropalomeridae was given by STEYSKAL (1957a) and DO PRADO (1965). I studied a preparation of *Rhytidops floridensis* (ALDRICH).

Description of male postabdomen and genitalia (fig. 94)

All postabdominal spiracles in membrane: 7th left spiracle near edge of 7th sternum, displaced dorsally in relation to preceding spiracles.

Postabdomen asymmetrical. 6th tergum well developed, about half as long as 5th tergum (slightly asymmetrical in *Rhytidops*, larger on right side). 6th sternum represented by asymmetrical, strongly sclerotized band which is fused with the narrow 7th sternum on the left side: 6th and 7th sterna fused dorsally with large 8th sternum. 7th and 8th terga absent.

Periandrium bearing discrete telomeres which are joined by a broad transverse interparameral sclerite. Cerci small, more or less discrete. Aedeagal apodeme laterally compressed, expanded anteriorly where it is fused with the hypandrium. Hypandrial arms broadly fused above base of aedeagus, where they bear a conspicuous upcurved process ('x' on fig. 94). Aedeagus sclerotized only at its base, with long membranous distal area, able to be swung through wide arc against aedeagal apodeme to anteriorly directed rest position. Large postgonites present. Ejaculatory apodeme not traced.

Characterization and discussion

The neotropical Ropalomeridae are characterized as the sister-group of the Sepsidae by the following apomorphous conditions with respect to the ground-plan of the Sepsidae family-group.
(1) Aedeagus (♂) with long, completely membranous distal area (fig. 94).
(2) Hypandrial arms (♂) broadly fused above base of aedeagus, where they bear a conspicuous upcurved process ('x' on fig. 94).

This process is what STEYSKAL (1957a) has called the 'spinus titillatorius', a synonym for the epiphallus of Calyptratae and some other groups of Muscoidea. However I do not think that the characteristic process found in Ropalomeridae should be called an epiphallus, since its sclerotization is confluent with the hypandrial arms, not with the base of the aedeagus.

(3) Frons impressed: eyes protruding (HENDEL 1936).
(4) Hind tibiae laterally compressed, strongly expanded, sabre-shaped (HENDEL 1936).

Relatively plesiomorphous conditions of the Ropalomeridae in comparison with the Sepsidae are: (1) the presence of discrete telomeres (♂), (2) the presence

183

of a clearly delimited 7th sternum (♂) on the left side, (3) 6th abdominal spiracles (♂) lying in membrane, and (4) incomplete fusion of the hypandrium and aedeagal apodeme (♂).

D. 6. Family Sepsidae

Sources of information

The male postabdomen and genitalia of the Sepsidae were treated by HENNIG (1937a, 1949). I studied preparations of the following: – *Orygma luctuosum* MEIGEN; *Nemopoda nitidula* (FALLÉN); *Sepsis vicaria* WALKER; *Themira putris* (L.).

Description of male postabdomen and genitalia (figs. 95-96)

6th abdominal spiracles within 6th tergum: in *Orygma* 7th spiracles asymmetrically situated in membrane (the left displaced dorsally in relation to preceding spiracles, the right displaced centrally): in other genera (Sepsini) 7th spiracles symmetrically situated close to 6th spiracles within sclerotized area formed by fusion of 6th tergum and 8th sternum.

Postabdomen slightly asymmetrical in *Orygma*, fully symmetrical in Sepsini. 6th tergum discrete (about half as long as 5th tergum) only in *Orygma:* in Sepsini 6th tergum represented by narrow band which is fused with the large 8th sternum, with boundary between 6th tergum and 8th sternum indicated by visible suture in some species, while in others the fusion is complete. 6th sternum large and slightly asymmetrical in *Orygma*, linked to 8th sternum on left side; in Sepsini represented by symmetrical ventral strip of sclerotization. 7th sternum not delimited (absent or fused with 8th sternum). 8th sternum well developed, especially in *Orygma*. 7th and 8th terga absent.

Telomeres represented by lobes of periandrium, hardly articulated (at most delimited by partial suture): in Sepsini (but not in *Orygma*) each telomere bears a slender apodeme extending from its inner corner which is linked to the hypandrial arms. Cerci small and discrete *(Orygma)*, or scarcely differentiated from membrane (Sepsini). Aedeagal apodeme elongate, forming trough which receives the aedeagus when at rest, fully fused with hypandrium anteriorly. Hypandrium bearing setose, posteriorly directed lobes (pregonites), which appear discrete in *Orygma*. Aedeagus able to be swung through wide arc against aedeagal apodeme to anteriorly directed rest position, in *Orygma* rather elongate with bilobed distal area, in Sepsini rather short and broad with complex sclerotization. Small postgonites present. Ejaculatory apodeme present.

Characterization and discussion

The Sepsidae are characterized in their groundplan by the following conditions which are apomorphous with respect to the groundplan of the Sepsidae family-group.

(1) 6th and 7th abdominal spiracles ($\female$) lying within their respective terga (HENNIG 1958).
(2) Hypandrium and aedeagal apodeme ($\male$) completely fused (fig. 96).
(3) Telomeres ($\male$) fused with periandrium, at most delimited by partial suture.
(4) 6th abdominal spiracles ($\male$) lying within 6th tergum.

In *Orygma* (fig. 95) the 6th spiracles lie within the large 6th tergum, while the 7th spiracles lie asymmetrically in membrane (the 7th right spiracle being slightly displaced centrally in relation to the preceding right spiracles, as postulated for the groundplan of the Muscoidea). In other Sepsidae the reduced 6th tergum has become fused with the inverted 8th sternum, and the 6th and 7th spiracles lie close together (symmetrically on both sides) within the margins of the resulting sclerotized area.

(5) 7th sternum ($\male$) not delimited (either lost or fused with 8th sternum).

In most Sepsidae full symmetry of the male postabdomen has been restored. However *Orygma*, the plesiomorphous sister-group (Orygmatini) of all other recent Sepsidae (Sepsini) according to HENNIG (1965b), retains an asymmetrical 6th sternum (contiguous with the inverted 8th sternum on the left side only) and a displaced 7th right spiracle. Thus the symmetrical postabdomen of most Sepsidae is secondary, derived from the asymmetrical pattern of sclerites characteristic of other Sciomyzoinea and the Muscoidea generally. I have also confirmed from serial sections of *Nemopoda* that the '*musculus hypandriotergalis*', linking the (inverted) 8th sternum and the hypandrium on the left side, is present, as in most other Muscoidea.

(6) Anal vein shortened, not reaching the wing margin (HENNIG 1958).

J. F. MCALPINE (1963) has given 'anal vein abbreviated' as a condition of the Sepsidae family-group, but it seems to me that this vein is more or less distinct to the wing margin in some Ropalomeridae, as is implied in the table given by HENNIG (1965b: 80).

The genus *Orygma* is interesting as it clearly represents the sister-group of all other Sepsidae, as stated by HENNIG (1965b). It is the only sepsid genus in which the males retain a large 6th tergum and large 6th sternum, with the 7th spiracles situated asymmetrically in membrane. The synapomorphies between *Orygma* and other Sepsidae (as indicated by my characterization above) are clear enough, and I see no morphological grounds for classifying *Orygma* in the Coelopidae, as it still sometimes done (for instance by EGGLISHAW 1960).

C. 30. Family Megamerinidae

Sources of information

HENNIG (1941c) described the male postabdomen and genitalia of *Megamerina*. I studied a preparation of *Megamerina dolium* (F.). My present treatment is based solely on *Megamerina*, since I was unable to obtain information on other genera.

Description of male postabdomen and genitalia (figs. 97-98)

6th and 7th left abdominal spiracles absent: 6th and 7th right spiracles in membrane, the latter displaced centrally in relation to preceding spiracles (to about centre-line of venter).

Postabdomen asymmetrical. 6th tergum represented by narrow band of dorsal sclerotization. 6th sternum very small, asymmetrical, linked to 7th sternum on left side. 7th sternum very small, asymmetrically developed on left side where it is fused dorsally with the large elongate 8th sternum. 7th tergum represented by small area of ventral sclerotization (near the 7th right spiracle). 8th tergum absent.

Periandrium bearing long discrete telomeres, which are joined by a weakly sclerotized U-shaped interparameral sclerite. Cerci discrete, rather large. Aedeagal apodeme rod-like, fused with hypandrium anteriorly. Aedeagus elongate, with small epiphallus arising from basal phallophore, supported along most of its length by slender paraphalli, but with complex lobed distal section which bears a pair of twisted filaments; aedeagus able to be swung through wide arc against aedeagal apodeme to anteriorly directed rest position. Postgonites bearing anteriorly directed lobe with comb of bristles. Large ejaculatory apodeme present.

Characterization and discussion

The Megamerinidae are characterized by the following conditions which are apomorphous with respect to the groundplan of the Sciomyzoinea and Muscoidea.

(1) Ocellar bristles lost (HENNIG 1958).
(2) Postverticals lost (HENNIG 1958).
(3) Only one upper fronto-orbital bristle *(ors)* retained (HENNIG 1958).
(4) Hind femur thickened and armed with two rows of spines on underside (HENNIG 1958).
(5) Abdominal segments elongate, except for male 6th and 7th segments.

In the male (fig. 97) the preabdominal segments, the 8th segment and the andrium are conspicuously elongate, while the sclerites of the 6th and 7th segments are contrastingly very small. In the female all the abdominal segments as far as the 8th (genital) segment are elongate.

(6) 6th tergum (♂) reduced to narrow band (fig. 97).
(7) 6th left and 7th left spiracles lost (♂).

Megamerina is unique among the Schizophora which I have examined in having lost the 6th and 7th left spiracles, while retaining the full complement of right spiracles. The 7th right spiracle is situated near the centre-line of the venter (fig. 97).

(8) 8th tergum vestige lost (♂).
(9) Aedeagal apodeme (♂) fused with hypandrium anteriorly (fig. 98).

(10) Aedeagus (♂) very long, with complex lobed distal section which bears a pair of twisted filaments; epiphallus present (fig. 98).

(11) Postgonites (♂) bearing anteriorly directed lobe with comb of bristles (fig. 98).

(12) Two elongate spermathecae (♀) (HENNIG 1958, fig. 27).

HENNIG (1958, 1965b) considered the Megamerinidae as closely related to the Nothybidae and Diopsidae, because of the common absence of certain bristles in these families. But the structure of the male postabdomen does not support the view that there is a close relationship between these groups. As far as I can judge, none of the other groups included in HENNIG's concept of Nothyboidea show any synapomorphous conditions with the Megamerinidae in respect of the structure of the male postabdomen and genitalia, and I therefore think that the loss of frontal bristles in the Megamerinidae was probably independent. The reduced 6th tergum and retained 7th right spiracle of *Megamerina* suggest to me that the genus is not closely related to the Tanypezidae. Since reduction of the 6th tergum is an apomorphous condition of the Sciomyzoinea, it seems possible that the Megamerinidae belong to this prefamily. However this classification is only tentative, since the characterization of the Sciomyzoinea is still too weak to allow confident interpretation of the limits of the group.

The neotropical genus *Syringogaster* has rightly been removed from the Megamerinidae by DO PRADO (1969). The structure of the male abdomen of *Syringogaster* is very different from that of *Megamerina*, resembling closely the condition shown by the Diopsidae. I therefore group the Syringogastridae with the Diopsidae in this work (in the prefamily Diopsioinea).

C. 31. Family Cremifaniidae status novus

Sources of information

The works of DELUCCHI & PSCHORN-WALCHER (1954) and J. F. MCALPINE (1963) are the main sources of information on the Cremifaniidae. Details of the male postabdomen and genitalia are included. I studied a preparation of *Cremifania nigrocellulata* CZERNY.

Description of male postabdomen and genitalia (figs. 99-100)

All postabdominal spiracles in membrane: 7th left spiracle displaced dorsally in relation to preceding spiracles, at margin of 7th sternum: 7th right spiracle displaced centrally.

Postabdomen asymmetrical. 6th tergum short, less than half as long as 5th tergum. 6th sternum asymmetrical, linked to 7th sternum on left side. 7th sternum asymmetrically developed on left side, where it is fused dorsally with the large 8th sternum. 7th tergum represented by small lateroventral sclerite on right side (near 7th right spiracle). 8th tergum absent.

Periandrium bearing discrete bifid telomeres. Interparameral sclerotization scarcely differentiated. Cerci discrete. Aedeagal apodeme rod-like, free from hypandrium. Aedeagus able to be swung through wide arc against aedeagal apodeme to anteriorly directed rest position, its basal phallophore supporting weakly sclerotized epiphallus: distally the aedeagus is upturned and bears several sclerotized lobes and processes. Small postgonites present. Ejaculatory apodeme present.

Characterization and discussion

In the Recent fauna the Cremifaniidae contain only the two known species of *Cremifania*. This genus has hitherto been classified in the Chamaemyiidae, although DELUCCHI & PSCHORN-WALCHER (1954) said that they doubted whether this was correct. J. F. McALPINE (1963) also emphasized the differences between *Cremifania* and true Chamaemyiidae, and proposed to classify *Cremifania* in a new subfamily of Chamaemyiidae (Cremifaniinae). He concluded that 'the genus *Cremifania* CZERNY combines characters of various taxa which HENNIG (1958) placed in two superfamilies, the Sciomyzoidea and the Lauxanioidea. In some ways it can be considered a taxonomic link between the two groups.' McALPINE's conclusion that *Cremifania* could be retained within the Chamaemyiidae despite many divergent characters was closely linked with his view that the Sciomyzoinea and Lauxanioidea are monophyletic. This interpretation allowed him to explain similarities between *Cremifania* and the Sciomyzoinea as representing plesiomorphous groundplan conditions of the Chamaemyiidae in his sense (including *Cremifania*).

On the basis of my present analysis I do not accept the view that the Lauxanioidea and Sciomyzoinea are monophyletic, since the structure of the male postabdomen and aedeagus in these groups is fundamentally different. All families of Sciomyzoinea show, at least in their groundplan, the asymmetrical arrangement of postabdominal sclerites characteristic of the Muscoidea; their aedeagus has well differentiated sclerites in its walls, and can be swung through a wide arc into an anteriorly directed rest position. The conditions of the male postabdomen shown by the Lauxanioidea do not seem derived from the muscoid pattern, and the aedeagus is of a simpler type, always retained in a more or less posteriorly directed position (without swinging mechanism). If my concept of Muscoidea is accepted as a probable monophyletic group, then the Sciomyzoinea cannot be considered more closely related to the Lauxanioidea than to other groups of Muscoidea. Consistent with this interpretation I do not accept the classification of *Cremifania* in the Chamaemyiidae. The postabdominal structure and type of aedeagus shown by this genus is typical of the Muscoidea. I refer *Cremifania* to the Sciomyzoinea, subject to the need for clarification of the limits of this prefamily, as previously discussed.

Cremifania shows many plesiomorphous conditions for the Muscoidea. As far as I can see, the structure of the female abdomen, as described by McALPINE (1963), cannot be characterized as apomorphous in any respect. Three spermathecare are retained, each with a seperate duct. The structure of the male post-abdomen and genitalia seems little removed from the postulated groundplan condition for the Muscoidea. The 7th tergum vestige is retained in its usual position near the 7th right spiracle.

McAlpine (1963, fig. 12) indicates a large sclerite on the right side of the abdomen of *Cremifania nearctica* McAlpine (♂) as representing the 7th tergum. I suspect that the figure is incorrect in this respect, since neither *Cremifania nigrocellulata* nor any other member of the Muscoidea has a large 7th tergum of the type indicated. The 7th tergum in *C. nigrocellulata* Czerny (fig. 99) is represented only by a lateroventral vestige, as normally in Muscoidea. Since *Cremifania nearctica* is known only from the type I was unable to see any material.

It is not clear to what extent the condition of the aedeagus shown by *Cremifania* is apomorphous. The relatively rigid condition of the aedeagus with strong paraphalli is close to the postulated groundplan condition for Muscoidea (and Sciomyzoinea), although the characteristic upturning of the distal part of the aedeagus of *Cremifania* may perhaps constitute some degree of apomorphy (fig. 100).

Since I have found no firm grounds for including *Cremifania* within any previously recognized family, I here accord the group full family rank. That the group is of some antiquity has been confirmed by Hennig's (1956b) description of the fossil *Procremifania* from Baltic amber.

The Cremifaniidae are characterized, like all the families here referred to the Sciomyzoinea, by reduction of the 6th tergum (which is less than half as long as the 5th tergum). Apomorphous conditions of the Cremifaniidae with respect to the groundplan of the Sciomyzoinea are as follows.
(1) Postverticals reduced.

Short divergent postvertical (or 'postocellar') bristles are reported to be present in some individuals of *Cremifania nigrocellulata* Czerny, but these are often absent.

(2) Propleural bristle absent (Hennig 1965b).
(3) Subcostal cell infuscated; vein r_1 bent opposite end of subcosta.
(4) Anal vein much shortened, not reaching wing margin.
(5) 8th tergum vestige (♂) lost.
(6) Epiphallus present (♂) (fig. 100).
(7) Telomeres (♂) bifid, with 'more or less rigid, blade-like anterior lobe, and a moveable, lobulate, posterior lobe' (J. F. McAlpine 1963).
(8) Larvae predators of Adelgidae (Homoptera), with their integument bearing branched spinules and flower-shaped papillae ('blumenförmige Papillen'): toothed 'accessory mouth hooks' present (Delucchi & Pschorn-Walcher 1954).

The known larvae of Chamaemyiidae are also predators of Homoptera (Aphidoidea and Coccoidea), but Delucchi & Pschorn-Walcher have reported that these larvae are morphologically very different from those of *Cremifania*. Consistent with my view of the relationships of the Cremifaniidae, I consider that the similar larval feeding habit was acquired independently by *Cremifania* and the Chamaemyiidae.

B. 11. Prefamily Anthomyzoinea

The families included in this group were mostly treated as families of uncertain relationship by HENNIG (1958), except for the Sphaeroceridae, which he classified in the 'Milichioidea'. In subsequent works HENNIG has tentatively proposed a superfamily Anthomyzoidea, for reception of (according to his latest treatment in the paper now in press) the Clusiidae, Acartophthalmidae, Anthomyzidae, Opomyzidae, Chyromyidae, Aulacigastridae, Asteiidae, Teratomyzidae and Periscelididae. On the basis of my studies of the male postabdomen and genitalia I conclude that the Teratomyzidae and Periscelididae should be excluded from this consideration, as they belong to the Nothyboidea. The remaining families are all referable to the Muscoidea in the sense of this work, and show many conditions in common. I have little doubt that most of these families are rather closely related and can be validly grouped together in a prefamily Anthomyzoinea. But the limits of this group are still very much in doubt. I propose to exclude the Clusiidae and Acartophthalmidae, since they are probably more closely related to families outside the Anthomyzoinea (see below under Agromyzoinea and Tephritoinea). On the other hand I propose to add to the Anthomyzoinea the Heleomyzidae, Rhinotoridae, Borboropsidae and Trixoscelididae (grouped together by some authors as 'Heleomyzidae' in the widest sense), all still treated by HENNIG (in press) as groups of uncertain relationship. These families show all the conditions which can be suggested as groundplan conditions of the Anthomyzoinea if they are excluded, with the exception of shortening of the anal vein; and, in particular, their possession of setose postgonites seems to me to provide evidence that they should be included in the Anthomyzoinea, since I have not found this condition in any other group of Muscoidea. The Sphaeroceridae should also be referred to the Anthomyzoinea, since the structure of their male genitalia indicates that they are closely related to the Chyromyidae and Aulacigastridae and were misplaced in HENNIG's 'Milichioidea' (included in Tephritoinea in this work). The complete list of families here referred to the Anthomyzoinea is as follows: Heleomyzidae, Rhinotoridae, Anthomyzidae, Borboropsidae, Trixoscelididae, Asteiidae, Opomyzidae, Sphaeroceridae, Chyromyidae and Aulacigastridae.

It is possible that the Asteiidae are misplaced in the Anthomyzoinea, since they show the following conditions differing from the postulated groundplan conditions of this prefamily: (1) postverticals parallel or divergent, (2) costa unbroken, and (3) postgonites not setose. On the other hand these conditions are possibly secondary (apomorphous with respect to the groundplan of the Anthomyzoinea). Pending clarification I follow HENNIG's classification of this family in the Anthomyzoinea.

If the Anthomyzoinea in my present sense are monophyletic, they may be characterized by the following conditions in their groundplan which are apomorphous with respect to the groundplan of the Muscoidea.

(1) Postverticals convergent.

The postverticals are convergent in the groundplan of most of the families here included in the Anthomyzoinea, and I therefore ascribe this condition to the groundplan of the Anthomyzoinea. But the postverticals are reduced or absent in the groundplans of the Opomyzidae, Asteiidae and Aulacigastridae. Long divergent postverticals are shown only by the problematical genus *Stenomicra* (see under Anthomyzidae).

(2) Vibrissae present.

Vibrissae are absent only in the Opomyzidae.

(3) Costa broken at end of subcosta.

Among Recent Anthomyzoinea the costal break is absent only in the Asteiidae. HENNIG (1965b, 1967) referred to the Anthomyzidae two species from Baltic amber in which the costa appears to be unbroken (*Protanthomyza* and *Xenanthomyza*), but it is evident from his remarks that the classification of these insects is problematical.

(4) 6th tergum ($\male$) somewhat shortened, at most two-thirds of length of 5th tergum.
(5) Postgonites ($\male$) setose.

This character has not previously been considered. I find that the postgonites bear several setulae in some or all the species studied of most of the families here included in the Anthomyzoinea, with the exception of the Asteiidae and Opomyzidae. In the latter the postgonites are absent. I did not find such setose postgonites in any other group of Muscoidea. I have no doubt that these structures are postgonites, since at least in some species they are turned outwards when the aedeagus is swung into the copulatory position, an action which can be observed, for instance, in respect of the postgonites of Agromyzidae.

HENNIG (in press) discusses the widespread occurrence of hook-shaped antennae ('Hakenfühlern') among the Anthomyzoinea. Clearly such conditions should be considered in future studies of the Anthomyzoinea, and they may assist in defining some of the included groups. But since there are still doubts about whether hook-shaped antennae should be ascribed to the groundplan of some of the families, I do not include such conditions in the characterization presented in this work.

Members of the Anthomyzoinea exhibit a variety of conditions of the aedeagus, and it is not now possible to define the groundplan condition. In the Anthomyzidae and most genera of Heleomyzidae the aedeagus is elongate. However in the Suillinae (Heleomyzidae) the distal part of the aedeagus is broad and flattened. An undoubtedly apomorphous condition is shown by the Rhinotoridae, in which the aedeagus is posteriorly directed, with the swinging mechanism lost. The Borboropsidae, Chyromyidae, Sphaeroceridae and Aulacigastridae are all characterized by a short aedeagus with very complex sclerotization. It is not clear whether the various types of broad aedeagus should be considered apomorphous or plesiomorphous in relation to the more elongate conditions shown by the Anthomyzidae and most Heleomyzidae.

In addition to the groups here recognized as families, HENNIG (in press) also refers to the Anthomyzoinea the South American genera *Schizostomyia*, *Gayomyia* and *Paraleucopis*. None of these is known to me. STEYSKAL kindly sent me a sketch of the male postabdomen of *Schizostomyia*, which indicates that this genus shows the asymmetrical arrangement of the postabdominal sterna characteristic of the Muscoidea and a reduced 6th tergum (represented only by a narrow band). The validity of HENNIG's exclusion of *Schizostomyia* from the Psilidae (Nothyboidea) is thus confirmed. The genus clearly belongs to the Muscoidea, but I am unable to give an independent opinion on whether it belongs to any family of Anthomyzoinea without further information.

C. 32. Family Heleomyzidae

Sources of information

No detailed treatment of the male postabdomen and genitalia is yet available for most of the genera included in Heleomyzidae. Information on some Australian genera has been given by D. K. MCALPINE (1967), and on the South American *Cephodapedon* by HENNIG (1969a). I studied preparations of the following: – *Fenwickia claripennis* MALLOCH; *Allophylopsis scutellata* (HUTTON); *Orbellia tokyoensis* CZERNY; *Heleomyza serrata* (L.); *Pseudoleria pectinata* (LOEW); *Acantholeria armipes* (LOEW); *Eccoptomera simplex* COQUILLETT; *Anorostoma marginatum* LOEW; *Lutomyia hemiptera* (CURRAN); *Allophyla laevis* LOEW; *Suillia plumata* (LOEW).

Description of male postabdomen and genitalia (figs. 101-103)

6th abdominal spiracles in membrane: 7th left spiracle present in all genera studied except *Allophylopsis*, displaced dorsally in relation to preceding spiracles, usually in membrane near edge of 7th sternum but within 7th sternum in *Allophyla:* 7th right spiracle absent in *Allophylopsis*, *Suillia* and *Allophyla*, when present displaced centrally in relation to preceding spiracles, usually in membrane but within sclerotized ring formed by sclerites of 7th segment in *Fenwickia*.

Postabdomen usually strongly asymmetrical, but scarcely so in *Fenwickia*. 6th tergum variably developed, usually represented by discrete narrow band (at most about half as long as 5th tergum in *Orbellia*); but absent in *Eccoptomera*, *Acantholeria* and *Allophyla*, and fused with 8th sternum in *Suillia*. 6th sternum asymmetrical except in *Fenwickia*, linked to 7th sternum on left side. 7th sternum usually asymmetrically developed on left side, where it is fused dorsally with the large 8th sternum: but in *Fenwickia* the 6th and 7th sterna form an almost symmetrical sclerotized ring around the venter of the postabdomen, and MCALPINE (1967) indicates that the 7th sternum forms a complete ventral ring in *Leriopsis* and *Austroleria*. 7th tergum usually absent, but represented by small lateroventral sclerite on right side in *Orbellia* and *Anorostoma*. 8th tergum absent except in *Orbellia*, in which it is represented by a small lateroventral sclerite on the right side.

192

Periandrium usually bearing discrete telomeres (except in *Pseudoleria*), which are double or furcate in some species: interparameral sclerotization variable, usually consisting of pair of discrete sclerites (processus longi) or single U-shaped sclerite extending from inner basal corners of telomeres. Cerci discrete or fused: in *Acantholeria, Heleomyza, Anorostoma* and *Fenwickia* a ventral link (point of articulation) between the cerci and telomeres is present, but this is poorly developed or absent in the other genera studied. Aedeagal apodeme usually rodlike (but rather short in *Allophylopsis*), linked to hypandrium by pair of ventral processes (e.g. *Orbellia, Anorostoma*) or single anteroventral process (e.g. *Lutomyia, Eccoptomera*, Suillinae). Aedeagus able to be swung through wide arc against aedeagal apodeme to anteriorly directed rest position, elongate, usually without epiphallus (except *Fenwickia*), with complex distal sclerotization only in *Cephodapedon* and genera from Australia and New Zealand; in most holarctic genera the distal part of the aedeagus is slender and somewhat flexible, but in Suillinae the aedeagus is very broad and flattened distally with striate pigmentation. Postgonites in most genera with distal setulae, which are sometimes large and conspicuous (setulae also present in *Heleomyza*, in which the postgonites themselves are scarcely differentiated): only in *Allophylopsis* was no trace of the postgonites found. Ejaculatory apodeme present (but small in Suillinae).

Characterization and discussion

The limits of the Heleomyzidae cannot be satisfactorily defined in the present state of knowledge. I attempt in this work to improve the classification by removing from the Heleomyzidae some groups which are probably heterogenous, such as the Heteromyzidae, Trixoscelididae and Notomyzidae. But this process is doubtless incomplete. Most of the South American genera referred by MALLOCH (1933) to the Heleomyzidae do not belong here. In this paper *Heloclusia* is referred to the Cypselosomatidae, *Notomyza* to a new family (Notomyzidae), *Prosopantrum* to the Cnemospathidae (Tephritoinea) and both *Anastomyza* and *Apophoneura* to the Rhinotoridae. The only one of MALLOCH's genera which I provisionally retain in the Heleomyzidae is *Cephodapedon*, on the evidence of the figures of the male postabdomen and genitalia given by HENNIG (1969a). The position of MALLOCH's other South American genera which are not known to me must be regarded as doubtful until they are more critically studied.

A crucial question affecting the delimitation of the Heleomyzidae is the relationship between the southern hemisphere and holarctic genera currently referred to this family. In most holarctic Heleomyzidae the aedeagus is slender to its apex, and shows little differentation of apical sclerites. The New Zealand genera studied by me *(Fenwickia* and *Allophylopsis)*, the Australian *Diplogeomyza* and its relatives (D. K. MCALPINE 1967), and the South American *Cephodapedon* (HENNIG 1969a) all differ from holarctic Heleomyzidae in that they show complex sclerotization of the apical area of the aedeagus. Their wing venation is apomorphous in respect of the course of the subcosta (which is approximated to or fused apically with vein r₁) and shortening of the anal vein. The possibility is thus indicated that there is a southern hemisphere group which is the sister-group of holarctic Heleomyzidae. However I cannot offer any conclusion at this time, as there are many relevant genera which I have not seen; and, even if the Heleomyzidae are divisible into two monophyletic groups, the possibility that these are not most closely related to one another must be considered, since it has not been demonstrated that the Heleomyzidae as presently understood are a monophyletic group. I agree with HARRISON (1959) and D. K. McALPINE (1967) that the Australian and New Zealand genera are best retained in the Heleomyzidae for the present. I do not see any grounds for referring them to the Trixoscelididae. The holarctic Suillinae *(Suillia* and *Allophyla)* show a unique condition of the aedeagus, which is very broad and flattened distally with characteristic striate pigmentation. I suspect that this condition was derived from an elongate type of aedeagus, since I note the development of similar striation also in *Anorostoma*, in which the aedeagus is elongate.

On the basis of existing analyses I can only suggest one common condition of the Heleomyzidae which is apomorphous with respect to the groundplan of the Anthomyzoinea, as follows.

(1) Aedeagal apodeme (♂) linked to hypandrium by pair of sclerotized ventral processes.

If the Heleomyzidae are a monophyletic group, I postulate that in their groundplan paired ventral processes of the aedeagal apodeme were present which linked the apodeme posteriorly (beyond its middle) to the hypandrial arms. This is the condition shown by *Orbellia* and *Anorostoma*, whose postabdominal morphology appears relatively plesiomorphous due to retention of the 7th tergum vestige *(Anorostoma)*, or of both the 7th and 8th tergum vestiges *(Orbellia)*. In some other Heleomyzidae (such as *Lutomyia* and *Eccoptomera*) the ventral process is single, linking the apodeme anteriorly with the apex of the hypandrium. This is probably a relatively apomorphous condition. Various intermediate conditions also occur, so that it seems possible that the variation in the position of the sclerotized link between the hypandrium and aedeagal apodeme is due to further modification of a single apomorphous groundplan condition. Some genera with a free aedeagal apodeme have been classified in the Heleomyzidae, but I think that these belong elsewhere (see under Heteromyzidae, Rhinotoridae and Trixoscelididae). The aedeagal apodeme is free, without any sclerotized link with the hypandrium, in most other families of Anthomyzoinea except Anthomyzidae and Opomyzidae.

Most Heleomyzidae retain three spermathecae (STURTEVANT 1926, HENNIG 1958), but only two are present in *Cephodapedon* (HENNIG 1969a).

In most Heleomyzidae the 7th and 8th tergum vestiges (♂) have been lost. But both are well developed in *Orbellia* (fig. 101), which thus shows the most plesiomorphous condition of the male postabdomen among those Heleomyzidae studied by me. I have noted the presence of the 7th tergum vestige also in *Anorostoma*.

The Heleomyzidae are often characterized in keys to identification by their possession of conspicuous costal bristles. The value of this character for the purposes of phylogenetic systematics is problematical, as a weak differentiation of costal bristles occurs very widely among the Schizophora. Long and conspicuous costal bristles occur in many disjunct groups, such as *Neottiophilum*

(Picphilidae *s.l.*), *Helcomyza* (Dryomyzidae), Helosciomyzidae, *Curtonotum* (Curtonotidae), *Heloclusia* (Cypselosomatidae) and many Calyptratae. Evidently we are dealing with a character which may vary readily between closely related groups. Within the Heleomyzidae the development of these bristles is variable, and there are some species in which they are only weakly differentiated.

Figures of the male postabdomen of two problematical genera, *Cinderella* and *Dichromyia*, have recently been given by HENNIG (1969a and in press). Neither genus is known to me. HENNIG's figures clearly indicate that these genera belong to the Muscoidea, but I am not able to give an independent opinion on whether they are likely to belong to the Heleomyzidae (where HENNIG provisionally classifies them) without further information. I definitely exclude from the Heleomyzidae *Apetaenus* and *Listriomastax* (Tethinidae) and *Chiropteromyza* (Chiropteromyzidae).

C. 33. *Family Rhinotoridae*

Sources of information

Information on the male postabdomen and genitalia of the Rhinotoridae can be found in the works of MALLOCH (1933) (figures of *Anastomyza* prepared by F. W. EDWARDS), STEYSKAL (1957a) (description and figures of *Rhinotora*) and D. K. MCALPINE (1968) (detailed treatment of *Cairnsimyia*, and comparisons with the South American genera). I studied preparations of the following: – *Anastomyza neglecta* EDWARDS; *Apophoneura inconspicua* MALLOCH; *Neorhinotora mutica* (SCHINER); *Tapeigaster marginifrons* BEZZI.

Description of male postabdomen and genitalia (figs. 105-106)

In *Tapeigaster* and *Anastomyza* postabdominal spiracles on left side both in membrane, with 7th near edge of 7th sternum; but both left spiracles within sclerotized area in *Neorhinotora* and *Apophoneura*. 6th and 7th right spiracles in membrane, except in *Apophoneura*.

Postabdomen asymmetrical (but only weakly so in *Neorhinotora*). In *Neorhinotora* and *Apophoneura* the postabdominal sclerites (6th tergum and 6th to 8th sterna) are extensively fused, with the 6th sternum forming a complete ventral band. In *Tapeigaster*, *Cairnsimyia* and *Anastomyza* the 6th tergum is narrow and usually discrete; the 6th sternum is asymmetrically developed on the left side, where it is linked to the 7th sternum *(Tapeigaster)* or fused with the 7th sternum *(Cairnsimyia, Anastomyza)*; and the 7th sternum is asymmetrically developed on the left side, where it is fused dorsally with the large 8th sternum. 7th and 8th terga absent.

Periandrium bearing large telomeres which are usually discrete (except in *Anastomyza*): in *Neorhinotora* rod-like interparameral sclerites (processus longi) extend from the inner basal corners of the telomeres, but conspicuous interparameral sclerites are not differentiated in the other genera examined. Cerci fused, very small, linked to or fused with adjacent margins of periandrium in *Neorhinotora* and *Anastomyza*, scarcely differentiated in *Apophoneura*. Aedeagal apodeme rod-like and free from hypandrium in *Tapeigaster* and *Neorhinotora* (rather short in the latter), but connected with hypandrium by posterior ventral process in *Anastomyza* and *Apophoneura*. Hypandrium extending well posteriorly from its articulation with periandrium; extending dorsally to form anterior bridge above aedeagal apodeme in *Tapeigaster*, but not in other genera. Aedeagus directed more or less posteriorly when at rest, not able to be swung through wide arc against aedeagal apodeme, with rather complex sclerotization in *Tapeigaster* and *Cairnsimyia;* various highly modified types of aedeagus are shown by the South American genera. Postgonites usually well developed (setose in some species), but not differentiated in *Apophoneura*. Slender ejaculatory apodeme present.

Characterization and discussion

The customary limits of the Rhinotoridae (including the neotropical genera *Rhinotora*, *Rhinotoroides* and *Neorhinotora*) have been widened by D. K. MCALPINE (1958, 1968) to include *Cairnsimyia*, distributed in Australia and New Guinea. I additionally refer to this family the South American genera *Anastomyza* and *Apophoneura*, which were classified in the Heleomyzidae by MALLOCH (1933), and the Australian *Tapeigaster*. My studies of the male genitalia confirm that HENNIG (1958) was right in rejecting the customary classification of the last genus in the 'Neottiophilidae' (Piophilidae *s.l.*). The family Rhinotoridae in the present widened sense is well characterized as a monophyletic group by the following groundplan conditions which are apomorphous with respect to the groundplan of the Anthomyzoinea.

(1) Peristomal opening large; haustellum (= Unterlippenbulbus) somewhat elongate, bearing only small labella.

Relevant figures have been given by MALLOCH (1933) for *Anastomyza* and *Apophoneura*, and by D. K. MCALPINE (1958, 1968) for *Cairnsimyia*. The specimens of *Neorhinotora* and *Tapeigaster* studied by me also show this characteristic condition.

(2) Front femora thickened, with spinous bristles on apical half of their ventral surface.

All the genera here referred to the Rhinotoridae are characterized by strong legs, with at least the front femora somewhat thickened. In some species of *Tapeigaster* the hind femora are also thickened. In *Cairnsimyia* the development of ventral spines on the femora is variable between the species. MCALPINE's (1968) key indicates that these are present on all femora in some species, but absent entirely in others. Since these spines are present in all other genera, I regard their absence in some species of *Cairnsimyia* as probably secondary.

(3) Anal vein fading distally, reaching wing margin at most as fold.
(4) Cerci of males small, fused below anus.

196

McALPINE (1968) states that in *Cairnsimyia* the small cerci may be fused or separate. Probably the latter condition is secondary, as the cerci are fused in all other genera.

(5) 7th and 8th tergum vestiges lost ($\circlearrowleft$).

(6) Aedeagus ($\circlearrowleft$) directed more or less posteriorly when at rest, not swung through wide arc against aedeagal apodeme (fig. 105).

Secondary loss of the mechanism for swinging the aedeagus against its apodeme is a distinctive autapomorphous condition of the Rhinotoridae. The aedeagus thus remains more or less posteriorly directed when at rest. In *Tapeigaster* and *Cairnsimyia* the aedeagus retains a fairly complex sclerotization, but in the neotropical genera this tends to become simplified. For instance in *Rhinotora* and *Neorhinotora* the aedeagus is broad and upcurved ('scimitar-shaped', STEYSKAL 1957a), showing a superficial resemblance to the aedeagus of some Lauxanioidea. But the muscoid affinities of the Rhinotoridae are clearly apparent from the arrangement of the postabdominal sclerites in the Australasian genera and some of the neotropical genera (especially *Anastomyza*). In these a typical muscoid pattern can be seen, with asymmetrical development of the 6th and 7th sterna (the latter fused dorsally with the inverted 8th sternum) and a discrete 6th tergum. In the *Rhinotora*-group (*Rhinotora, Neorhinotora* and *Rhinotoroides*) further modifications of the sclerotization have obscured the previous muscoid pattern. In this group the 6th tergum is more or less fused with the 8th sternum, and a complete ventral band (rightly interpreted by McALPINE as secondary) is developed from the 6th sternum; even the asymmetry of the 7th sternum is reduced, for this extends back across the centre-line towards the right side.

(7) Hypandrium ($\circlearrowleft$) extending well posteriorly from its articulation with periandrium (fig. 105).

This character is probably correlated with the loss of the mechanism for swinging the aedeagus.

(8) Two spermathecae ($\female$).

HENNIG (1958) has reported that both *Tapeigaster* and *Rhinotora* have only two spermathecae.

Two unusual charcters which occur frequently among the Rhinotoridae, although not in all species, are: (1) the development of supernumerary cross-veins between the costa and r_{2+3} (recorded for some species of each of the neotropical genera), and (2) a tendency for the 5th sternum of the male to become split into two halves (fig. 106), as observed in all the preparations studied by me. The frequent occurrence of these two conditions in Rhinotoridae are perhaps exemples of homoiology.

The groundplan condition of the wing venation in the Rhinotoridae is probably indicated by the condition of *Tapeigaster*, in which the subcosta is complete and well separated from r_1 apically.

The affinities of *Anastomyza, Apophoneura* and *Cairnsimyia* with the *Rhinotora*-group (Rhinotoridae in the old sense) were well discussed by D. K. McALPINE (1958, 1968), and most of the synapomorphous conditions listed above were already stated in his 1968 paper. The only part of McALPINE's conclusions which I regard as doubtful is his view that *Cairnsimyia* is more closely related to the *Rhinotora*-group than are the remaining neotropical genera (*Anastomyza* and *Apophoneura*). It seems to me that *Anastomyza* and *Apophoneura* are synapomorphous with the *Rhinotora*-group in respect of: (1) the presence of a strong postgenal bristle situated well above the peristomal margin; and (2) cerci of male fused laterally with the edges of the periandrium. Neither of the Australasian genera show these conditions. MCALPINE's view that *Cairnsimyia* is more closely related to the *Rhinotora*-group was based primarily on modifications of the head (especially the excavated vertex and very broad cheeks). I think it more probable that a slight tendency to depression of the vertex was already present in the groundplan of the family, and that this has become more pronounced independently in the Australasian *Cairnsimyia* and the neotropical *Rhinotora*-group.

In my opinion the Rhinotoridae are better treated as a distinct family, not included in the Heleomyzidae as D. K. MCALPINE (1958) has proposed, since I can see no firm grounds for postulating that the Rhinotoridae and Heleomyzdae are monophyletic. I do not think that the presence of a sclerotized link between the aedeagal apodeme and hypandrium, as in the postulated groundplan of the Heleomyzidae, can be ascribed to the groundplan of the Rhinotoridae. In *Tapeigaster*, whose postabdominal morphology seems to be the most plesiomorphous among the species studied by me, the aedeagal apodeme is rod-like and bears no ventral processes; it meets the sclerotization of the hypandrium only at the base of the aedeagus.

C. 34. Family Anthomyzidae

Sources of information

Information on the male postabdomen and genitalia of Anthomyzidae was given by HENNIG (1939a). I studied preparations of *Mumetopia occipitalis* MELANDER and an *Anthomyza* species.

Description of male postabdomen and genitalia (fig. 104)

6th abdominal spiracles in membrane: 7th left spiracle within 7th sternum, displaced dorsally in relation to preceding spiracles: 7th right spiracle in membrane, displaced centrally in relation to preceding spiracles.

Postabdomen asymmetrical. 6th tergum variably developed, over half as long as 5th tergum in some *Anthomyza* species, but represented only by narrow band in *Mumetopia*. 6th sternum asymmetrical, fused with 7th sternum on left side. 7th sternum asymmetrically developed, fused with large 8th sternum on left side. 7th and 8th terga absent.

Periandrium bearing discrete telomeres whose inner margins are confluent with a broad interparameral sclerite. Cerci discrete. Aedeagal apodeme rod-like, linked to hypandrium anteriorly by ventral process. Hypandrium quadrate anteriorly, with its arms fused posteriorly to form bridge above base of aedeagus. Aedeagus able to be swung through wide arc against aedeagal apodeme to anteriorly directed rest position, with very complex structure which includes an expanded medial area ('x' on fig. 104) from which arises a slender distal section (furcate in *Mumetopia*): in *Anthomyza* an expanded epiphallus is present, but this is absent in *Mumetopia*. Large pectinate postgonites present, discrete in *Mumetopia* but fused basally with hypandrium in *Anthomyza*. Ejaculatory apodeme minute.

Characterization and discussion

My present treatment is based primarily on *Anthomyza* and *Mumetopia*, since the limits of this family are still in doubt. The structure of these genera indicates that the Anthomyzidae (at least in a narrow sense) are characterized by the following conditions which are apomorphous with respect to the groundplan of the Anthomyzoinea.

(1) Face (prefrons) with broad central membranous area (HENNIG, in press).
(2) Subcosta reduced, fading before reaching wing margin (HENNIG 1958).
(3) Anal vein not reaching wing margin (HENNIG 1958).
(4) Only one postalar bristle *(pa)* present (HENNIG, in press).
(5) 7th abdominal spiracles (♀) lying within 7th tergum (HENNIG 1958).
(6) Two spermathecae (♀) (STURTEVANT 1926, HENNIG 1958).
(7) Aedeagal apodeme (♂) linked to hypandrium anteriorly by single ventral process (fig. 104).
(8) 7th and 8th tergum vestiges lost (♂).
(9) 7th left spiracle (♂) lying within 7th sternum.
(10) Aedeagus (♂) with expanded medial area ('x' on fig. 104), from which arises a slender distal section.
(11) Ejaculatory apodeme (♂) minute.

HENNIG (1939a) suggested that the expanded medial area of the aedeagus of *Anthomyza* and *Mumetopia* might be homologous with the ejaculatory bulb ('Samenspritze'). But this is not correct, as a reduced ejaculatory apodeme is present in the normal position (free from the body wall) in the preparations studied by me.

I also studied a preparation of a species of *Stenomicra*, a genus which HENNIG (1958) referred to the Anthomyzidae but now thinks closely related to the Aulacigastridae (HENNIG, in press). I doubt whether the latter interpretation is correct. In *Stenomicra* the aedeagus is long and slender, very different from the broad type of aedeagus shown by the Aulacigastridae, and the aedeagal apodeme is linked to the hypandrium anteriorly by a ventral process (as in Anthomyzidae but not in Aulacigastridae). The structure of the male genitalia of *Stenomicra* is thus more compatible with the view that the genus belongs to the Anthomyzidae. But I do not think the evidence conclusive, in particular because in *Stenomicra* the expanded medial area of the aedeagus characteristic of *Anthomyza* and *Mumetopia* is absent, the postverticals are divergent and the face does not have a membranous central area (HENNIG, in press). The only definite conclusion about *Stenomicra* which I venture to draw from present information is that the genus certainly belongs to the Muscoidea, not to the Drosophiloidea as has sometimes been suggested. I defer judgement on whether it belongs to the Anthomyzidae until more detailed information is available. Since the distribution of *Stenomicra* is extensive (including Australia and South America) it is possible that the genus as presently understood is a group which should be accorded higher rank in the classification system.

The South American genus *Melanthomyza* seems unlikely to belong to the Anthomyzidae according to the information presented by HENNIG (in press). The same applies to the Australian *Waterhouseia*, which according to COLLESS & D. K. McALPINE (1970) has a symmetrical 6th sternum in the male. Neither genus is known to me.

C. 35. Borboropsidae familia nova (type-genus Borboropsis CZERNY 1902)

Sources of information

I studied preparations of the sole known species, *Borboropsis puberula* (ZETTERSTEDT), whose male genitalia have also been figured by HACKMAN & ANDERSSON (1969b).

Description of male postabdomen and genitalia (figs. 110-111)

6th abdominal spiracles in membrane: 7th left spiracle within margin of 7th sternum, displaced dorsally in relation to preceding spiracles: 7th right spiracle in membrane, displaced centrally in relation to preceding spiracles.

Postabdomen asymmetrical. 6th tergum represented by narrow band of dorsal sclerotization. 6th sternum asymmetrical, linked to 7th sternum on left side. 7th sternum asymmetrically developed on left side, where it is fused dorsally with the large 8th sternum. 7th tergum represented by narrow lateroventral sclerite on right side. 8th tergum represented by trace of ventral sclerotization on right side.

Periandrium bearing discrete telomeres, whose inner basal corners articulate with an ill-defined U-shaped area of interparameral sclerotization. Cerci discrete, very small. Aedeagal apodeme rod-like, free from hypandrium. Aedeagus able to be swung through wide arc against aedeagal apodeme to anteriorly directed rest position, with very complex sclerotization which includes a slender epiphallus arising from the basal phallophore. Large setose postgonites present. Slender ejaculatory apodeme present.

Characterization and discussion

This family is proposed for the single holarctic species *Borboropsis puberula*

200

(ZETTERSTEDT) (= *fulviceps* STROBL), which has been hitherto classified in the Heleomyzidae. However *Borboropsis* does not show the conditions of the male genitalia characteristic of the Heleomyzidae. In *Borboropsis* the aedeagal apodeme is rod-like, without any sclerotized link with the hypandrium; and the aedeagus is short, with very complex sclerotization (fig. 110). This type of aedeagus is more similar to that shown, for instance, by the Chyromyidae, Aulacigastridae and Sphaeroceridae. However it is not clear to what extent such resemblances in the form of the aedeagus are due to synapomorphy, and I therefore do not offer any conclusion on which other group of Anthomyzoinea may be most closely related to the Borboropsidae. Many plesiomorphous conditions are retained by *Borboropsis*. The subcosta and anal vein both reach the wing margin as distinct veins. The 7th tergum vestige (♂) is retained (fig. 111), and there seems also to be a trace of the 8th tergum vestige. Three spermathecae are present (♀) in a preparation studied by me. Apomorphous conditions of *Borboropsis* with respect to the groundplan of the Anthomyzoinea are as follows.

(1) 6th tergum (♂) much shortened, less than one quarter of length of 5th tergum.
(2) 7th left spiracle (♂) lying within margin of 7th sternum.
(3) 8th tergum vestige (♂) reduced.
(4) Epiphallus (♂) well developed (fig. 110).

The epiphallus in *Borboropsis* is a slender posterior process of the phallophore, similar to that of the Calyptratae.

The combination of apomorphous and plesiomorphous conditions shown by *Borboropsis* provides no basis for including the genus in any other family of Anthomyzoinea, at least as currently defined. I therefore propose a new family.

C. 36. Family Trixoscelididae

Sources of information

HACKMAN (1970) has given information on the male postabdomen and genitalia of *Trixoscelis*. I studied a preparation of a *Trixoscelis* species (British Columbia).

Description of male postabdomen and genitalia (fig. 107)

6th abdominal spiracles in membrane: 7th spiracles absent.
Postabdomen asymmetrical. 6th tergum represented by narrow band of dorsal sclerotization. 6th sternum asymmetrical, fused with 7th sternum on left side. 7th sternum asymmetrically developed on left side, where it is fused dorsally with the large 8th sternum. 7th and 8th terga absent.

Periandrium bearing discrete telomeres, whose inner basal corners are linked to a conspicuous transverse interparameral sclerite. Cerci minute, fused. Aedeagal apodeme rod-like, free from hypandrium. Aedeagus able to be swung through wide arc against aedeagal apodeme to anteriorly directed rest position, with variable complex sclerotization; in most species the distal part of the aedeagus is expanded (see HACKMAN 1970 for figures illustrating the range of variation). Postgonites well developed, conspicuously setose. Slender ejaculatory apodeme present.

Characterization and discussion

A group based on *Trixoscelis* has usually been recognized either as a subfamily of the Heleomyzidae or as a distinct family. But the limits of this group have never been satisfactorily defined for the purposes of phylogenetic systematics (HENNIG 1958). The available taxonomic treatments are based only on a limited range of external characters, and need to be supplemented by studies of the postabdomen and genitalia of both sexes. Since only *Trixoscelis* is known to me, my present treatment is based solely on this genus and I can only leave open the question of what additional genera can be associated with this in a monophyletic group. *Trixoscelis* shows the following conditions which are apomorphous with respect to the groundplan of the Anthomyzoinea.
(1) Subcosta becoming fused distally with vein r_1 (CZERNY 1927).
(2) Anal vein shortened, not reaching wing margin (CZERNY 1927).
(3) 7th abdominal spiracles lost in both sexes.

This condition is also shown by the Opomyzidae and Asteiidae. The possibility of these families being closely related to the Trixoscelididae thus merits further investigation.

(4) 7th abdominal tergum and sternum (♀) fused, forming complete ring (HENNIG 1958, fig. 243).
(5) 6th tergum (♂) much shortened, less than one quarter of length of 5th tergum.
(6) 7th and 8th tergum vestiges lost (♂).
(7) Cerci (♂) minute, fused.
The male genitalia of *Trixoscelis* appear more plesiomorphous in some respects than those of the Heleomyzidae. The aedeagal apodeme is rod-like, without any sclerotized link with the hypandrium (fig. 107). The aedeagus bears a cylindrical phallophore, a pair of long paraphalli and in most species an expanded distal area with complex sclerotization ('glans').

Trixoscelis retains three spermathecae, as in the groundplan of the Anthomyzoinea (STURTEVANT 1926). In a preparation of *T. fumipennis* MELANDER (♀) studied by me these are pear-shaped.

202

C. 37. Family Asteiidae

Sources of information

I studied preparations of *Leiomyza laevigata* (MEIGEN) and *Asteia amoena* MEIGEN. No detailed treatment of the male postabdomen and genitalia of Asteiidae was found in the literature.

Description of male postabdomen and genitalia (fig. 108)

6th abdominal spiracles in membrane or within 6th tergum (*Leiomyza*): 7th spiracles absent.

Postabdomen asymmetrical. 6th tergum well developed, over half as long as 5th tergum. 6th sternum asymmetrical, fused with 7th sternum or synsternum (7 + 8) on left side. 7th sternum asymmetrically developed on left side, where it is fused dorsally with the large 8th sternum (*Leiomyza*): in *Asteia* 7th and 8th sterna fully fused, without any visible suture between them. 7th and 8th terga absent.

Telomeres fused with periandrium, hardly articulated; in *Leiomyza* the telomeres are double, the inner pair joined by a sclerotized bridge. Cerci discrete. Aedeagal apodeme rod-like, free from hypandrium. Aedeagus able to be swung through wide arc against aedeagal apodeme to anteriorly directed rest position, with complex sclerotization. Small postgonites present (not setose). Large ejaculatory apodeme present.

Characterization and discussion

The Asteiidae are characterized as a monophyletic group by the following groundplan conditions which are apomorphous with respect to the groundplan of the Anthomyzoinea.

(1)   Postverticals reduced, parallel or divergent.

If the Asteiidae are correctly included in the Anthomyzoinea, the parallel or divergent direction of the small postverticals retained by some species should probably be regarded as secondary. Probably not much significance can be attached to the direction of bristles which have been much reduced. In many Asteiidae the postverticals have been completely lost.

(2)   Subcosta fading distally, reaching wing margin at most as fold, not as true vein (HENNIG 1958).

(3)   Costa unbroken, only attenuated near end of r_1.

Since the costal break is present in all other families of Anthomyzoinea, the unbroken costa of the Asteiidae should probably be considered secondary, if the Asteiidae are correctly included in the Anthomyzoinea.

(4)   Anal vein and anal cell absent, or at most indicated by folds (in *Phlebosotera*); base of m_4 (= 'tb', lower cross-vein) lost (HENNIG 1958).

(5)   7th abdominal spiracles lost in both sexes.

The same condition is also shown by the Trixoscelididae and Opomyzidae. Whether these resemblances are due to synapomorphy or homoplasy is not clear.

(6)   Two spermathecae ($\female$) (HENNIG 1958).
(7)   7th and 8th tergum vestiges lost ($\male$).
(8)   Telomeres ($\male$) more or less fused with periandrium, at most delimited by partial suture (see the figures given by SABROSKY 1957).
(9)   Postgonites ($\male$) small, without setulae.

See the comment on the postgonites in my characterization of the Anthomyzoinea.

An additional apomorphous condition listed by HENNIG (1958) ('only 1 *ors* present') should be deleted in the light of his subsequent discussion (HENNIG 1969b). The latter work contains an important discussion of the classification of the Asteiidae.

The Asteiidae retain some noteworthy plesiomorphous conditions. The aedeagal apodeme ($\male$) is rod-like and free from the hypandrium. The 6th tergum ($\male$) is relatively large, over half as long as the 5th tergum (as in the groundplan of the Anthomyzoinea).

The aedeagus of *Asteia* (fig. 108) includes a basal area supported by the paraphalli and a heavily sclerotized distal area. In *Leiomyza* the basal area is shorter, probably a relatively apomorphous condition. It is not possible to characterize the type of aedeagus shown by *Asteia* as apomorphous with respect to the groundplan of the Anthomyzoinea on the basis of my present analysis. The morphological gap between the aedeagus of *Asteia* and that, for instance, of *Trixoscelis* does not seem great.

The structure of the male postabdomen and genitalia of the Asteiidae seems to me to indicate that HENNIG (1965b) was justified in rejecting previous suggestions of a close relationship of this family to the Drosophilidae or Chloropidae. It is possible that the Asteiidae, Trixoscelididae and Opomyzidae form a monophyletic group, characterized by loss of the 7th abdominal spiracles in both sexes. However more detailed studies of these families are needed before any firm judgement on this question can be made.

C. 38. *Family Opomyzidae*

Sources of information

HENNIG (1939a) and VOCKEROTH (1961) gave information on the male postabdomen and genitalia of Opomyzidae. I studied preparations of *Opomyza germinationis* (L.) and *Geomyza tripunctata* FALLÉN.

Description of male postabdomen and genitalia (fig. 109)

6th abdominal spiracles in membrane: 7th spiracles absent.

Postabdomen asymmetrical. 6th tergum represented by narrow band in *Anomalochaeta*, absent in *Opomyza* and *Geomyza*. 6th sternum asymmetrical, linked to 7th sternum on left side. 7th sternum asymmetrically developed on left side, where it is fused dorsally with the large 8th sternum. 7th and 8th terga absent.

Periandrium without articulated telomeres. Hypandrial arms fused to form bridge above base of aedeagus. Cerci large, discrete, elongate, with basal apodemes which are linked to the hypandrial bridge. Aedeagal apodeme rod-like, linked to hypandrium by paired ventral processes from about its middle. Aedeagus able to be swung through wide arc against aedeagal apodeme to anteriorly directed rest position, with very complex sclerotization and large epiphallus arising from basal phallophore. Postgonites absent. Ejaculatory apodeme minute.

Characterization and discussion

The Opomyzidae are well characterized as a monophyletic group by the following groundplan conditions which are apomorphous with respect to the groundplan of the Anthomyzoinea.

(1) Postverticals reduced.

Weak divergent postverticals are present in *Anomalochaeta*. In *Opomyza* and *Geomyza* the postverticals have been completely lost.

(2) Vibrissae reduced.

Since vibrissae are present in all other families of Anthomyzoinea, their absence in Opomyzidae is probably due to secondary loss (see HENNIG 1958: 637).

(3) Only one fronto-orbital bristle (*ors*) present.
(4) Face (prefrons) with membranous central area (HENNIG, in press).
(5) Subcosta not reaching wing margin as distinct vein, becoming fused with vein r_1 distally except for weak spur opposite costal break (HENNIG 1958, VOCKEROTH 1961).
(6) Anal vein not reaching wing margin (HENNIG 1958).
(7) 7th abdominal spiracles lost in both sexes.
(8) 6th tergum (♂) reduced to narrow band.

In the species studied by me the 6th tergum is completely absent. But the retention of a narrow discrete 6th tergum should clearly be ascribed to the groundplan of the Opomyzidae, as VOCKEROTH (1961) reports such as sclerite in *Anomalochaeta*.

(9) 7th and 8th tergum vestiges lost (♂).
(10) Telomeres (♂) not differentiated from periandrium.
(11) Aedeagal apodeme (♂) linked to hypandrium by paired ventral processes.
(12) Hypandrial arms (♂) fused to form bridge above base of aedeagus.
(13) Cerci (♂) large, elongate, with basal apodemes which are linked to hypandrial bridge.

(14) Aedeagus (♂) with very complex sclerotization (fig. 109), with large epiphallus arising from basal phallophore.

(15) Postgonites (♂) absent.

(16) Ejaculatory apodeme (♂) minute.

(17) Two spermathecae (♀) (STURTEVANT 1926, HENNIG 1958).

(18) 7th tergum and sternum (♀) enlarged, together forming a kind of laterally compressed ovipositor sheath (but not fused); terminal segments of postabdomen somewhat blade-like, with fused cerci (HENNIG 1958 and in press).

C. 39. Family Sphaeroceridae (= Borboridae)

Sources of information

HACKMAN's (1969a) review of the classification of the Sphaeroceridae gives some details of variation in the male genitalia. Further information can be found in the many papers on this group by O. W. RICHARDS, of which I here cite two (RICHARDS 1961 and 1968) as of particular interest to me because they give details on the genera *Frutillaria* and *Scutelliseta*. KIM & COOK (1966) have attempted a detailed morphological treatment of the male postabdomen and genitalia, but this is unfortunately misleading in some respects (see below). I studied preparations of the following: – *Sphaerocera curvipes* LATREILLE; *Copromyza equina* FALLÉN; *Leptocera* sp.

Description of male postabdomen and genitalia (figs. 112-114)

6th and 7th left abdominal spiracles both within the 6th sternum, the latter displaced dorsally in relation to preceding spiracles: 6th right spiracle, when present, in membrane (*Sphaerocera, Copromyza*), absent in *Leptocera:* 7th right spiracle always absent.

Postabdomen asymmetrical. 6th tergum absent. 6th sternum asymmetrical, expanded on left side where it is linked to or fused with 7th sternum: in *Leptocera* the ventral area of the 6th sternum is also expanded and there is a transverse cleft in its sclerotization. 7th sternum asymmetrically developed on left side where it is fused dorsally with the well developed 8th sternum. 7th and 8th terga absent.

Periandrium narrow, bearing large discrete telomeres: in *Leptocera* the periandrium is fused with the 8th sternum and hypandrium, and the telomeres are double. Inner basal corners of telomeres linked by V-shaped or H-shaped interparameral sclerite. Cerci variably developed, sometimes discrete (e.g. *Scutelliseta*), fused in *Frutillaria*, weakly differentiated in *Sphaerocera*, absent in *Leptocera*. Aedeagal apodeme rod-like, free from hypandrium. Vertical sections of hypandrial arms expanded (*Sphaerocera, Copromyza*), or fused with periandrium and 8th sternum (*Leptocera*). Aedeagus able to be swung through wide

206

arc against aedeagal apodeme to anteriorly directed rest position, its sclerotization consisting of short cylindrical phallophore (bearing small epiphallus in *Archiborborus* and *Frutillaria*) and expanded complex distal area. Postgonites well developed, setose in some groups (e.g. *Sphaerocera*). Ejaculatory apodeme minute (*Sphaerocera*) or absent.

Characterization and discussion

In HENNIG's (1958) classification the Sphaeroceridae were doubtfully included in the Milichioidea on the basis of 'comparatively trivial indices', namely the nearly straight course of the anal cross-vein and the presence of two rows of interfrontal bristles. In my opinion this classification was incorrect. The Milichiidae are referable to a wider group for which I use the name Tephritoinea in this work. The type of aedeagus shown by the Sphaeroceridae is very different, and does not seem derived from the condition ascribed to the groundplan of the Tephritoinea. SPEIGHT (1969) has also suggested that the Sphaeroceridae were misplaced in the Milichioidea, because they show a reduced prosternal basisternum. He suggests that the family should be classified in a 'superfamily including the Heleomyzidae'. My conclusion that the Sphaeroceridae should be classified in the Anthomyzoinea is thus in agreement with SPEIGHT's view. The aedeagus of Sphaeroceridae is short, with very complex sclerotization, comparable with the types of aedeagus shown by the Borboropsidae, Chyromyidae and Aulacigastridae. In my opinion the most closely related group to the Sphaeroceridae is probably one or other of these families.

The Sphaeroceridae are well characterized as a monophyletic group by the following groundplan conditions which are apomorphous with respect to the groundplan of the Anthomyzoinea

(1)   Metatarsus of hind legs shortened and thickened (HENNIG 1958).

A few little-known genera which do not show this condition have been classified in the Sphaeroceridae, but these are probably all misplaced (see HACKMAN 1969a). One such genus, the New Zealand *Protoborborus*, was transferred to the 'Pseudopomyzidae' (included in Cypselosomatidae in this work) by D. K. MCALPINE (1966).

(2)   Subcosta fading distally, reaching wing margin at most as fold (HENNIG 1958).

(3)   Second median (m_4) abruptly cut off apically, not reaching wing margin.

(4)   Anal vein (cu_{1b} + 1a) shortened, not reaching wing margin (HENNIG 1958).

(5)   Anal lobe of wing secondarily broadened (HENNIG 1958).

(6)   Peristomal opening large, with 'im Verhältnis zum Kopf gewaltig entwickelten Mundteilen' (FREY 1921).

(7)   Interfrontal bristles present in two rows on interfrontal ridges (HENNIG 1958).

(8)   6th tergum ($\male$) completely lost.

(9)   7th and 8th tergum vestiges lost ($\male$).

(10) 7th right spiracle ($\male$) lost; 6th and 7th left spiracles both lying within 6th sternum (fig. 113).

This characteristic arrangement of postabdominal spiracles is probably a groundplan condition of the Sphaeroceridae, though additional genera need to be examined to confirm this. The anterior displacement of the 7th left spiracle onto the 6th sternum has led to confusion in the homologies of the sclerites in the paper by KIM & COOK (1968) (see below). The displacement of the 7th left spiracle in Sphaeroceridae may possibly represent synapomorphy with the Chyromyidae, in which this spiracle also lies on the 6th segment.

(11) Vertical sections of hypandrial arms (♂) expanded (fig. 114).

(12) Ejaculatory apodeme (♂) minute.

The structure of the male postabdomen of *Sphaerocera* (figs. 112-114) and *Copromyza* is readily comparable with that of other families of Muscoidea and poses no special problems of homology. These genera also have relatively plesiomorphous wing venation. It seems reasonable to postulate that such genera as *Sphaerocera*, *Copromyza* and *Archiborborus* are closer to the groundplan of the family in most respects than *Leptocera* and its relatives, which show many highly modified conditions. The male postabdomen of *Leptocera* and its relatives has been modified by a considerable expansion of the 6th sternum (as also of the preabdominal sterna) and fusion of the 8th sternum with the periandrium and hypandrium. However, I can see no difficulty in interpreting the conditions shown by these genera as derived from the conditions shown by *Sphaerocera* and *Copromyza*. KIM & COOK's (1966) homology of part of the postabdominal sclerotization of *Leptocera* with the 7th tergum seems to me unwarranted, since according to my present analysis this tergum was already reduced to a vestige situated lateroventrally on the right side in the groundplan of the Muscoidea.

Further comments on KIM & COOK's (1966) paper are as follows. The presence of two spiracles on the left side of the 6th sternum has led these authors to suppose that part of this sclerite represents the tergum, which has become 'lateroverted' and carried the 6th right spiracle onto the left side. This is untenable, because the true 6th right spiracle is situated on the right side in the membrane of the genital pouch where it would be expected (fig. 113) and where it is in fact shown in their own figure of *Copromyza* (fig. 36B). The areas shown as '6th tergite' and '6th sternite' on their figures are not discrete sclerites. It appears that they have considered the internal ridge (serving for muscle attachment) as indicating the boundary between two sclerites. A ridge of this type on the 6th sternum is of widespread occurrence among muscoid families. Usually the ridge more or less follows the anterior margin of the sternum. The extension of sclerotization anterior to the ridge in Sphaeroceridae is unusual, but not unparalleled. Since there is no break in the external sclerotization along the ridge, I see no reason for thinking that it indicates a line of fusion between separate areas of sclerotization. As for the position of the spiracles, there are several other examples among the Schizophora of the movement of postabdominal spiracles onto segments to which they did not originally belong (for instance the movement of the 7th right spiracle onto the 8th sternum in Calyptratae). The simplest explanation of the postabdominal morphology of Sphaeroceridae is to postulate a forward movement of the 7th left spiracle onto the 6th sternum, since this involves no unusual assumption about the homologies of the sclerites, which are arranged in the typical muscoid pattern in many genera (such as *Sphaerocera* and *Copromyza*). A further point is that KIM & COOK appear to

208

consider the periandrium of *Sphaerocera* as part of the proctiger, since they call it the '10th tergite'. They say that in *Sphaerocera* and *Copromyza* the 'ninth tergite is represented by a very narrow strip-like sclerite between the eighth and ninth tergites and by a triangular sclerite on each side'. On checking the species figured (*Sphaerocera curvipes* LATREILLE) I find that the 'triangular sclerites' are part of the hypandrium (expanded vertical sections of the hypandrial arms), and I can see no 'narrow strip-like sclerite'. What KIM & COOK call the '10th tergite' is evidently the periandrium (= 'epandrium' of most recent authors) and it is to this that the telomeres ('valvulae laterales') are attached, as would be expected. Furthermore KIM & COOK homologize the postabdomen of *Leptocera* inconsistently with their interpretation of *Sphaerocera*. In *Leptocera* they attribute part of the sclerotization of the postabdomen to the '7th tergite', which they consider to be absent in *Sphaerocera*. On the other hand they consider the '10th tergite' of *Sphaerocera* to be absent in *Leptocera*. I see no grounds for supposing that the sclerotization of the male postabdomen in different genera of Sphaeroceridae is of different segmental origin.

C. 40. Family Chyromyidae

Sources of information

I studied preparations of *Chyromya flava* (L.) and *Aphaniosoma occulicauda* COLLIN. No detailed treatment of the male postabdomen and genitalia of Chyromyidae was available in the literature.

Description of male postabdomen and genitalia (figs. 115 and 117)

In *Chyromya* the 6th and 7th pairs of abdominal spiracles are both situated within the 6th tergum (symmetrically on either side) with the 7th pair displaced dorsally in relation to the 6th pair: the position of the spiracles in *Aphaniosoma* could not be established on account of the extremely small size of these insects.

Postabdomen fully symmetrical. 6th tergum well developed in *Chyromya*, about half as long as 5th tergum, with lateroventral extensions on either side: in *Aphaniosoma* the 6th tergum is divided into two fragments, which bear lateroventral extensions as in *Chyromya*. 6th and 7th sterna absent. 7th and 8th terga absent. 8th sternum fused centrally with periandrium, but partly delimited from this by lateral sutures (*Chyromya*): in *Aphaniosoma* the 8th sternum is not differentiated from membrane.

Periandrium in *Chyromya* bearing slender, more or less fused telomeres, whose inner basal corners are linked to an H-shaped area of interparameral sclerotization: cerci small, discrete. In *Aphaniosoma* the periandrium and cerci are scarcely differentiated from membrane. Aedeagal apodeme fused with body wall posteriorly, where it is expanded into a broad area of sclerotization which shields the aedeagus when at rest: this sclerotization is linked anteriorly to the hypandrium (appearing as a ventral process of the apodeme in lateral view). Aedeagus able to be swung through wide arc against aedeagal apodeme to anteriorly directed rest position, its sclerotization consisting of short cylindrical phallophore and expanded complex distal area. Postgonites well developed, setose in *Chyromya*. Ejaculatory apodeme not traced.

Characterization and discussion

The Chyromyidae are characterized by the following groundplan conditions which are apomorphous with respect to the groundplan of the Anthomyzoinea.

(1) Face (prefrons) with membranous central area (HENNIG 1958 and in press).
(2) Palpi very short (FREY 1921).

I have confirmed that the palpi are also reduced in *Aphaniosoma*, as well as in *Chyromya* which was studied by FREY.

(3) Subcosta closely approximated to vein r_1.
(4) Anal vein shortened, not reaching wing margin (HENNIG 1958).
(5) 6th tergum ($\male$) with lateroventral extensions on either side (fig. 115); 6th sternum lost.

The presence of these characteristic lateroventral extensions of the 6th tergum is probably an autapomorphous condition of the Chyromyidae. No discrete 6th sternum is present, although it is perhaps possible that this has been involved in the formation of the apparent extensions of the tergum through splitting and fusion of the resulting sternum halves with the tergum. But since no break in the sclerotization can be seen in the preparations studied by me, I regard loss of the 6th sternum as the more probable explanation of the observed structure. In *Aphaniosoma* the 6th tergum has become split into two fragments, but the lateroventral extensions are still well developed.

(6) 6th and 7th abdominal spiracles ($\male$) symmetrically situated on either side within 6th tergum, with 7th pair displaced dorsally in relation to 6th pair.

It is possible that the forward movement of the 7th spiracles onto the 6th segment in Chyromyidae may be due to synapomorphy with the Sphaeroceridae, in which the 7th left spiracle is situated within the 6th sternum.

(7) 7th sternum ($\male$) lost.
(8) 7th and 8th tergum vestiges lost ($\male$).
(9) 8th sternum ($\male$) fused centrally with periandrium, only partly delimited by lateral sutures (fig. 115).

This condition is shown by *Chyromya*. In the minute *Aphaniosoma* the 8th sternum and periandrium are not differentiated from membrane, so that the 6th tergum fragments are the last dorsal sclerites of the male abdomen.

(10) Aedeagal apodeme ($\male$) fused with body wall posteriorly, where it is expanded into a broad area of sclerotization which shields the aedeagus when at rest; this sclerotization is linked anteriorly to the hypandrium (fig. 117).

210

(11) Telomeres ($\circlearrowright$) more or less fused with periandrium.

(12) Two spermathecae ($\circleddash$) (STURTEVANT 1926, HENNIG 1958).

STURTEVANT comments on similarity between the spermathecae of *Chyromya* and *Aulacigaster*. However the latter retains three spermathecae (as also in Borboropsidae and some Sphaeroceridae).

It is doubtful whether the inward direction of the anterior fronto-orbital bristle (*ors*) shown by *Chyromya* can be considered a groundplan condition of the Chyromyidae, since in *Aphaniosoma* all the *ors* are posteriorly directed (see HENNIG 1965b). It is also improbable that absence of the propleural bristle can be ascribed to the groundplan of the Chyromyidae (as suggested by HENNIG 1965b), since this bristle is present in *Aphaniosoma*.

The aedeagus of Chyromyidae (fig. 117) is short, with very complex sclerotization, similar to the conditions of the aedeagus shown by the Borboropsidae, Sphaeroceridae and Aulacigastridae. I think it possible that one of the last two families is the sister-group of the Chyromyidae, although the evidence is not conclusive.

C. 41. Family Aulacigastridae

Sources of information

The limits of the Aulacigastridae have been revised by HENNIG (1969a), who gives some information on the male postabdomen and genitalia. I studied preparations of *Aulacigaster leucopeza* (MEIGEN) and *Cyamops nebulosus* MELANDER.

Description of male postabdomen and genitalia (fig. 116)

6th abdominal spiracles in membrane: 7th left spiracle in membrane near margin of 7th sternum (*Cyamops*) or within sclerotized area (*Aulacigaster*): 7th right spiracle in membrane.

Postabdomen asymmetrical (but only weakly so in *Aulacigaster*). 6th tergum represented by discrete narrow band (*Cyamops*) or fused with 6th to 8th sterna (in *Aulacigaster*; and also in *Schizochroa* according to HENNIG 1969a). 6th sternum asymmetrically developed on left side where it is fused with the 7th sternum: the suture between the 6th and 7th sterna is conspicuous in *Cyamops*, but in *Aulacigaster* the 6th to 8th sterna (and 6th tergum) are not clearly delimited from each other. 7th sternum asymmetrically developed on left side, clearly delimited in *Cyamops* in which it is relatively large in comparison with the 8th sternum. In *Cyamops* the 8th sternum is asymmetrical, well developed only on the left side, reduced to a slender strip on the right side; however the fused area which appears to represent the 8th sternum in *Aulacigaster* is relatively large and more or less symmetrical. 7th and 8th terga absent: in *Cyamops* a strip of secondary sclerotization is present on the right side of the venter posterior to the 6th sternum.

Periandrium with small fused telomeres in *Aulacigaster*; in *Cyamops* the telomeres are discrete, but asymmetrically developed (the left elongate, the right short and broad). A pair of slender interparameral sclerites (processus longi) connects the inner basal corners of the telomeres with the bridge formed by the hypandrial arms in *Aulacigaster*: but no interparameral sclerites were found in *Cyamops*. Cerci weakly sclerotized, poorly differentiated from membrane, more or less discrete. Aedeagal apodeme rod-like, free from hypandrium, with posterior lateral expansions which shield the base of the aedeagus ('x' on fig. 116); these are asymmetrically developed (the larger on the right side) in *Cyamops*. Hypandrium concave centrally, where it receives the aedeagus when at rest, with its arms fused to form bridge above base of aedeagus in *Aulacigaster*; but in *Cyamops* the hypandrium is asymmetrical and no such bridge is formed. Aedeagus able to be swung through wide arc against aedeagal apodeme to anteriorly directed rest position, rather short, with very complex sclerotization. Postgonites absent in *Aulacigaster*, but single setose postgonite present in *Cyamops*. Large ejaculatory apodeme present.

Characterization and discussion

The Aulacigastridae are characterized by the following groundplan conditions which are apomorphous with respect to the groundplan of the Anthomyzoinea.
(1)   Subcosta fused with r_1 for a certain distance before its end (HENNIG 1969a).
(2)   Anal vein shortened, not reaching wing margin (HENNIG 1958).
(3)   Peristomal opening large (HENNIG 1969a).
(4)   Of the two orbital bristles (*ors*) the anterior pair are directed inwards (HENNIG 1969a).
(5)   Postvertical bristles lost; ocellar bristles reduced, at most represented by fine hairs (HENNIG 1969a).
(6)   Only one postalar bristle (*pa*) present (HENNIG 1969a).
(7)   7th and 8th tergum vestiges lost (♂).

A strip of sclerotization posterior to the 6th sternum is present on the right side of the venter in *Cyamops*. I am doubtful whether this represents the 7th tergum vestige, as the 7th right spiracle lies to the right of it. More probably this weak sclerotization is secondary.

(8)   Aedeagal apodeme (♂) with posterior lateral expansions which shield the base of the aedeagus (fig. 116).
The genus *Stenomicra*, which Hennig (in press) interprets as closely related to the Aulacigastridae, is discussed above under Anthomyzidae.

B. 12. Prefamily Agromyzoinea

In this group I include the Agromyzidae and the Clusiidae. The relationships of the Agromyzidae have been variously interpreted. Most authors have suggested that they are closely related to one or other family of Tephritoinea, in particular to the Tephritidae, Odiniidae or Milichiidae. HENNIG (1965b) accepted the view that the Odiniidae and Agromyzidae are sister-groups, although admitting that there are no decisive grounds for this interpretation. In my opinion these families are unlikely to be sister-groups, because the Odiniidae show a condition of the aedeagus which seems derived from that which I ascribe to the groundplan of the Tephritoinea. The only apomorphous conditions which have been put forward to support the hypothesis of a sister-group relationship between the Odiniidae and Agromyzidae are either widespread (such as inwardly directed *ori*, costal break present, anal vein shortened, two spermathecae) or proportional changes (namely the presence of a long aedeagal apodeme and hypandrium in both families, to which SPENCER 1969 refers). I suggest that the sister-group of the Agromyzidae is more probably the Clusiidae. The agreement in frontal chaetotaxy and wing venation between the Agromyzidae and Clusiidae is no less than that between the Agromyzidae and Odiniidae; and the Clusiidae show a type of aedeagus which is similar to that of the Agromyzidae in respect of length and complexity of sclerotization. In respect of most character sequences analysed the Clusiidae are more plesiomorphous than the Agromyzidae. The latter family is highly divergent in some respects from all other Muscoidea, which is the reason why judgement of its relationship is difficult.

If the Clusiidae and Agromyzidae are correctly grouped in the Agromyzoinea, then this prefamily seems characterized in its groundplan by the following conditions which are apomorphous with respect to the groundplan of the Muscoidea. (1) Aedeagus (♂) extremely elongate, with complex sclerotization including phallophore, paired paraphalli and distal sclerites of varying form; epiphallus well developed (figs. 118 and 121).

The term paraphalli in the above characterization refers to the sclerites which have been called the 'arms of the basiphallus' by workers on Agromyzidae. The terminology in current use for that family is unfortunately inconsistent with that used for other groups of Schizophora (see later discussion under Agromyzidae). In the Agromyzidae and most Clusiidae the mechanism for swinging the aedeagus against its apodeme is well developed, as in the groundplan of the Muscoidea, so that the aedeagus is anteriorly directed when at rest. But in *Clusiodes* (Clusiidae) this mechanism has been lost, and the aedeagus remains more or less posteriorly directed when at rest. The latter condition is clearly apomorphous, not belonging to the groundplan of the Clusiidae. A detailed statement of the groundplan condition of the aedeagus of Agromyzoinea is hindered by insufficient analysis of the variation within the Agromyzidae, and it is possible that some of the conditions stated below as characterizing the Agromyzidae may eventually prove to be groundplan conditions of the Agromyzoinea. Members of the Agromyzoinea show some of the largest and most complexly sclerotized aedeagi among the whole of the Diptera. In some Clusiidae the basal area of the aedeagus is somewhat flexible, becoming coiled when at rest, a condition approaching that shown by Tephritoinea. However such a condition is not shown by the Agromyzidae, in which the basal section is always supported by rigid paraphalli; it seems unlikely that such coiling can be ascribed to the groundplan of the Agromyzoinea.

An epiphallus was clearly present in the groundplan of the Agromyzoinea. According to D. K. McALPINE (1960) such a structure is absent among the Clusiidae only in *Clusiodes*, which has highly modified genitalia due to loss of the swinging mechanism. In *Clusia* (fig. 118) the epiphallus is a large upcurved posterior process of the phallophore, which latter is open below (not fully cylindrical as in Agromyzidae). In Agromyzidae the epiphallus is discrete from the phallophore and has a rather complex structure (see below under Agromyzidae).

(2)   8th tergum vestige lost (♂).

The male postabdomen of Clusiidae shows the asymmetrical pattern of sclerites typical of the Muscoidea. Of the sclerites present in the groundplan of the Muscoidea only the 8th tergum vestige has been lost.

(3)   8th abdominal segment (♀) elongate and largely membranous, forming retractile ovipositor which is retracted within the subconical 7th segment when at rest.

The condition of the female postabdomen shown by the Clusiidae is clearly more plesio-morphous than the characteristic boring ovipositor of the Agromyzidae. A retractile ovipositor occurs in other groups of Muscoidea (such as the Calyptratae, Tanypezoinea and Tephritoinea), and it is possible that some of these resemblances may be due to synapo-morphy. But no firm judgement can be made until more extensive comparative studies on the female postabdomen of Muscoidea are available. I have noted that in a preparation of *Clusio-des melanostoma* (Loew) the small 8th sternum is entire, not divided as in the Tephritoinea and Calyptratae.

(4)   Two spermathecae (♀) (Sturtevant 1926, Hennig 1958, Saskawa 1958).
(5)   Vibrissae present (Hennig 1958).
(6)   Inwardly directed lower fronto-orbital bristle (*ori*) present.

The number of fronto-orbital bristles is variable in the Clusiidae, and I find no clear state-ment of the groundplan condition in the literature. However the presence of a pair of in-wardly directed *ori*, in addition to two or more upwardly directed orbitals, is common to the majority of species (see Hennig 1958, figs 190-194). The groundplan condition for the Agro-myzidae is clearly one in which two pairs of upwardly directed *ors* are present, and one or two pairs of inwardly directed *ori*. Deviations from this pattern occur only in a small minority of species.

(7)   Costa broken at end of subcosta.

Hennig (1958) postulated that the costa was unbroken in the groundplan of the Clusiidae, as shown by *Heteromeringia*. However D. K. McAlpine (1960) states that the costa was incised in the material of *Heteromeringia* seen by him, as in all other Clusiidae know to him.

C. 42. Family Clusiidae

Sources of information

Information on the male postabdomen and genitalia of Clusiidae was given by D. K. McAlpine (1960) and Sasakawa (1966). I studied preparations of *Clusia lateralis* (Walker) and *Clusiodes melanostoma* (Loew).

214

Description of male postabdomen and genitalia (figs. 118-119)

6th abdominal spiracles in membrane: 7th left spiracle usually displaced dorsally in relation to preceding spiracles (except *Clusiodes*), in membrane near margin of 7th sternum or within 7th sternum: 7th right spiracle in membrane, displaced centrally in relation to preceding spiracles.

Postabdomen asymmetrical. 6th tergum relatively large, at least half as long as 5th tergum (in some species almost fully as long). 6th sternum asymmetrical, linked to or fused with 7th sternum on left side. 7th sternum usually asymmetrically developed on left side and discrete from 8th sternum except for dorsal point of fusion (except *Clusiodes*). 7th tergum represented by small lateroventral sclerite on right side (near 7th right spiracle) in *Clusia* and *Tetrameringia*. In *Clusiodes* 7th sternum and 7th tergum not discrete, but fused to form ventral annulus which is fully fused (without visible suture) with the 8th sternum on the left side. 8th sternum large. 8th tergum absent.

Periandrium bearing large discrete telomeres (sometimes bilobed), which in *Clusia* are linked to the hypandrial bridge by a pair of rod-like interparameral sclerites (processus longi) which are fused with each other anteriorly: however no such interparameral sclerites are present in *Clusiodes*. Cerci discrete. Aedeagal apodeme rod-like, free from hypandrium in *Clusia*, linked to hypandrium by ventral process in *Clusiodes*. Hypandrial arms fused to form bridge above base of aedeagus. Aedeagus elongate, with variable (often very complex) sclerotization, sometimes divided distally or bearing distal processes (e.g. *Paraclusia*, *Heteromeringia*), sometimes with somewhat flexible basal section which can be coiled when at rest (*Heteromeringia*, *Tetrameringia*); usually a conspicuous epiphallus arises from the basal phallophore (except *Clusiodes*); aedeagus usually able to be swung through wide arc against aedeagal apodeme to anteriorly directed rest position (except *Clusiodes*, in which it remains posteriorly directed when at rest). Postgonites usually well developed, but scarcely differentiated in *Paraclusia* and *Clusia*. Ejaculatory apodeme present.

Characterization and discussion

For most of the character sequences analysed the groundplan condition of the Clusiidae is that ascribed to the groundplan of the Agromyzoinea as a whole. The view that the family is monophyletic is supported only by the following conditions which are apomorphous with respect to the groundplan of the Agromyzoinea.
(1) 2nd antennal article with angular projection on its outer side (HENNIG 1958).
(2) Face (prefons) with membranous central area (HENNIG, in press).

HENNIG (1958) referred to the inclusion of spiracles 6 and 7 (♀) within the terga in *Clusia*, but this is probably not a groundplan condition of the Clusiidae since these spiracles are situated in membrane in *Clusiodes*. The structure of the female postabdomen in the Clusiidae is more plesiomorphous than in the Agromyzidae. The 7th tergum and sternum are discrete, and the proctiger remains short. No 'egg-guides' have been reported.

C. 43. Family Agromyzidae

Sources of information

The first detailed treatments of the male genitalia of Agromyzidae were those of SPEIJER (1934), DE MEIJERE (1938) and FRICK (1952). Further progress was made by NOWAKOWSKI (1959, 1964) and his works probably provide the best detailed morphological treatment currently available. IPE's (1967) work on *Melanagromyza obtusa* (MALLOCH) is also helpful, and includes details of the internal reproductive system. I am well acquainted with the range of variation within this family as a result of my previous work on it, and have studied preparations of the male postabdomen and genitalia of large numbers of species in all major genera.

Description of male postabdomen and genitalia (figs. 120-121)

6th and 7th abdominal spiracles symmetrically situated, in most species in membrane below 6th tergum, but within this tergum in *Selachops*.

Postabdomen fully symmetrical. 6th tergum large, of similar length to 5th tergum. 6th and 7th sterna absent, the last ventral sclerite before the hypopygium being the large 5th sternum. 7th and 8th terga absent. Between the 6th tergum and the hypopygium lies a narrow strip of sclerotization which represents the 8th sternum; this is more or less discrete in some species, but more often fused with the 6th tergum (differentiated only by a change in the intensity of sclerotization).

Periandrium with or without discrete telomeres: a discrete well-defined interparameral sclerite or pair of sclerites ('bacilliform sclerites') is linked anteriorly to the vertical sections of the hypandrial arms. Cerci discrete. Aedeagal apodeme long and rod-like, free from hypandrium, extending anteriorly to near base of abdomen (at least into 4th abdominal segment). Hypandrium extending anteriorly above 5th sternum, forming roof of genital pouch which receives the aedeagus when at rest. Aedeagus consisting of two sections, with complex sclerotization which includes cylindrical phallophore, pair of paraphalli (of which the right is joined basally to the phallophore), very complex sclerotization of distal section (which is divided apically in many groups), and often additional sclerites in a dorsal lobe or lobes (the so-called 'hypophallus') ('x' on fig. 121); from the base of the aedeagus arises a fold (aedeagal hood) which contains a discrete epiphallus (consisting of paired basal sclerites and an unpaired apical sclerite) and also in some groups a pair of lateral sclerites which articulate with the apex of the epiphallus; aedeagus able to be swung through wide arc against aedeagal apodeme to anteriorly directed rest position within genital pouch. Postgonites well developed. Ejaculatory apodeme usually large and fan-shaped (but minute in *Phytomyza milii* group).

Characterization and discussion

Subsequent research on the Agromyzidae has confirmed that the group was correctly delimited in HENDEL's (1931-36) monograph of the Palaearctic fauna. The Agromyzidae are well characterized as a monophyletic group by the following apomorphous conditions with respect to the groundplan of the Agromyzoinea.

(1) 7th segment (♀) completely sclerotized, forming conical non-retractile ovipositor sheath ('basal cone' of ovipositor), with dorsal apodeme extending anteriorly into 6th segment.

(2) 8th segment (♀) bearing numerous anteriorly directed denticles.

(3) Gonopore (♀) flanked by pair of serrated 'egg-guides'.

(4) Proctiger (♀) elongate, so that the anus is situated well posterior to the gonopore.

(Note on characters 1-4).
Detailed treatments of the female postabdomen of Agromyzidae have been given by SASA-KAWA (1958) and IPE (1966). It is evident that all female Agromyzidae show a characteristic complex of modifications of the postabdomen which enable them to insert their eggs into living plant tissues. The most outstanding apomorphous conditions of this complex have been summarized under the four headings above. No other Schizophora are known to show this combination of conditions. The apparent similarity between the 'ovipositor' of the Agromyzidae and those of the Tephritidae family-group and Lonchaeidae s.l. ('Otitoidea') is only superficial, as pointed out by HENNIG (1958) with respect to the Tephritidae s.l. ('Otitoidea'). In these groups the ovipositor sheath (7th segment) does not bear an anterior apodeme extending into the 6th segment, there are no egg-guides, and the proctiger is not elongate (so that the gonopore lies near the tip of the abdomen). I regard these structural differences as confirming that a piercing ovipositor was evolved independently by the ancestors of the Agromyzidae, as implied by my classification.
The homologies of some parts of the agromyzid ovipositor have not been resolved. In all Agromyzidae a pair of 'egg-guides' or 'valvulae inferiores' flank the gonopore. These are usually serrated, blade-like structures (except in certain groups of Agromyza, see SASAKAWA 1958). I think it improbable that these egg-guides can be homologized with the valvulae of other insect orders (as the term 'valvulae inferiores' implies), since no comparable structures have been reported in other families of Diptera. From study of HERTING's (1957) work, I suggest two possible explanations of their origin: (1) they may be derivatives of the 8th sternum (like the 'vaginal plates' of Drosophila), or (2) they may be derived from rudiments of the 9th sternum, as HERTING has suggested for the 'lingulae' of the Calyptratae. A further problem of homology is posed by the elongate segment which lies posterior to the gonopore. SASAKAWA (1958) accepted this as the ninth segment, and homologized its dorsal and ventral sclerites as the 9th tergum and 9th sternum. This interpretation is improbable, since to the best of my knowledge the 9th tergum and 9th sternum are never developed as typical dorsal and ventral sclerites in any other female Schizophora. I think it more probable that these sclerites represent the 10th tergum and sternum, and that the postgenital segment consequently represents an elongated proctiger. The morphology of intersexes, if such can be found, might be helpful in clarifying this point, since they may be expected to show conditions intermediate between the male and female proctiger.

(5) 6th and 7th sterna (♂) lost; 8th sternum reduced to narrow band (fig. 120).

The arrangement of the male abdominal sclerites is uniform throughout the Agromyzidae. I can find no trace of asymmetry in the postabdominal sclerotization, notwithstanding FRICK's (1952) statement that 'the sixth, seventh and eighth sternites are greatly reduced and consist of narrow sclerotized strips stretching along the left side of the sixth tergite'. To the best of my knowledge the 6th and 7th sterna have been completely lost in all Agromyzidae. There are only five abdominal sterna, of which the fifth is the largest. The last large dorsal sclerite is the unreduced 6th tergum, which is followed by a small band-like sclerite which had

217

apparently been overlooked before I recently drew attention to its presence (GRIFFITHS 1964). I now regard this as representing a vestige of the inverted 8th sternum. The homology with the 8th tergum which I suggested in 1964 has become untenable, since it now seems clear that the 8th segment is inverted in all Cyclorrhapha. I reported that this vestige of the 8th sternum is discrete in certain species of three disjunct genera (*Melanagromyza, Cerodontha* and *Phytomyza*), and it thus seems probable that it was present as a discrete sclerite in the groundplan of the Agromyzidae (GRIFFITHS 1964). However in many species this sclerite has become fused with the 6th tergum, differentiated only by a change in the intensity of sclerotization.

HENNIG (1958) suggested that the postabdominal spiracles are situated within the 6th tergum in male Agromyzidae. The spiracles in Agromyzidae are small and often difficult to locate. However I have carefully checked preparations of four species, *Agromyza phragmitidis* HENDEL, *Liriomyza graminicola* DE MEIJERE, *Phytomyza cineracea* HENDEL and *Cerodontha (Poemyza) pygmaea* (MEIGEN), and found that in every case both pairs of postabdominal spiracles are situated in membrane (symmetrically on both sides of the abdomen). Thus there is no reason to think that any of the male abdominal spiracles were situated within terga in the groundplan of the Agromyzidae.

(6) Aedeagus (♂) (fig. 121) with large distal section and dorsal lobe or lobes arising from area of articulation between basal and distal sections: its sclerotization consisting of cylindrical phallophore, paired paraphalli (of which the right is joined basally to the phallophore), and variable sclerotization of distal section and dorsal lobe(s): when at rest the aedeagus lies within a genital pouch formed above the 5th sternum (with the hypandrium forming the roof of this pouch).

FRICK (1952) considered that the aedeagus of Agromyzidae was divisible into three sections (basal, median and distal). A division between the median and distal sections in FRICK's sense is not shown by all groups of Agromyzidae, and it is not clear whether such a division was present in the groundplan of the family. But a division into two articulated sections, a basal section containing the phallophore and paraphalli (= 'arms of basiphallus') and a distal section containing such sclerites as the distiphallus and mesophallus, can be observed in all genera and is clearly attributable to the groundplan of the family. The form of the sclerites of the basal section of the aedeagus (the phallophore and paraphalli) shows little variation, except in a few groups in which the aedeagus is highly modified (such as the *Phytomyza obscurella* group). But the sclerotization of the distal section is highly variable both between genera and species, and I do not attempt here to define the groundplan condition precisely. The terminal portion of the ejaculatory duct is bifid in many Agromyzidae (as probably also in some Clusiidae), and SPENCER (1969: 13) has suggested that a divided distiphallus (containing the twin ducts) belongs to the groundplan of the Agromyzidae. This interpretation seems to me probable, as the condition in question occurs in many genera. However I do not think that a definitive judgement can be made at present, as the relationships between the various groups included in the Phytomyzinae (a group whose monophyly has not been demonstrated) are still poorly understood. The dorsal lobe mentioned in the above characterization is currently known as the 'hypophallus'. This normally contains two or more small sclerites ('x' on fig. 121).

The terminology which is currently applied to the aedeagus of the Agromyzidae and its sclerites (based largely on the proposals of NOWAKOWSKI 1959) poses certain problems when considered in the light of the application of these terms to other families of Schizophora. First, it is clear that the position of the aedeagus which is comparable with the position of this organ in groups which lack a swinging mechanism (that is all Diptera except the Nothyboidea and Muscoidea) is the posteriorly directed position adopted for copulation. The customary use by writers on Agromyzidae (including myself) of the terms 'dorsal' and 'ventral' refers to the rest position of the aedeagus, thus reversing their significance for purposes of comparative morphology. This difficulty has already been recognized by NOWAKOWSKI (1964: 181, footnote 7). The term 'hypophallus' is thus a misnomer when applied to lobes whose position is dorsal with reference to the copulatory position of the aedeagus. (The hypophallus of Calyptratae is truly ventral in position and cannot be homologous with the so-called hypophallus of Agromyzidae.) A second difficulty concerns the use of the term 'paraphalli' for sclerites of the distal section of the aedeagus of Agromyzidae. The term was originally proposed by LOWNE (1893) for sclerites in the aedeagus of *Calliphora* which are joined basally to the phallophore. NOWAKOWSKI (1959) was able to interpret these distally situated sclerites as 'paraphalli' because he postulated that the phallophore ('Phallotheca') had become secondarily divided in Agromyzidae, so that the whole basal section of the aedeagus was homologous with the phallophore of the Calyptratae. However I do not see any good evidence for this interpretation, and use the term paraphalli in this paper for the paired sclerites of the basal section of the aedeagus which are of widespread occurrence among the Muscoidea. These are what workers on Agromyzidae have called the 'arms of the basiphallus'. I propose that the sclerites of the distal section of the aedeagus in Agromyzidae which have been called 'paraphalli' should be renamed **paramesophalli**.

(7) Aedeagal apodeme ($\male$) and hypandrium extending anteriorly.

The aedeagal apodeme is very large in Agromyzidae, extending anteriorly to near the base of the abdomen (at least into the 4th abdominal segment). The hypandrium is likewise enlarged, and extends anteriorly above the 5th sternum, forming the roof of the genital pouch. The apomorphy of these structures lies primarily in their increased size; their shape is in no way unusual, and there is no sclerotized connection between them. The extension of the aedeagal apodeme and hypandrium has been associated with corresponding increases in the length of the muscles which originate on these sclerites. The musculature of the genitalia has thus become extremely powerful. DE MEIJERE (1938) reported that in *Amauromyza abnormalis* (MALLOCH) not only the aedeagus but also part of the aedeagal apodeme penetrates the female's genital opening during copulation.

(8) Aedeagal hood containing discrete epiphallus and paired lateral sclerites ($\male$).

In Agromyzidae the fold of the integument posterior to the base of the adeagus is large and conspicuous. FRICK (1952) proposed for this the term aedeagal hood (= Peniskappe, NOWAKOWSKI 1964). The epiphallus is no longer a process of the phallophore, but a discrete area of sclerotization which appears to be composite, consisting of paired basal sclerites and an unpaired apical sclerite (NOWAKOWSKI 1964). In many groups there is also a pair of lateral sclerites of the aedeagal hood which articulate with the apex of the epiphallus. As the aedeagus is swung into the copulatory position these sclerites are turned outwards about their articulation with the epiphallus. I provisionally ascribe the presence of these lateral sclerites to the groundplan of the Agromyzidae, although this requires to be checked by further comparative studies.

(9) Orbital setulae present; centre of frons almost bare.

It is characteristic of the Agromyzidae that the centre of the frons bears no setulae, or at most a few fine scattered setulae visible only at high magnification. However there are usually one or more rows of distinct setulae on the orbits, between the large orbital bristles and the eye-margin. The loss of these orbital setulae in some groups of *Paraphytomyza* is doubtless secondary.

(10) Pseudotracheae with ventral main channel ('Sammlungsrohr') (FREY 1921).

(11) Subcosta fading distally, closely approximated to vein r_1.

The groundplan condition for the Agromyzidae is probably that shown by the Phytomyzinae, in which the subcosta fades distally, but its course can be followed to the wing margin (at the costal break). The Agromyzinae are characterized by a more strongly apomorphic condition in which the subcosta becomes fused with r_1 distally (see HENDEL 1931: 16). In Clusiidae the subcosta is complete.

(12) Anal vein shortened, not reaching wing margin.

(13) Larvae feeding inside living plant tissue, with their mouth-hooks directed obliquely or almost vertically in relation to labial sclerite (Halsstück);

front spiracles lying close together on dorsal surface of prothorax (see HENNIG 1952a).

B 13. Prefamily Tephritoinea

I include in this prefamily the Otitoidea, Pallopteroidea (except Lonchaeidae) and part of the Milichioidea of HENNIG's (1958) classification, as well as the families Chloropidae, Chiropteromyzidae, Mormotomyiidae, Cnemospathidae, Acartophthalmidae and Odiniidae, which were unplaced in HENNIG's classification. I do not include in this prefamily the Braulidae, Canacidae and Sphaeroceridae, which were tentatively placed by HENNIG in the 'Milichioidea'. In this work I refer the Sphaeroceridae to the Anthomyzoinea, and treat the Braulidae and Canacidae as families of uncertain relationship. The complete list of families here referred to the Tephritoinea is as follows: Chiropteromyzidae, Mormotomyiidae, Cnemospathidae, Odiniidae, Tethinidae, Acartophthalmidae, Carnidae, Milichiidae, Chloropidae, Conopidae, Eurygnathomyiidae, Richardiidae, Piophilidae *s.l.* and Tephritidae *s.l.*

SPEIGHT (1969) and COLLESS & D. K. MCALPINE (1970) have proposed to classify the Tethinidae, Milichiidae and Chloropidae in the Drosophiloidea, but I am convinced that this is incorrect. The Chloropidae include some genera, such as *Lasiopleura*, in which the asymmetrical structure of the male postabdomen characteristic of Muscoidea is retained. The simpler symmetrical structure shown by the closely related Milichiidae thus cannot be derived from the groundplan condition of the Drosophiloidea. The structure of the male postabdomen and genitalia in the Carnidae, which probably belong to a monophyletic group containing also the Milichiidae, Chloropidae and Acartophthalmidae (called the Chloropidae family-group below), is hardly separable from that of some of the groups included by HENNIG in the Otitoidea and Pallopteroidea. It seems to me most improbable that the complex and highly modified type of aedeagus concerned was evolved independently in different lineages, and no incompatibility with the evidence of other characters arises from postulating that this type of aedeagus indicates common ancestry in every case. The impossibility of drawing any clear distinction in respect of the structure of the male postabdomen and genitalia between the Chloropidae family-group and the families included in Hennig's Pallopteroidea and Otitoidea is further emphasized by the structure of the Chiropteromyzidae and Cnemospathidae. These show the same characteristic type of aedeagus, but would be excluded from HENNIG's Pallopteroidea and Otitoidea, as well as from the Chloropidae family-group, on account of the less apomorphous condition of the female postabdomen. In my opinion a wider group is needed to embrace these families, as well as the Chloropidae group and HENNIG's Pallopteroidea (except Lonchaeidae) and Otitoidea. The name Tephritoinea is here proposed in a wide sense for a group of this kind.

The Tephritoinea are characterized as a monophyletic group by the following groundplan conditions which are apomorphous with respect to the groundplan of the Muscoidea.

220

(1) Aedeagus (♂) extremely elongate, flexible, coiled when at rest, extended by pressure of body fluid, bearing numerous fine cuticular processes ('pubescence') (figs. 123, 130, 131, 134, 138 and 145).

In other Muscoidea the aedeagus is usually relatively rigid, and is extruded primarily by contraction of muscles originating on the aedeagal apodeme, which causes the aedeagus to swing about its attachment to the apodeme. This swinging mechanism is clearly retained in most Tephritoinea (except Odiniidae and Conopidae), since the base of the aedeagus articulates with the aedeagal apodeme in the normal manner and a distinct phallophore is usually present. But some form of pumping action is clearly also required to produce stiffening and straightening of the extremely long aedeagus, which is flexible and coiled when at rest. HANNA (1938) suggested that erection of the aedeagus in *Ceratitis* (Tephritidae) is achieved by pressure of the seminal fluid in the ejaculatory duct produced by contraction of the muscles of the ejaculatory bulb. But it is not clear from his account to what extent his conclusion was based on observation, and MUNRO (1947: 73) has questioned the accuracy of HANNA's work. I doubt whether the volume of seminal fluid would be sufficient to stiffen and straighten the aedeagus; more probably this result is achieved by pumping of body fluid. In many Tephritoinea a pair of sclerotized rods may be seen in the wall of the aedeagus. I suggest that these represent the paraphalli of other Muscoidea, and that the coiled aedeagus of Tephritoinea has been derived from a less modified muscoid type of aedeagus by progressive increase in the length and elasticity of this organ. When at rest the aedeagus is either coiled within a genital pouch on the right side of the postabdomen, or is looped around the dorsum of the postabdomen from the right side. The latter condition is shown by the Chiropteromyzidae, Cnemospathidae, and some of the groups included in HENNIG's Otitoidea and Pallopteroidea. The aedeagus is usually clothed with fine 'pubescence', a characteristic condition which should probably be ascribed to the groundplan of the Tephritoinea on account of its wide distribution. In a few groups the cuticular processes are strongly sclerotized, and are more appropriately called 'spinules' rather than 'pubescence'.

(2) 7th and 8th abdominal segments (♀) elongate, forming slender retractile ovipositor: 8th sternum divided into paired longitudinal sclerites.

It has been the frequent practice in classifications of the Schizophora to define a group partly coinciding with the Tephritoinea in my present sense on the basis of the structure of the ovipositor. For instance HENDEL (1936) characterizes his 'Trypetides' with the words 'Basalstück des Ovipositors zu einem chitinisierten Tubus verwachsen'. HENNIG's Otitoidea were similarly characterized ('Beim Weibchen 7. Abdominalsegment zu einer Legrohrscheide umgebildet'). However such characterization applies to only part of the Tephritoinea in the present sense. The condition which I ascribe to the groundplan of the Tephritoinea is in my opinion an earlier stage in the character sequence leading to the type of ovipositor described in the above quotations.

Some degree of retractility of each segment into the preceding segment is a general feature of the abdomen in Diptera. The formation of a 'retractile ovipositor' involves lengthening of the genital segment (segment 8) and strengthening of the retraction mechanism, so that the genital segment can be fully retracted despite its increased length. The groundplan condition for the Tephritoinea appears to be one in which both the 7th and 8th segments are elongate, with the distal part of the 7th segment being largely membranous and becoming infolded as the 8th segment is retracted; the sclerites of the 7th segment remain more or less unmodified as dorsal and ventral plates on the anterior part of the segment. Such a condition is shown, for instance, by the Chiropteromyzidae and Cnemospathidae. Two different trends of further modification from the groundplan can be seen. In some groups (notably the Odiniidae, Carnidae and Chloropidae) the 7th sternum or both the sclerites of the 7th segment are somewhat reduced, so that the preceding segments become the effective sheath for the ovipositor. In other groups the sclerites of 7th segment are enlarged and sometimes fuse to form a conspicuous ovipositor sheath (or 'basal cone' of the ovipositor), which is the condition used by HENNIG (1958) to characterize his group Otitoidea.

Secondary shortening of the ovipositor occurs at least in *Meromyza* (Chloropidae) and in *Pelomyia* (Tethinidae). I did not undertake a survey of the structure of the ovipositor, and undoubtedly much useful information for purposes of classification could be obtained from further studies.

(3) *'Musculus hypandriotergalis'* lost (?).

In my discussion above of the characterization of the Muscoidea I have put forward the hypothesis that this unpaired muscle (linking the hypandrium with the inverted 8th sternum on the left side) was present in the groundplan of the Muscoidea. This muscle has not been reported in any member of the Tephritoinea. It may be definitely recorded as absent in the tephritid *Urophora* (= *Euribia*) (HENNIG 1936a), in *Thecophora* (Conopidae) and in *Chlorops* (Chloropidae); the latter two records are based on serial sections in my possession. Of course my ascription of loss of this muscle to the groundplan of the Tephritoinea is only tentative in view of the limited information.

(4) Costa broken at end of subcosta (or near end of r_1 if the subcosta has been reduced).

Different conditions of the costa occur only among the Acartophthalmidae, Conopidae and Tephritidae *s.l.* The almost universal presence of a single costal break at the end of the subcosta in all other families indicates that this is probably the groundplan condition for the Tephritoinea.

I agree with HENNIG (1958) that the presence of a spur ('Zipfel') on the anal cell in some groups of Tephritidae *s.l.* should be regarded as apomorphous. A recurved anal cross-vein is shown by most other Tephritoinea which retain a long anal vein, including Chiropteromyzidae, Cnemospathidae, *Hemeromyia* (Carnidae), Acartophthalmidae and Eurygnathomyiidae.

The conditions of the male postabdominal sclerites shown by the various families of Tephritoinea can be arranged in a character sequence leading from the groundplan condition of the Muscoidea (with the 6th tergum and vestiges of the 7th and 8th terga retained) to the much simpler types of structure shown, for instance, by the Tethinidae, Milichiidae, Conopidae and Acartophthalmidae.

I provisionally omit the vibrissa character from my characterization of the families of Tephritoinea. Some of the included families lack distinct vibrissae (such as Conopidae, Eurygnathomyiidae and most Tephritidae *s.l.*), others show well differentiated vibrissae (such as Chiropteromyzidae, Cnemospathidae, Odiniidae, Piophilidae and Acartophthalmidae), and there are others (such as the Milichiidae, Carnidae and Tethinidae) in which the differentiation of vibrissae from the other peristomal bristles is variable. It seems to me that the phylogenetic interpretation of this character requires clarification.

The Tephritoinea contain two subordinate groups which are dominant in the Recent fauna, each containing thousands of species. There are the Tephritidae family-group (equivalent to the Pallopteroidea and Otitoidea of HENNIG 1958) and the Chloropidae family-group. However there are in addition several smaller families of Tephritoinea not referable to these groups, and it does not seem possible to make a dichotomy of the Tephritoinea on present information.

222

C. 44. Family Chiropteromyzidae

Sources of information

This treatment is based on study of the male syntype of *Chiropteromyza wegelii* FREY.

Description of male postabdomen and genitalia (figs. 122-123)

6th abdominal spiracles in membrane: 7th left spiracle within 7th sternum, displaced dorsally in relation to preceding spiracles: 7th right spiracle in membrane, displaced centrally in relation to preceding spiracles.

Postabdomen asymmetrical. 6th tergum narrow, less than half as long as 5th tergum. 6th sternum asymmetrical, extending dorsally on left side where it is adjacent to the 7th sternum. 7th sternum asymmetrically developed on left side, where it is fused dorsally with the large 8th sternum. 7th tergum represented by narrow lateroventral sclerite on right side. 8th tergum absent.

Periandrium bearing discrete furcate telomeres. Cerci discrete. Aedeagal apodeme long and rod-like, linked to hypandrium posteriorly by rather indistinct ventral processes. Aedeagus extremely elongate, flexible, finely pubescent, coiled when at rest in fold below posterior margin of 6th tergum; base of aedeagus able to be swung through wide arc against aedeagal apodeme. Condition of postgonites not established. Slender ejaculatory apodeme present.

Characterization and discussion

This family contains the single known species *Chiropteromyza wegelii* FREY, bred from a bat roost in Finland. The syntype figured is the only known male, and the figures of the male postabdomen and genitalia given by HENNIG (in press) thus refer to the same specimen. HENNIG's figures portray the abdomen in the condition in which I received it (mounted in balsam). Owing to telescoping of the terminal segments he did not observe some details of the structure, and makes the incorrect statement that the 'inner copulatory apparatus is strongly reduced'. I removed the abdomen from its balsam mount into my temporary mounting medium (see section 1), and was able to restore its elasticity sufficiently to extend the terminal segments. As a result I am able to offer a more detailed description and figures.

The aedeagus of the male syntype of *Chiropteromyza wegelii* is elongate, with conspicuous pubescence, looped around the dorsum of the postabdomen from the right side in the fold below the posterior margin of the 6th tergum: the structure of its postabdominal sclerites is not much removed from the postulated groundplan condition for the Muscoidea, with both a discrete 6th tergum in normal dorsal position and the 7th tergum vestige retained. The female of this species has a retractile ovipositor, agreeing with the condition ascribed above to the groundplan of the Tephritoinea. In my opinion these conditions of the male and female postabdomen indicate that *Chiropteromyza* does not belong to the Heleomyzidae, where HENNIG (1958 and in press) tentatively refers it.

The following conditions shown by *Chiropteromyza* I consider to be apomorphous with respect to the groundplan of the Tephritoinea.

(1) 8th sternum ($\male$) with four outstanding marginal bristles (of which the outer two are longer) (fig. 123).

(2) Telomeres ($\male$) strongly furcate (figs. 122 and 123).

(3) Aedeagal apodeme ($\male$) largely free and rod-like, but linked to hypandrium posteriorly by rather indistinct ventral processes.

A completely free aedeagal apodeme (without any sclerotized connection with the hypandrium) should probably be attributed to the groundplan of the Tephritoinea. In *Chiropteromyza* the apodeme is well developed and little removed from this postulated groundplan condition, but rather ill-defined sclerotized processes linking it to the hypandrium are visible posteriorly (near the base of the aedeagus).

(4) 6th tergum ($\male$) less than half as long as 5th tergum.

The 6th tergum is less reduced in *Chiropteromyza* than in most Tephritoinea, remaining as a discrete sclerite in normal dorsal position. Nevertheless, if the large 6th tergum shown by Mormotomyiidae is indicative of the groundplan condition for the Tephritoinea, then the condition shown by the Chiropteromyzidae may be apomorphous to some degree.

(5) 8th tergum vestige ($\male$) lost.

(6) 7th left spiracle ($\male$) lying within 7th sternum (fig. 122).

(7) Two spermathecae ($\female$).

This characterization is based on my examination of a female syntype of *Chiropteromyza wegelii* FREY.

(8) Postverticals convergent.

While many Tephritoinea have long divergent postverticals, interpreted by HENNIG (1958) as a plesiomorphous condition, the Chiropteromyzidae, Cnemospathidae and many members of the Milichiidae family-group have convergent postverticals.

(9) Subcosta becoming fused with vein r_1 distally (HENNIG, in press).

(10) Anal vein shortened, not reaching wing margin.

The anal vein (cu_{1b} + ia) is long in *Chiropteromyza*, although not reaching the wing margin (see HENNIG, in press).

C. 45. Family Mormotomyiidae

Sources of information

The morphology of the single known species, *Mormotomyia hirsuta* AUSTEN, has been treated by AUSTEN (1936), VAN EMDEN (1950) and HENNIG (in press). I also studied preparations of this species.

224

Description of male postabdomen and genitalia (figs. 124-126)

6th abdominal spiracles in membrane near ventral corner of 6th tergum: 7th spiracles absent.

Postabdomen slightly asymmetrical. 6th tergum large, not much shorter than 5th tergum. 6th sternum asymmetrically developed (expanded on left side), fused on both sides with large pregenital sclerite (synsternum 7 + 8): no suture between the 7th and 8th sternum is visible on the latter. 7th tergum not discrete (absent or involved in formation of pregenital sclerite). 8th tergum absent.

Periandrium without telomeres: a small interparameral sclerite lies below the proctiger. Cerci discrete. Aedeagal apodeme long and rod-like, free from hypandrium. Hypandrium forming pubescent trough-like structure, above which lies a complex group of sclerites ('x' on fig. 126) with which the postgonites are articulated. Aedeagus elongate, flexible, pubescent, divided distally, with large basal phallophore which bears a small epiphallus; phallophore able to be swung through wide arc against aedeagal apodeme. Postgonites large, each with single seta. Fan-shaped ejaculatory apodeme present.

Characterization and discussion

This family contains the single known species *Mormotomyia hirsuta* AUSTEN, found in a bat cave in Kenya. VAN EMDEN (1950) discussed the relationships of *Mormotomyia* and concluded that 'the genus represents a well-founded family intermediate between the Cordyluridae and the Acalyptrata'. VAN EMDEN's work has given rise to the opinion that *Mormotomyia* belongs to the Calyptratae, although he did not in fact state this as a definite conclusion. The basis of VAN EMDEN's statement that 'the key characters of the adult are those of Calyptrata' was the presence in *Mormotomyia* of a cleft on the second antennal article. HENNIG (in press) also concludes that the form of the antennae remains the only useful indication that *Mormotomyia* belongs to the Calyptratae. However it should be appreciated that a conspicuous cleft on the second antennal article is present in several disjunct groups of Schizophora, apart from the Calyptratae; such a condition is shown in the groundplan of the Lonchaeoidea and Drosophiloidea, as well as in some genera of Tephritidae *s.l.*, Megamerinidae and Psilidae. Clearly no species can be referred to the Calyptratae on the basis of this condition alone. The Calyptratae show many additional apomorphous conditions, especially of the mouthparts and male genitalia, as listed in my treatment of the group. VAN EMDEN found that *Mormotomyia* does not show the apomorphous conditions of the mouthparts characteristic of the Calyptratae, and my present analysis of the male postabdomen and genitalia also provides no basis for classifying *Mormotomyia* in the Calyptratae.

The male genitalia of *Mormotomyia* (figs. 124-126) show several autapomorphous conditions which do not help in clarifying the relationship of the genus. But the flexible pubescent condition of the aedeagus seems to me to be evidence for classifying *Mormotomyia* in the Tephritoinea. The condition of the female postabdomen, redescribed by HENNIG (in press), is also well compatible with this interpretation. What HENNIG has described is a retractile ovipositor with the 8th sternum divided into a pair of sclerites, as I postulate for the groundplan of the Tephritoinea. This condition is less modified than that shown by the Calyptratae (with the possible exception of the Hippoboscidae family-group), and indicates that *Mormotomyia* cannot be an aberrant member of any particular family of Calyptratae. If *Mormotomyia* belongs to the Calyptratae, then it can only be classified as the sister-group of all other Calyptratae, as HENNIG (in press) concludes. However this conclusion depends on the premise that *Mormotomyia* is a calyptrate, which is not supported by any evidence which I think convincing. The interpretation that *Mormotomyia* is an aberrant member of the Tephritoinea seems to me to rest on firmer foundations.

The Tephritoinea include numerous species whose larvae develop in faeces, and it is interesting that both *Chiropteromyza* and *Mormotomyia* are found in bat roosts. However *Chiropteromyza* does not show any cavernicolous modifications and there do not seem to be any firm morphological grounds for supposing that *Mormotomyia* is more closely related to *Chiropteromyza* than to other groups of Tephritoinea. Consequently the Mormotomyiidae are retained as a monobasic family in my present treatment.

The Mormotomyiidae show the following apomorphous conditions with respect to the groundplan of the Tephritoinea.

(1) Entire body, including legs and wing vestiges, clothed with ochraceous pubescence, which is conspicuously longer and denser in the male; macrochaetae not differentiated from this pubescence, except for the peristomal bristles.
(2) Eyes very small; ocelli lost.
(3) Ocellar plate enlarged (especially in male), occupying most of frons.
(4) 3rd antennal article small, partly concealed by enlarged 2nd article which bears a longitudinal suture externally.
(5) Wings vestigial, represented by linear strips; halteres reduced.
(6) 7th abdominal spiracles lost in both sexes.

Parallel loss of the 7th spiracles in both sexes is also characteristic of the South American Cnemospathidae, but I am doubtful whether this resemblance is due to synapomorphy in the absence of confirmation from other characters.

(7) Two very small spermethecae ($♀$) (HENNIG, in press).
(8) 7th tergum ($♀$) divided along centre-line of dorsum; 7th sternum ill-defined, weakly sclerotized (HENNIG, in press).
(9) 7th sternum ($♂$) fully fused with 8th sternum, forming pregenital sclerite (synsternum 7 + 8) with which the 6th sternum is linked on both sides (fig. 124): 7th tergum not discrete.

The 6th tergum in *Mormotomyia* is large and unmodified, with the spiracles of the 6th segment lying in membrane near its edges. However the 6th sternum is fused with the pregenital sclerite on both sides. The area of fusion is broader on the left side, a condition readily interpretable as derived from the normal asymmetrical morphology of the 6th sternum in Musco-

226

idea The origin of the area of fusion on the right side is not so clear: either the 7th tergum vestige is fused with adjacent areas of the 6th sternum and 8th sternum, or, if the 7th tergum has been lost, the 6th sternum is extended and fused directly with the 8th sternum. There is no delimited area on the left side identifiable as the 7th sternum, which is presumably fully fused with the 8th sternum. The 6th tergum overlaps the anterior edge of the pregenital sclerite in the preparations studied by me, the membrane which separates them being infolded. VAN EMDEN (1950, fig. 4) indicated the 6th tergum in a position well anterior to the pregenital sclerite and 6th sternum; I think this potentially misleading, as I do not see how the sclerites can be brought into such a position without tearing the membrane between the 6th tergum and sternum.

(10) 8th tergum vestige (♂) lost.
(11) Telomeres (♂) lost.
(12) Hypandrium (♂) modified to form pubescent trough-like structure; above this lies a complex group of sclerites ('x' on fig. 126) with which the postgonites are articulated.

VAN EMDEN (1950) referred to the trough-like structure as the fulcrum. He homologized the anteriorly directed processes of the sclerite above this with the pregonites ('anterior parameris'), but this seems to me doubtful as pregonites are not usually thus oriented. The ventral sclerotization of the genital segment in *Mormotomyia* is evidently highly modified, and some of the structures may be neomorphous. No similar modifications have been reported in other families.

Mormotomyia is the only group here referred to the Tephritoinea in which the 6th tergum (♂) is virtually as long as the 5th tergum. On this account I postulate in my analysis that the 6th tergum was unreduced in the groundplan of the Tephritoinea.

C. 46. Family Cnemospathidae

Sources of information

F. W. EDWARDS (in MALLOCH 1933) gave descriptions and figures of the male postabdomen and genitalia of six species of *Prosopantrum*. I studied a preparation of *Prosopantrum inerme* MALLOCH.

Description of male postabdomen and genitalia (fig. 127)

5th abdominal spiracles in membrane: 7th spiracles absent.

Postabdomen asymmetrical. 6th tergum represented by narrow band of dorsal sclerotization. 6th sternum asymmetrical, linked to 7th sternum on left side. 7th sternum asymmetrically developed on left side where it is linked dorsally to the large 8th sternum. 7th and 8th terga absent.

Periandrium with pair of lateral notches on its anterior margin ('x' on fig. 127), with telomeres represented by finger-like lobes (not articulated): inner sides of these lobes confluent with pair of interparameral sclerites which are linked anteriorly to the hypandrium. Cerci large, discrete. Aedeagal apodeme short, linked to hypandrium posteriorly by conspicuous ventral process. Hypandrium bearing two pairs of posteriorly directed processes. Aedeagus elongate, flexible, coiled when at rest, with numerous spiniform cuticular processes: at the base of the aedeagus lies a cylindrical phallophore which can be swung through a wide arc against the aedeagal apodeme. Postgonites long. Ejaculatory apodeme absent.

Characterization and discussion

The name Cnemospathidae is available for this South American group following ENDERLEIN's (1938) proposal. The type-genus *Cnemospathis* was synonymized with *Prosopantrum* by HENNIG (1948c). This group can scarcely belong to the Heleomyzidae where it was classified by MALLOCH (1933), since its members show the coiled pubescent type of aedeagus and retractile ovipositor characteristic of the Tephritoinea. At present I refer to this family only the genus *Prosopantrum* (including synonyms), but it is possible that other South American genera which I have not seen also belong here.

The Cnemospathidae are characterized by the following apomorphous conditions with respect to the groundplan of the Tephritoinea.

(1) Periandrium (♂) with pair of lateral notches on its anterior margin ('x' on fig. 127); telomeres not discrete, represented by finger-like lobes of periandrium.

The figures given by F. W. EDWARDS (in MALLOCH 1933) indicate that this characterization applies to all species of *Prosopantrum*.

(2) Aedeagal apodeme (♂) short, linked to hypandrium posteriorly by conspicuous ventral process (fig. 127).

(3) 6th tergum (♂) much shortened, less than one quarter of length of 5th tergum.

In Cnemospathidae the 6th tergum remains as a discrete sclerite in normal dorsal position, but is much shorter than the preceding terga.

(4) 7th and 8th tergum vestiges (♂) lost.

(5) Ejaculatory apodeme (♂) lost.

(6) Spermathecae (♀) extremely long and slender (about equal in length to sclerites of 7th segment).

This condition, shown in a preparation of *Prosopantrum inerme* MALLOCH studied by me, is unique to the best of my knowledge. The spermathecae are three in number, as in the groundplan of the Tephritoinea.

(7) 7th abdominal spiracles lost in both sexes.

This characterization also applies to the Mormotomyiidae (see above).

228

(8) Postverticals convergent.
(9) Subcosta weak, becoming fused with vein r_1 distally.
(10) Anal vein shortened, not reaching wing margin.

MALLOCH (1933: 182) states that the anal vein ends 'abruptly about two-thirds of the distance to the wing margin'.

C. 47. Family Odiniidae

Sources of information

Information on the male postabdomen and genitalia of the Odiniidae was given by HENNIG (1938a, 1938e). I studied preparations of the following: – *Neoalticomerus formosus* (LOEW); *Odinia boletina* (ZETTERSTEDT); *Odinia* sp.

Description of male postabdomen and genitalia (fig. 128)

6th and 7th abdominal spiracles all in membrane, the left near margin of 7th sternum.

Postabdomen asymmetrical. 6th tergum short, about one-third of length of 5th tergum in *Neoalticomerus;* represented by narrow strip of sclerotization or absent in *Odinia.* 6th sternum asymmetrical, linked to 7th sternum on left side. 7th sternum asymmetrically developed on left side where it is fused dorsally with the 8th sternum. 8th sternum relatively large and more or less symmetrical in *Neoalticomerus,* but becoming narrow on right side in *Odinia.* 7th and 8th terga absent.

In *Neoalticomerus* the periandrium bears discrete telomeres which are connected with the vertical sections of the hypandrial arms by a broad interparameral sclerite, and a pair of slender posteriorly directed processes arises from this interparameral sclerite (lying between the telomeres): in *Odinia* the telomeres are scarcely differentiated from the periandrium and the interparameral sclerite and its processes are absent. Cerci small and discrete *(Odinia),* scarcely differentiated in *Neoalticomerus.* Aedeagal apodeme long and rod-like, free from hypandrium. Hypandrium long and narrow anteriorly. Aedeagus consisting of strongly sclerotized phallophore which is directed in the same direction as the aedeagal apodeme and cannot be swung through a wide arc against this, and flexible pubescent distal area which is folded anteriorly against the hypandrium when at rest. Postgonites very small or absent. Ejaculatory apodeme absent.

Characterization and discussion

The Odiniidae have often been considered the sister-group of the Agromyzidae. HENNIG (1965 and in press) has accepted this view, although admitting that no decisive grounds for it can be put forward. I have concluded that this interpretation of the relationships of the Odiniidae is probably mistaken, because the condition of the aedeagus shown by this family seems to be a modification of the condition ascribed to the groundplan of the Tephritoinea. The sister-group of the Agromyzidae is in my opinion more probably the Clusiidae (see under Agromyzoinea).

The Odiniidae are characterized by the following groundplan conditions which are apomorphous with respect to the groundplan of the Tephritoinea.

(1)   Aedeagus (♂) somewhat shortened: phallophore not able to be swung against aedeagal apodeme (fig. 128).

> The aedeagus of the Odiniidae is flexible and finely pubescent beyond the basal phallophore, as in most Tephritoinea. But its condition is clearly apomorphous in respect of loss of the articulation between the phallophore and the aedeagal apodeme. At rest only the flexible distal area of the aedeagus is directed anteriorly, while the phallophore remains directed in the same direction as the apodeme (posteriorly to posteroventrally). The swinging mechanism has been similarly lost in the Conopidae.

(2)   Hypandrium (♂) long and narrow anteriorly (fig. 128).
(3)   6th tergum (♂) shortened, only about one-third of length of 5th tergum.
(4)   7th and 8th tergum vestiges lost (♂).
(5)   Postgonites (♂) reduced.
(6)   Ejaculatory apodeme (♂) lost.
(7)   Sclerites of 7th abdominal segment (♀) reduced.

> HENNIG's (1958) and STEYSKAL's (1963) figures of the female postabdomen of *Neoalticomerus* and *Traginops* respectively indicate thath the 7th segment is largely membranous with only small sclerites. Both the 7th and 8th segments are retractile within the preceding segments. This condition may be considered a further development of the 'retractile ovipositor' condition which I ascribe to the groundplan of the Tephritoinea.

(8)   Two spermathecae (♀) (STURTEVANT 1926, HENNIG 1958, STEYSKAL 1963).
(9)   Of the three pairs of orbital bristles *(ors)* the most anterior pair are directed inwards (HENNIG 1958).
(10) Subcosta closely approximated to vein r_1.
(11) Anal vein shortened, not reaching wing margin (HENNIG 1958).

FREY (1921) classified the Odiniidae as a subfamily of Carnidae on the basis of similarities in the mouthparts. But I do not find any clear demonstration of synapomorphy in the characterization which he gives. The retention of sclerotized spermathecae by the Odiniidae excludes them from the Chloropidae family-group, in which I include the Carnidae. If my delimitation of the latter group is correct, then the Odiniidae cannot be the sister-group of the Carnidae, although it is possible that they are the sister-group of the Chloropidae family-group as a whole.

C. 48. Family Tethinidae

Sources of information

Information on the male postabdomen and genitalia of the Tethinidae has been given by HENNIG (1939a and in press). I studied preparations of *Tethina parvula* (LOEW), *Pelomyiella melanderi* (STURTEVANT) and *Macrocanace littorea* (HUTTON).

Description of male postabdomen and genitalia (figs. 129-130)

6th and 7th abdominal spiracles symmetrically situated on either side, 6th pair in membrane, 7th pair within sclerotized area before hypopygium (fused 6th tergum + 8th sternum).

Postabdomen fully symmetrical. Only a single dorsal sclerotized area (fused 6th tergum + 8th sternum) lies between the 5th segment and the hypopygium, with the boundary between the fused sclerites indicated by a conspicuous suture. 6th and 7th sterna absent, the last ventral sclerite before the hypopygium being the 5th sternum. 7th and 8th terga absent.

Periandrium bearing well developed telomeres, which are discrete in *Apetaenus*, *Macrocanace*, *Pelomyia* and *Pelomyiella* (shielded in lateral view by lobes of periandrium in the latter two genera), but partly fused with the periandrium in *Tethina* and *Rhicnoessa:* inner basal corners of telomeres linked to broad interparameral sclerite. Cerci discrete. Aedeagal apodeme long and rodlike: hypandrium much shorter than aedeagal apodeme, with its anterior apex linked to posterior part of apodeme. Aedeagus with basal phallophore (bearing epiphallus in *Pelomyiella*, but not as far as known in other genera) and slender, flexible, pubescent distal area: phallophore able to be swung through wide arc against aedeagal apodeme. Postgonites represented by pubescent lobes or apparently absent. Large ejaculatory apodeme present.

Characterization and discussion

I do not share the doubts on the monophyly of the Tethinidae raised by MALLOCH (1934). It seems to me that the Tethinidae are well characterized as a monophyletic group by the following groundplan conditions which are apomorphous with respect to the groundplan of the Tephritoinea.

(1) Postabdomen (♂) fully symmetrical (fig. 129); only a single dorsal sclerotized area (fused 6th tergum + 8th sternum) lies between the 5th segment and the hypopygium; 6th and 7th sterna lost; 7th and 8th tergum vestiges lost.

The structure of the male postabdomen in Tethinidae has been radically simplified, with complete loss of the asymmetry characteristic of most Muscoidea. The posterior area of the symmetrical protandrial sclerite clearly represents the inverted 8th sternum, but there is also

231

an anterior band delimited by a conspicuous suture. The presence of the 6th spiracles near the lateral corners of this anterior band indicates that it probably represents the 6th tergum. All seven pairs of abdominal spiracles are present in the two species studied by me, and I therefore reject HENNIG's (1958) suggestion that loss of the 7th spiracles may be a groundplan condition of the Tethinidae.

(2) Hypandrium (δ) short, with its anterior apex linked to posterior part of aedeagal apodeme.

It is characteristic of the Tethinidae that the aedeagal apodeme is long and rod-like (a plesiomorphous condition), in contrast with the short hypandrium. I have not found a similar condition in the other families of Tephritoinea.

(3) Two spermathecae ($\female$) (STURTEVANT 1926; HENNIG 1958 and in press).
(4) Labella elongate (HENNIG 1958, after FREY 1921).
(5) Postverticals convergent (HENNIG 1958).
(6) Subcosta fading distally, closely approximated to vein r_1.

I propose to widen the limits of the Tethinidae to include the genera *Listriomastax*, *Apetaenus* and *Macrocanace*, distributed in certain subantarctic islands (Kerguelen, Possession, Marion, Macquarie, Antipodes, Bounty and Campbell islands). The figures of the male postabdomen and genitalia of *Apetaenus* given by HENNIG (in press) agree well with the preparation of *Macrocanace* which I studied. This subantarctic group shows all the apomorphous conditions of the male postabdomen which I ascribe to the groundplan of the Tethinidae. In fact the only modification to the complete list of apomorphous conditions of the Tethinidae necessitated by the inclusion of this group is that reduction of the anal vein can no longer be ascribed to the groundplan of the family, since this vein is retained in *Listriomastax* and *Macrocanace*. The genus *Macrocanace* was previously misplaced in the Canacidae.

HENNIG's (in press) figures of the female postabdomen of *Apetaenus* portray a retractile ovipositor with divided 8th sternum, as I ascribe to the groundplan of the Tephritoinea. In *Pelomyia* the 8th segment is rather short. Probably the latter condition is the result of secondary shortening, as has also occurred in *Meromyza* (Chloropidae).

C. 49. Chloropidae family-group

I include in this group the four families Acartophthalmidae, Carnidae, Milichiidae and Chloropidae. The close relationship of the last two families was emphasized by STURTEVANT (1926: 15), who stated that 'the rudimentary seminal receptacles with long fine ducts, and the pocket-like ventral receptacle indicate that these two groups are close to each other...' Both HENNIG (in press) and I accept STURTEVANT's conclusion. Sclerotized spermathecae are absent also in the Acartophthalmidae (HENNIG 1958: 623) and Carnidae, according to my examination of a preparation of *Meoneura obscurella* (FALLÉN). The delimitation of the Chloropidae family-group here proposed is based partly on the

232

premise that synapomorphous modifications of the female reproductive system are shown by all four families. Clearly it is desirable that the morphological basis for this premise should be checked by study of fresh specimens of Acartophthalmidae and Carnidae (which were unfortunately not available to STURTEVANT), since the nature of the spermathecal ducts cannot be determined precisely from mounts of dried material. However an additional probably synapomorphous condition of all four families is provided by the structure of the male genitalia (character 2 below); I therefore think it reasonable to postulate the limits of this family-group, at least provisionally.

The following apomorphous conditions with respect to the groundplan of the Tephritoinea are put forward in support of the view that the Chloropidae family-group in the sense here proposed is monophyletic.

(1) Two rudimentary spermathecae (♀); spermathecal ducts long and fine (see STURTEVANT 1926).

This characterization requires to be checked for the Acartophthalmidae and Carnidae, as stated above.

(2) Aedeagal apodeme (♂) free anteriorly, but becoming fused with body wall posteriorly so that a sclerotized area is formed anterior to the aedeagus base between the hypandrial arms (fig. 136).

The sclerotized area anterior to the aedeagus base is discrete from the hypandrium in *Hemeromyia* (Carnidae) and Chloropidae. Probably this is the groundplan condition. In *Meoneura* (Carnidae) and Acartophthalmidae the sclerotized area becomes confluent laterally with the hypandrium. I regard the extensive fusion of the aedeagal apodeme and hypandrium shown by the Milichiidae as the most apomorphous condition in this character sequence.

(3) 6th tergum (♂) reduced.

Only in *Hemeromyia* (Carnidae) is the 6th tergum retained. In that genus it is reduced to a narrow band developed asymmetrically mainly on the right side.

(4) 8th tergum vestige (♂) lost.
(5) Anal vein shortened, not reaching wing margin.

The most complete wing venation among the Chloropidae family-group is shown by the Acartophthalmidae and *Hemeromyia* (Carnidae). In these groups the anal vein is well developed, but cut off before reaching the wing margin.

D. 7. Family Acartophthalmidae

Sources of information

The male postabdomen and genitalia of the Recent genus *Acartophthalmus* were treated by HENNIG (1938b). I studied preparations of *Acartophthalmus nigrinus* (ZETTERSTEDT).

Description of male postabdomen and genitalia (figs. 131-132)

6th abdominal spiracles in membrane: 7th spiracles absent.

Postabdomen almost symmetrical. 6th tergum absent. 6th sternum represented by narrow ventral band which is joined to the dorsal pregenital sclerite on both sides. 7th and 8th sterna fully fused (without any visible suture between them), forming almost symmetrical pregenital sclerite. 7th and 8th terga absent.

Hypopygium bearing only single pair of discrete articulated lobes (telomeres or enlarged cerci) whose inner basal corners bear apodemes linked anteriorly to the hypandrial arms. Aedeagal apodeme free anteriorly, but becoming fused with body wall posteriorly, thus forming sclerotized area ('x' on fig. 131) which is confluent laterally with the hypandrium. Aedeagus with small basal phallophore and slender, flexible, pubescent distal area: phallophore able to be swung through wide arc against aedeagal apodeme. Postgonites present. Ejaculatory apodeme minute.

Characterization and discussion

Acartophthalmus is more plesiomorphous than all other members of the Chloropidae family-group in respect of its retention of a well developed subcosta (separated from vein r_1 distally), and possibly also in respect of its frontal chaetotaxy (including the presence of long divergent postverticals). Therefore the Acartophthalmidae should be recognized as a monobasic family, unless family rank is accorded to the Chloropidae family-group as a whole. HENNIG's (1965b) remark that 'die Acartophthalmidae haben den deutlichen Charakter einer Reliktgruppe' seems appropriate.

The Acartophthalmidae are characterized by the following apomorphous conditions with respect to the postulated groundplan of the Chloropidae family-group.
(1) Costal break situated shortly beyond humeral cross-vein, not at end of subcosta (HENNIG 1958).

This characterization applies to the Recent species, but there is no costal break in the Baltic amber fossil *Acartophthalmites* (HENNIG 1965b).

(2) 7th abdominal spiracles lost in both sexes (HENNIG 1958).
(3) 6th tergum (♂) completely lost.
(4) 7th and 8th sterna (♂) fully fused, forming almost symmetrical pregenital sclerite with which the 6th sternum is joined on both sides (fig. 132); 7th tergum vestige lost.

In *Acartophthalmus* symmetry of the male postabdomen has been almost restored through loss and fusion of sclerites, the left side of the pregenital sclerite being only slightly more expanded anteriorly than the right. These conditions are strongly apomorphous. In the groundplan of the Chloropidae family-group the asymmetrical arrangement of postabdominal sclerites characteristic of most Muscoidea was retained, as shown by the Carnidae (especially *Hemeromyia*) and some Chloropidae (such as *Lasiopleura*).

234

(5)   Hypopygium (♂) bearing only single pair of articulated lobes (telomeres or enlarged cerci?).

(6)   Ejaculatory apodeme (♂) minute.

Among the members of the Chloropidae family-group the ejaculatory apodeme is well developed only in the Chloropidae.

(7)   Cerci (♀) long and slender (HENNIG 1965b).

Acartophthalmus shows a flexible pubescent aedeagus typical of the Tephritoinea (as also in Carnidae).

D. 8.   Family Carnidae

Sources of information

Information on the male postabdomen and genitalia of the Carnidae was given by HENNIG (1937b). I studied preparations of the following: – *Meoneura pteropleuralis* SABROSKY; *Meoneura flavifacies* COLLIN; *Hemeromyia* sp. (California).

Description of male postabdomen and genitalia (fig. 134)

6th abdominal spiracles in membrane: in *Meoneura* the 7th left spiracle lies within the 7th sternum, displaced dorsally in relation to the preceding spiracles, and the 7th right spiracle is strongly displaced beyond the centre-line onto the left side of the insect (lying in membrane near the 7th tergum vestige): in *Hemeromyia* both 7th spiracles are absent.

Postabdomen asymmetrical. 6th tergum represented by asymmetrical narrow band of sclerotization developed mainly on right side *(Hemeromyia)*; absent in *Meoneura*. 6th sternum asymmetrical, linked to or fused with 7th sternum on left side. 7th sternum asymmetrically developed on left side, where it is fused dorsally with the large 8th sternum. 7th tergum represented by small sclerite on centre-line of venter (near apex of hypandrium). 8th tergum absent.

Periandrium bearing discrete telomeres which are sometimes double or furcate: in *Meoneura pteropleuralis* SABROSKY the inner pair of the double telomeres is fused to form a conspicuous process below the proctiger: a broad interparameral sclerite or pair of sclerites connects the inner basal corners of the telomeres with the hypandrial arms. Cerci small, poorly differentiated from membrane, discrete or more or less fused. Aedeagal apodeme free anteriorly, but becoming fused with body wall posteriorly; the sclerotized area thus formed is discrete in *Hemeromyia*, but confluent laterally with the hypandrium in *Meoneura*. Aedeagus with small basal phallophore and slender, flexible, pubescent distal area (coiled when at rest): phallophore able to be swung through wide arc against aedeagal apodeme. Small postgonites present. Ejaculatory apodeme minute.

Characterization and discussion

FREY (1921) separated the Carnidae from the Milichiidae on the basis of differences in the mouthparts. HENNIG (1958) formerly rejected this separation, but now accepts (HENNIG 1965b) that the Carnidae should be recognized as a separate family, subject to the exclusion of the genera *Horaismoptera* and *Risa* which he transfers to the Milichiidae. My studies of the male postabdomen confirm that there is a marked distinction between the Carnidae and the Milichiidae. Since there are certain characters which suggest that the Milichiidae may be more closely related to the Chloropidae than to the Carnidae (see below), the validity of including the Carnidae and Milichiidae in a group (Milichiidae *sensu lato*) which does not also include the Chloropidae is in doubt. I therefore treat the Carnidae as a separate family in this work.

The Carnidae show the following groundplan conditions which are apomorphous with respect to the groundplan of the Chloropidae family-group.

(1) Mentum enlarged, enclosing sides of haustellum (FREY 1921).

FREY (1921) studied the mouthparts of *Meoneura* and *Carnus*; detailed information on *Hemeromyia* is not available. The above condition seems to be the only substantive apomorphous condition demonstrated by FREY's studies. In other respects (such as the development of the maxillae and labella) the Carnidae show more plesiomorphous conditions than the Milichiidae and Chloropidae. When FREY speaks of resemblance between the Carnidae and the Odiniidae and Piophilidae, I think he is referring to symplesiomorphous conditions.

(2) Upper fronto-orbital bristles *(ors)* directed outwards over eye-margin; lower fronto-orbitals *(ori)* directed inwards (HENNIG 1958).
(3) Second costal break present (shortly beyond humeral cross-vein) (HENNIG 1958).
(4) Subcosta weak, becoming fused with vein r_1 distally.
(5) 7th tergum vestige ($\male$) lying on centre-line of venter (near apex of hypandrium); 7th right abdominal spiracle displaced beyond centre-line onto left side of insect; 7th left spiracle lying within 7th sternum.

Both *Hemeromyia* and *Meoneura* show the asymmetrical development of the 6th and 7th sterna characteristic of the Muscoidea, and are the only members of the Chloropidae family-group which retain the 7th tergum vestige. However the position of this tergum vestige is unusual in that it has been displaced to the centre-line of the venter. In *Meoneura* the 7th right spiracle retains its normal position relative to the 7th tergum vestige, thus lying on the left side of the insect, and the true 7th left spiracle lies within the 7th sternum. In *Hemeromyia* both the 7th spiracles have been lost. It is not clear whether the unusual position of the 7th tergum vestige shown by the Carnidae can be ascribed to the groundplan of the Chloropidae family-group. Consequently I list this condition as an apomorphous condition of the Carnidae only.

(6) Ejaculatory apodeme ($\male$) minute.
(7) Cerci ($\female$) fused to form stylus-like structure (see HENNIG 1937b and 1965b).
(8) Sclerites of 7th abdominal segment ($\female$) small (see HENNIG 1965b).

236

The above two characters have not been confirmed for *Hemeromyia*.

The Carnidae, like the Acartophthalmidae, retain a flexible pubescent aedeagus typical of the Tephritoinea. This is particularly long and coiled when at rest in *Hemeromyia*. As far as I am aware, the latter is the only genus of the Chloropidae family-group in which the 6th tergum (♂) is retained.

D. 9. Family Milichiidae

Sources of information

The male postabdomen and genitalia of the Milichiidae were treated by HENNIG (1937b, 1939a). I studied preparations of the following: – *Pholeomyia indecora* (LOEW); *Leptometopa latipes* (MEIGEN); *Milichia speciosa* MEIGEN; *Phyllomyza hirtipalpis* MALLOCH; *Desmometopa m-nigrum* (ZETTERSTEDT).

Description of male postabdomen and genitalia (figs. 133 and 135)

6th and 7th abdominal spiracles symmetrically situated, usually within pregenital sclerite but in membrane in *Desmometopa* (in which the pregenital sclerite is reduced).

Postabdomen fully symmetrical. The 5th tergum and 5th sternum are the last large sclerites before the hypopygium. In most genera an additional narrow dorsal sclerite (pregenital sclerite) which is slightly expanded laterally lies between the 5th tergum and the hypopygium ('x' on fig. 133); but in *Desmometopa* this sclerite is represented only by a pair of lateral strips of sclerotization.

Periandrium bearing telomeres which are either discrete *(Leptometopa, Phyllomyza, Pholeomyia)* or partly fused with the periandrium *(Milichia, Desmometopa)*: inner basal corners of telomeres connected with hypandrial arms by broad interparameral sclerite or pair of sclerites. Cerci variably developed, convergent ventrally. Aedeagal apodeme and hypandrium extensively fused. Aedeagal sclerotization usually consisting mainly of basal phallophore which can be swung through a wide arc against the aedeagal apodeme, with distal area of aedeagus more or less membranous and showing at most weak traces of pubescence *(Milichia, Leptometopa, Desmometopa)*: in *Pholeomyia* and *Phyllomyza* the aedeagus is vestigial, scarcely differentiated from the aedeagal apodeme. Postgonites absent. Ejaculatory apodeme very small.

Characterization and discussion

The Milichiidae are characterized by the following groundplan conditions which are apomorphous with respect to the groundplan of the Chloropidae family-group.

237

(1) Labella elongate; maxillary lacinia rudimentary (see FREY 1921).

According to FREY reduction of the lacinia ('Galea') is characteristic of the Chloropidae and Milichiidae, but the Carnidae retain a normally developed lacinia.

(2) Upper fronto-orbital bristles *(ors)* directed outwards over eye-margin; lower fronto-orbitals *(ori)* directed inwards (HENNIG 1958).
(3) Postverticals convergent (HENNIG 1958).
(4) Second costal break present (shortly beyond humeral cross-vein) (HENNIG 1958).
(5) Subcosta weak, fading or becoming fused with vein r_1 distally.
(6) Anal vein lost (HENNIG 1958).
(7) 6th tergum (♂) completely lost.
(8) Postabdomen symmetrical (♂); only single narrow dorsal sclerite between 5th tergum and hypopygium ('x' on fig. 133); the 6th and 7th abdominal spiracles lie symmetrically on either side within the lateral areas of this sclerite; 7th tergum vestige lost.

This condition represents a fundamental simplification of the asymmetrical pattern of postabdominal sclerotization characteristic of the Muscoidea, as retained by the Carnidae and in the groundplan of the Chloropidae. The homology of the narrow pregenital sclerite is not clear. Probably the 8th sternum is the major element in its formation. No differentiated areas representing the 6th and 7th sterna can be discerned, although it is possible that these sclerites are involved in the formation of its lateroventral areas, where the 6th and 7th spiracles are located. However the position of the postabdominal spiracles does not always indicate the homology of areas of sclerotization in which they lie, as emphasized elsewhere in this work.

HENNIG (1937b, fig. 2) has indicated a ventral area of sclerotization as 'Sternit 6 + x' in *Milichia speciosa* MEIGEN. But from comparison of this species with the other Milichiidae before me I am of the opinion that the ill-defined sclerotization of this ventral area in that species is a secondary partial sclerotization of membrane, not a true sternum.

(9) Hypandrium (♂) and aedeagal apodeme extensively fused (fig. 135).

I have interpreted the groundplan condition for the Milichiidae family-group as one in which the aedeagal apodeme has become fused with the body wall posteriorly so that a sclerotized area is formed between the hypandrial arms (see above). In Milichiidae this sclerotized area is fully confluent with the hypandrium, so that the aedeagal apodeme and hypandrium are extensively fused.

(10) Aedeagus (♂) reduced (fig. 135).

The Milichiidae show a strong tendency to reduction of the aedeagus, which in some groups (such as *Pholeomyia*) has become rudimentary. The least modified condition among the species before me appears to be that shown by *Milichia speciosa* MEIGEN, in which weak pubescence of the aedeagus is retained. Even in this species the aedeagus is much shorter than in the Carnidae and Acartophthalmidae.

(11) Postgonites (♂) absent.
(12) Ejaculatory apodeme (♂) very small.

HENNIG (1958) gives as an additional apomorphous condition of the Milichiidae 'Interfrontalbörstchen auf Interfrontalleisten vorhanden'. I am not con-

238

vinced that this characterization should be applied to the groundplan of the Milichiidae, as distinct interfrontal ridges ('Interfrontalleisten') are not shown by all genera.

In my opinion the sister-group of the Milichiidae is more probably the Chloropidae than the Carnidae. Possibly synapomorphous conditions shown by the Milichiidae and Chloropidae but not by the Carnidae (at least in their groundplan) are: (1) reduction of the aedeagus ($\male$), (2) reduction of the maxillary lacinia, (3) loss of the 6th tergum ($\male$), (4) loss of the 7th tergum vestige ($\male$), and (5) postverticals convergent. As far as I am aware the only possibly synapomorphous conditions shown by the Carnidae and Milichiidae but not by the Chloropidae are: (1) the presence of inwardly directed *ori*, (2) the presence of two costal breaks, and (3) ejaculatory apodeme ($\male$) very small. However further studies are needed to clarify this question.

SPEIGHT (1969) has argued that *Australimyza, Neophyllomyza, Phyllomyza* and *Paramyia* are probably wrongly classified in the Milichiidae because they show a different type of prosternal basisternum. As far as *Australimyza* is concerned, my studies confirm that this is not a milichiid. I have proposed above a new family and prefamily based on this genus. However the morphology of the male postabdomen of *Phyllomyza* is similar to that of other Milichiidae. Presumably the same applies to *Neophyllomyza* and *Paramyia* (neither seen by me), since they are generally considered closely related to *Phyllomyza*. Since the postabdominal morphology of male Milichiidae is strongly apomorphous, I must reject SPEIGHT's suggestion that *Phyllomyza* and its allies may be more closely related to the Anthomyzidae than to the Milichiidae. Doubtless the difference of prosternal type to which SPEIGHT has drawn attention will prove useful for classification within the Milichiidae, but I do not see that it provides sufficient grounds for doubting the monophyly of *Phyllomyza* and its allies with other Milichiidae.

D. 10. Family Chloropidae (including Mindidae)

Sources of information

I did not find any detailed treatment of the male postabdomen and genitalia of the Chloropidae in the literature. I studied preparations of the following: – *Rhodesiella subditica* LAMB; *Lipara lucens* MEIGEN; *Chlorops sulphureus* LOEW; *Meromyza americana* FITCH; *Siphonellopsis lacteibasis* STROBL; *Lasiopleura shewelli* SABROSKY.

Description of male postabdomen and genitalia (figs. 136-137)

6th and 7th abdominal spiracles usually more or less symmetrically situated (except in *Lasiopleura*), variable in position in relation to sclerites: in *Lasiopleura* all these postabdominal spiracles lie in membrane with those on the left side lying above the 7th sternum: in *Siphonellopsis* both pairs of postabdominal spiracles lie within the pregenital sclerite (synsternum 7 + 8): in the other genera studied the left spiracles lie in membrane or semisclerotized area below the pregenital sclerite (like those on the right side).

Postabdomen obviously asymmetrical in *Lasiopleura*, but only weakly so or virtually symmetrical in other genera studied. 6th tergum absent. 5th sternum small in *Siphonellopsis* and absent in *Lasiopleura*; but well developed in other genera studied. 6th sternum, when present, asymmetrical, linked to 7th sternum on left side: among the genera studied the 6th sternum is present as a conspicuous sclerite in *Lasiopleura*, and it is also weakly developed in *Siphonellopsis*: in the other genera studied no 6th sternum is present. 7th sternum asymmetrically developed on left side where it is linked to the large 8th sternum *(Lasiopleura)*; in the other genera studied the 7th sternum is not delimited from the 8th sternum. 7th and 8th terga absent.

Periandrium bearing discrete telomeres, whose inner basal corners are connected to the hypandrial arms by an interparameral sclerite or pair of sclerites. Cerci small, usually fully fused (but discrete apically in *Lasiopleura*). Aedeagal apodeme free anteriorly, but becoming fused with body wall posteriorly, thus forming discrete sclerotized plate: between this plate and the hypandrial arms lies a pair of obliquely directed sclerites ('x' on fig. 136). Aedeagus short, its sclerotization consisting mainly of basal phallophore which can be swung through a wide arc against the aedeagal apodeme to an anteriorly directed rest position, usually with membranous distal area beyond phallophore. Small postgonites present in *Lasiopleura*, but not in other genera studied. Small ejaculatory apodeme present.

Characterization and discussion

The Chloropidae show the following groundplan conditions which are apomorphous with respect to the groundplan of the Chloropidae family-group.
(1)   Maxillary lacinia ('Galea') rudimentary (FREY 1921).
(2)   Postverticals convergent (see HENNIG 1958).
(3)   Ocellar plate large (HENNIG 1958).
(4)   Propleuron margined at anterior edge of its lateral exposure (MALLOCH 1934: 396).

The full characterization given by MALLOCH was: 'Prosternal plate with a sharp anterior margin, in front of which the surface is abruptly precipitous, the propleura similarly margined at anterior edge of its lateral exposure'. SPEIGHT (1969) accepts MALLOCH's characterization of the propleuron as indicating an apomorphous condition, but comments that 'MALLOCH's description of the chloropid prosternum includes nothing that does not also apply to precoxal bridges in other Acalypterates, and the sclerite itself does not seem in any way peculiar'. The first part of MALLOCH's characterization is accordingly omitted on SPEIGHT's authority.

(5)   Subcosta developed as distinct vein only at its base, from there to costal break visible at most as fold (HENNIG 1958).

The course of the subcosta diverges from r_1 apically in those chloropids in which the complete course is visible (as a fold). Thus the costal break in the groundplan of the family is distinctly separated from the apex of r_1. The course of the subcosta in Chloropidae thus appears more plesiomorphous than the conditions shown by the Milichiidae and Carnidae, but more apomorphous than the condition shown by *Acartophthalmus*.

(6) Anal cell not evident, with veins forming its posterior and distal boundaries (sections of 1a and cu_{1b}) absent or largely reduced to folds (at most with basal stub of 1a sclerotized) (see HENNIG 1958); anal vein lost.

(7) Base of m_4 ($=$ 'tb', lower cross-vein) lost (HENNIG 1958).

(8) 6th tergum ($\male$) completely lost.

(9) 7th tergum vestige ($\male$) lost.

(10) Aedeagus ($\male$) much reduced (fig. 136).

In Chloropidae the aedeagus is very short and no longer has the 'pubescent' appearance characteristic of most Tephritoinea. The aedeagi of *Lasiopleura*, *Siphonellopsis* and *Lipara* are of rather similar appearance, consisting of a cylindrical sclerotized basal area (phallophore) and a short membranous distal area. Probably this is close to the groundplan condition for the family.

(11) 7th tergum ($\female$) divided along centre-line of dorsum; 7th sternum small.

Some degree of reduction of the sclerites of the 7th segment is apparent in all the preparations of the female postabdomen of Chloropidae before me (*Lasiopleura*, *Siphonellopsis*, *Meromyza* and *Hippelates*). However my characterization is provisional only, as the structure of the female postabdomen in Chloropidae has not been studied from a comparative standpoint.

It is probable that additional autapomorphous conditions of the Chloropidae will be revealed by comparative study of the chaetotaxy of the head and thorax, which shows a tendency towards reduction. I have omitted such characterization for the present, as I am not able to judge the groundplan condition for the family from the available treatments in the literature.

PARAMONOW's (1956) new genus *Minda*, on which he based his proposed family 'Mindidae', has been reduced to synonymy with the chloropid genus *Pemphigonotus* by HENNIG (1958) on the authority of C. W. SABROSKY. HENNIG considers the absence of a costal break in this genus to be secondary (apomorphous).

C. 50. Family Conopidae (including Stylogastridae)

Sources of information

Information on the male postabdomen and genitalia of the Conopidae was given by STREIFF (1906) and STEYSKAL (1957a). I studied preparations of the following: – *Physocephala furcillata* (WILLISTON); *Zodion fulvifrons* SAY; *Thecophora modesta* (WILLISTON); *Dalmannia nigriceps* LOEW; *Stylogaster neglecta* WILLISTON.

Description of male postabdomen and genitalia (figs. 138-139)

6th and 7th abdominal spiracles symmetrically situated, usually within pregenital sclerite but in membrane in *Stylogaster*.

Postabdomen more or less symmetrical. 6th tergum absent. 6th sternum usually forming complete ventral band joined to dorsal pregenital sclerite on both sides (except *Stylogaster*). 7th and 8th sterna fully fused, usually forming more or less symmetrical pregenital sclerite, without any visible suture between them: in *Myopa* and *Thecophora* a strip of secondary sclerotization is visible on the venter of the 7th segment. 7th and 8th terga absent. In *Stylogaster* there are no clearly defined sclerites between the 5th segment and the hypopygium.

Periandrium with or without discrete telomeres (unusually large telomeres are shown by *Stylogaster*). Interparameral sclerites weakly developed or absent. Cerci discrete or fused (usually small). Aedeagal apodeme free from hypandrium, usually long and rod-like (but dorsoventrally flattened in *Stylogaster*). Hypandrium usually large, extending posteriorly to form sheath around posterior part of aedeagal apodeme (except *Stylogaster*, in which the hypandrium is modified to form a pair of large spinous lobes): in *Zodion* the anterior part of the hypandrium is represented by a pair of slender rods. Aedeagus variably developed; in *Dalmannia*, *Parazodion* and *Paramyopa* long, flexible and pubesent (coiled when at rest), supported by long paraphalli; in *Stylogaster* long and smooth, with conspicuous glans which bears paired terminal filaments; very short and pubescent, more or less membranous, in other genera studied by me: base of aedeagus not able to be swung through wide arc against aedeagal apodeme. Postgonites absent. Ejaculatory apodeme variably developed.

Characterization and discussion

My reference of this family to the Tephritoinea was foreshadowed by the comments of HENNIG (1952a), although he did not follow up this question in later works. Formerly the Conopidae were sometimes thought closely related to the Syrphidae, and were still classified as Syrphoidea in STONE *et al.* (1965). However such a classification is untenable, since it has been known at least since the 1880's that the Conopidae possess a ptilinum. Recently the most widely held opinion has been that the Conopidae are the sister-group ('Archischiza') of all other Schizophora ('Muscaria'). This classification has generally been credited to ENDERLEIN (1936), although in fact the same group concepts were proposed under different names by BRAUER (1890). ENDERLEIN's names have been tentatively accepted in recent works of HENNIG (1958, 1966b). The grounds for ENDERLEIN's view were not clearly stated. He merely remarked that the Conopidae 'Besitz der Stirnspalte mit Charakteren der Aschiza verbinden', without specifying which characters of the 'Aschiza' he meant. The resemblances in wing venation between the Syrphidae and some Conopidae (notably the elongate anal cell) are not shown by all members of the Conopidae and are probably secondary. The reduction or absence of frontal bristles in most Recent Conopidae is also clearly secondary, since both *Myopa* and the Baltic amber fossil *Palaeomyopa* retain an arrangement of frontal bristles typical of the Schizophora. HENNIG (1958) suggested that the short length of the ptilinal suture might be a plesiomorphous condition of the Conopidae, an interpretation originally suggested by BRAUER (1890). But this interpretation is also doubtful, since HENNIG has subsequently reported (HENNIG 1966b) that in some genera of Conopidae, such as *Dalmannia*, the suture is no shorter than in most other Schizophora. As for the structure of the ptilinum itself, STRICKLAND (1953) reported that the ptilina of Conopidae 'are voluminous and represent the most rugged type of ptilinum which we have encountered'. Thus no relatively plesiomorphous conditions have been found in the Conopidae which support the classification of this family as the sister-group of all other Schizophora. This opinion must surely be abandoned.

In my opinion the Conopidae are probably referable to the Tephritoinea, since some genera show the type of aedeagus which I ascribe to the groundplan of this prefamily. HENNIG (1966b) was led by his acceptance of ENDERLEIN's classification to postulate that the long pubescent aedeagus of the conopid genera *Dalmannia*, *Parazodion* and *Paramyopa* was the result of convergence with the condition shown by Tephritoinea. The need to postulate convergence is removed, if it is accepted that ENDERLEIN's division of the Schizophora into 'Archischiza' and 'Muscaria' was unwarranted. The aedeagus of *Dalmannia* (fig. 138) is long and flexible, with conspicuous pubescence, coiled when at rest; the paraphalli are retained as a pair of lateral strips, and apically the aedeagus is somewhat expanded to form a weak 'glans'. This structure is typical of the Tephritoinea, and I doubt whether such a complex condition was evolved independently in the Conopidae. The short, largely membranous aedeagus of the Conopinae may be due to secondary reduction. Similar considerations apply to the structure of the female ovipositor. In *Dalmannia* the 7th and 8th abdominal segments form an elongate ovipositor (STREIFF 1906: 199), as also in *Stylogaster*. But in Conopinae the 8th segment is short. I suggest that a secondary shortening of the ovipositor has occurred in Conopinae, correlated with reduction of the aedeagus.

The interpretation of some character sequences within the Conopidae is problematical. Certain conditions of *Dalmannia* which HENNIG (1966b) suggests may be apomorphous, namely (1) the short anal cell, (2) the long ptilinal suture, (3) the absence of 'sc$_2$', and (4) the absence of the 'unpaired organ' on the 5th sternum ($\female$), may rather be groundplan conditions of the Conopidae since they correspond with the conditions found in other families of Tephritoinea. HENNIG's interpretation of these characters was of course influenced by his provisional acceptance of the view that the Conopidae are the sister-group of all other Schizophora. Consequently I do not think it would be warranted to ascribe to the groundplan of the Conopidae the following conditions, although they are shown by most Recent Conopidae: (1) an elongate anal cell, (2) a short ptilinal suture, (3) 'sc$_2$' present, and (4) 5th sternum ($\female$) with 'unpaired organ'. The conditions in respect of which the Conopidae may with confidence be characterized as apomorphous in their groundplan with respect to the groundplan of the Tephritoinea are the following.

(1) 7th and 8th sterna (♂) fully fused, forming more or less symmetrical pregenital sclerite, with which the 6th sternum is joined on both sides (fig. 139); both the 6th and 7th abdominal spiracles lie symmetrically on either side within this pregenital sclerite.

In all Conopidae the 6th sternum is narrow, joined to the dorsal pregenital sclerite on both sides. *Dalmannia* does not differ in this respect from other Conopidae, for the large sclerite which STEYSKAL (1957a) labelled as the 'sixth sternite' is the 5th sternum. STEYSKAL referred to the presence of the '7th sternite' as a ventral strip in *Myopa*, and I have noted the same condition in a preparation of *Thecophora*. In view of its limited distribution I doubt whether this condition belongs to the groundplan of the Conopidae. More probably this strip is a secondary sclerotization of the venter of the 7th segment. The lateral area of the 7th sternum on the left side present in the groundplan of the Muscoidea (and Tephritoinea) seems fully fused with the inverted 8th sternum in all Conopidae.

(2) 6th tergum (♂) lost.

STEYSKAL (1957a) suggested that the 6th tergum is involved in the formation of the pregenital sclerite in Conopidae, and interpreted an 'impressed line' found on the posterior part of this sclerite in some genera as representing the demarcation between the 6th tergum and 8th sternum (which he called '8th tergite'). I doubt whether this interpretation is correct. The 'impressed line' is very feeble, and there is no break in the sclerotization along it. The 6th tergum is reduced or absent in many other Tephritoinea, and I think it probable that it has also been lost in the Conopidae.

(3) 7th and 8th tergum vestiges (♂) lost.
(4) Hypandrium (♂) extending posteriorly to form sheath around posterior part of aedeagal apodeme; base of the aedeagus not able to be swung through wide arc against aedeagal apodeme.

The aedeagus of *Dalmannia* (fig. 138) seems little removed from the condition postulated for the groundplan of the Tephritoinea, except that the swinging mechanism has been lost, so that little movement of the base of the aedeagus against the aedeagal apodeme is possible. Among other Tephritoinea loss of the swinging mechanism has occurred also in the Odiniidae.

(5) Postgonites (♂) absent.
(6) Ovipositor (♀) (segments 7 and following) directed obliquely in relation to longitudinal axis of abdomen (STREIFF 1906, HENNIG 1966b); 7th segment conical, largely covered by large 7th tergum.

This characterization may not wholly apply to the groundplan of the Conopidae, if *Stylogaster* (whose slender ovipositor is not directed obliquely) is the sister-group of all other Conopidae (see HENNIG 1966b: 11). However this question requires clarification. The sclerotization of the 7th segment is not complete in Conopidae, since the small 7th sternum is narrowly separated by membrane from the tergum (STREIFF 1906).

(7) Two spermathecae (♀).

Although four spermathecae (on two ducts) are reported for several genera of Conopidae (see STURTEVANT 1925, HENNIG 1966b), only two spermathecae are present in *Stylogaster* and *Dalmannia* (HENNIG 1966b, fig. 22). The latter condition occurs widely in other Tephritoinea, and probably represents the groundplan condition from which the four-spermathecae condition, known only in *Seioptera* (Tephritidae *s.l.*) among other families of Tephritoinea, has been derived.

244

(8) 2nd antennal article elongate.

(9) Haustellum elongate.

Both the above characters have been well discussed by HENNIG (1966b). The haustellum is not elongate in the Baltic amber fossil *Palaeomyopa*.

(10) Prosternal basisternum with its posterodorsal corners elongate, forming 'tails' (SPEIGHT 1969).

(11) Costa secondarily unbroken.

I think it probable that the costal break has been secondarily lost in Conopidae because there is no other family in the Tephritoinea which has a completely unbroken costa in its groundplan.

(12) Larvae developing as internal parasitoids of adult insects (see HENNIG 1952a).

I do not attempt to suggest a morphological characterization of the larvae of Conopidae, as the larvae of the critical genera *Dalmannia* and *Stylogaster* have not to my knowledge been described. There is little doubt that at least *Stylogaster* also has parasitoid larvae, since its eggs are known to be deposited on insects (see K. G. V. SMITH 1967).

ROHDENDORF (1964) proposed to consider *Stylogaster* as representing a separate family, Stylogastridae. I prefer not to follow this proposal, since it is not clear whether *Stylogaster* is the sister-group of other Recent Conopidae. As previously mentioned, some confusion has arisen in the interpretation of certain character sequences in Conopidae, due to acceptance of the unwarranted opinion that the Conopidae are the sister-group of all other Schizophora. The interrelationships between the genera of Conopidae therefore require to be re-analysed.

My reference of the Conopidae to the Tephritoinea does not imply that I think the Conopidae are monophyletic with the 'Pyrgotidae', as has sometimes been suggested. The structure of the male postabdomen and genitalia of the 'Pyrgotidae' clearly indicates that they belong to the Tephritidae *s.l.*, in accordance with the conclusions of HENNIG (1936c). I therefore consider that the Conopidae and 'Pyrgotidae' evolved parasitoid larval development independently.

C. 51. *Tephritidae family-group*

I include in this group the Eurygnathomyiidae, Richardiidae, Piophilidae *sensu lato* and the several so-called families here grouped together as the Tephritidae *sensu lato*. This group corresponds to the Otitoidea and Pallopteroidea of HENNIG's (1958) classification, except that I exclude the Lonchaeidae (see my previous discussion under Lonchaeoidea). The Tephritidae family-group is characterized by the following apomorphous conditions with respect to the groundplan of the Tephritoinea.

245

(1) 7th segment (♀) extremely elongate (at least twice as long as 6th segment), subconical in shape (broad anteriorly but tapering posteriorly); 8th segment extremely elongate (retractile within 7th segment); cerci more or less fused.

The characteristic condition of the female postabdomen provides the principal evidence for inferring that the Tephritidae family-group is monophyletic. I consider that this condition is a further development of the 'retractile ovipositor' condition which I ascribe to the ground-plan of the Tephritoinea. The 7th tergum and sternum were probably discrete in the ground-plan of the Tephritidae family-group, as in *Eurygnathomyia* and *Aenigmatomyia*.

(2) 8th sternum (♂) asymmetrically developed on left side (figs. 140, 142 and 144).

A result of this asymmetry is that the hypopygium is turned towards the right side of the insect, at least to a slight degree. In most species this turning is rather well marked.

In many respects the Tephritidae family-group (or at least some members of this group) show plesiomorphous conditions for the Tephritoinea. Three orbicular spermathecae are retained by *Eurygnathomyia* and some species of the Tephritidae *s.l.* The 7th tergum vestige (♂) is often well developed in Richardiidea, Piophilidae and Tephritidae *s.l.*, and in some Richardiidae the 8th tergum vestige (♂) is also retained. A long pubescent aedeagus of the type ascribed to the groundplan of the Tephritoinea may be ascribed to the groundplan of all the groups here treated as families. A plesiomorphous wing venation may also be ascribed to the groundplan of the Tephritidae family-group, with the anal vein and subcosta both reaching the wing margin as distinct veins. The postvertical bristles, when present, are with few exceptions divergent, which according to HENNIG (1958) is a plesiomorphous condition. There is much variation in the development of orbital bristles among members of the Tephritidae family-group, but probably short vertical plates bearing one, or at most two, orbital bristles should be ascribed to the groundplan of the group. Whether such a condition is apomorphous with respect to the groundplan of the Tephritoinea is not clear. The development of lower fronto-orbital bristles among some members of the Tephritidae *s.l.* has clearly proceeded independently of the similar development shown by the Chloropidae family-group and the Odiniidae.

The best developed 6th tergum among members of the Tephritidae family-group is shown by *Eurygnathomyia*, in which two fragments are present. But this condition probably did not belong to the groundplan of the Tephritidae family-group, since in *Heloparia* the 6th tergum extends across the dorsum as a narrow band. More probably the conditions shown by *Heloparia* and *Eurygnathomyia* are due to independent reduction of a relatively large 6th tergum in the groundplan of the family-group.

D. 11. Family Eurygnathomyiidae status novus

Sources of information

The male genitalia of *Eurygnathomyia bicolor* (ZETTERSTEDT) were described by MORGE (1967). I also studied a preparation of this species.

Description of male postabdomen and genitalia (figs. 140-141)

6th abdominal spiracles in membrane: 7th spiracles absent.

Postabdomen asymmetrical. 6th tergum represented by two fragments, one (well defined) situated dorsally on left side and one (rather ill-defined) situated lateroventrally on right side. 6th sternum large, asymmetrical, linked to 7th sternum on left side, with expanded area on right side of venter. 7th sternum asymmetrically developed on left side where it is fused dorsally with the 8th sternum. 8th sternum fairly large, asymmetrically developed on left side. 7th and 8th terga absent.

Periandrium bearing discrete telomeres whose inner basal corners are linked to a Y-shaped interparameral sclerite. Cerci fully fused. Aedeagal apodeme fused broadly with hypandrium and body wall from about its middle, so that its posterior part forms the roof of a trough which protects the base of the aedeagus when at rest. Aedeagus long, flexible, pubescent, coiled on right side of postabdominal venter when at rest, with row of conspicuous spinules distally; the small phallophore at the base of the aedeagus bears a small epiphallus, and can be swung through a wide arc against the aedeagal apodeme. Postgonites absent. Small ejaculatory apodeme present.

Characterization and discussion

This family is based on the single species *Eurygnathomyia bicolor* (ZETTER-STEDT). This has generally been treated as representing a monobasic subfamily of the 'Pallopteridae', but I do not see any evidence that it is more closely related to *Palloptera* than to other members of the Tephritidae family-group. I refer *Palloptera* to the Tephritidae *s.l.* (see below). The combination of apomorphous and plesiomorphous conditions shown by *Eurygnathomyia* is such that I cannot include it in any of the other groups here recognized as families within the Tephritidae family-group. Consequently I think that *Eurygnathomyia* must be recognized as constituting a separate family, if family rank is accorded below the level of the Tephritidae family-group.

Apomorphous conditions of *Eurygnathomyia* with respect to the groundplan of the Tephritidae family-group are as follows.

(1) 7th abdominal spiracles ($\male$) lost in both sexes.

In the Piophilidae and Tephritidae *s.l.* the 6th and 7th pairs of spiracles have been lost in the male, but the female retains the normal complement of seven pairs of abdominal spiracles.

Eurygnathomyia shows parallel loss of the 7th pair of spiracles in both sexes, while retaining the 6th pair. The absence of sex linkage in the loss of spiracles in *Eurygnathomyia* suggests that the genetic changes involved were different from those which led to the loss of spiracles in the male only of the Piophilidae and Tephritidae. I therefore regard the loss of the 7th spiracles in *Eurygnathomyia* as an autapomorphous condition.

(2) 6th tergum (♂) reduced to two fragments, one (well defined) situated dorsally on left side above 6th left spiracle, and one (rather ill-defined) situated lateroventrally on right side (fig. 140).

No other species of the Tephritidae family-group retains a fragment of the 6th tergum on the left side of the abdomen. The aedeagus passes below this left fragment of the 6th tergum when coiled in its rest position.

(3) 7th and 8th tergum vestiges lost (♂).
(4) Aedeagus (♂) with row of spinules distally (fig. 141).

These spinules have probably been formed by modification of the cuticular processes, which are developed as fine pubescence (as usually in Tephritoinea) along most of the length of the aedeagus.

(5) Aedeagal apodeme (♂) fused broadly with hypandrium and body wall from about its middle (fig. 141).

In *Eurygnathomyia* the aedeagal apodeme forms the roof of a trough which protects the base of the aedeagus when at rest. This condition appears to have arisen through fusion of the apodeme with the anterior end of the hypandrium and the membrane posterior to this along the centre-line. Similar conditions are shown by many members of the Piophilidae and Tephritidae *s.l.*, but in the Richardiidae the apodeme is free from the hypandrium.

(6) Postgonites (♂) lost.

D. 12. Family Richardiidae

Sources of information

Information on the male postabdomen and genitalia of the Richardiidae was given by STEYSKAL (1958c). I studied preparations of the following: – *Richardia* sp.; *Odontomera ferruginea* MACQUART; *Epiplatea erosa* LOEW.

Description of male postabdomen and genitalia (figs. 142-143)

6th left abdominal spiracle in membrane (*Richardia, Odontomera, Sepsisoma*) or within the 7th sternum (*Epiplatea*): 7th left spiracle within 7th sternum except possibly in *Odontomera* in which the limits of the 7th sternum are not clearly defined. 6th and 7th right spiracles in semisclerotized area on right side in *Epiplatea* (with 7th spiracle displaced centrally in relation to those preceding); but both these right spiracles are absent in *Richardia* and *Odontomera*.

248

Postabdomen asymmetrical. 6th tergum absent. 6th sternum large, asymmetrical, linked to or fused with 7th sternum on left side. 7th sternum asymmetrically developed on left side, where it is fused dorsally with the 8th sternum; usually well defined, but in *Odontomera* only weakly differentiated from membrane so that its limits are ill defined. 8th sternum fairly large, asymmetrically developed on left side. 7th tergum represented by small lateroventral sclerite on right side *(Richardia, Odontomera)* or absent *(Epiplatea)*. 8th tergum represented by small ventral sclerite on right side *(Richardia, Odontomera)* or absent *(Epiplatea)*.

Periandrium bearing discrete telomeres, which are usually double (except *Epiplatea*): in *Epiplatea* the inner sides of the telomeres are confluent with a pair of interparameral sclerites, but the latter are absent in *Richardia* and *Odontomera*. Cerci fused. Aedeagal apodeme rod-like, free from hypandrium in *Richardia* and *Odontomera*, but fused anteriorly with hypandrial arms in *Epiplatea:* the hypandrium of *Richardia* and *Odontomera* consists of a pair of free-ending laterally directed arms which do not meet below the aedeagal apodeme but are fused above it. Aedeagus long, flexible (coiled when at rest), at least partly pubescent, in some genera with conspicuous distal glans (e.g. *Richardia*): phallophore consisting of sclerotized ring which can be swung through a wide arc against the aedeagal apodeme: in some genera (e.g. *Richardia*) a characteristic process ('x' on fig. 143) lies above the base of the aedeagus, but this is absent in *Epiplatea* and *Odontomera*. Postgonites absent. Fan-shaped ejaculatory apodeme present.

Characterization and discussion

The primarily neotropical Richardiidae should not be considered a subfamily of 'Otitidae', as STEYSKAL (1958c) has rightly concluded, since they show plesiomorphous conditions of the male postabdomen which exclude them from the Tephritidae *s.l.* in the sense here proposed. These relatively plesiomorphous conditions are: (1) the retention in their groundplan of the full complement of seven pairs of abdominal spiracles, (2) the free aedeagal apodeme, without sclerotized connections with the hypandrium, and (3) the retention of the 8th tergum vestige. The latter sclerite has not been reported in any other family of Tephritoinea. The 7th tergum vestige (♂) is well developed in some Richardiidae, as well as in some species of Piophilidae and Tephritidae *s.l.* (but not in Eurygnathomyiidae).

Apomorphous conditions of the Richardiidae with respect to the groundplan of the Tephritidae family-group are as follows.
(1)   7th tergum and sternum (♀) fused, forming completely sclerotized oviscape (basal cone of ovipositor); 7th spiracles lying within 7th sternum.

The condition of the female postabdomen in Richardiidae appears identical with that shown by many members of the Tephritidae *s.l.* The suture between the 7th tergum and sternum is visible, and the 7th spiracles lie below this suture within the sternum.

(2) Two spermathecae ($\female$).

Two orbicular spermathecae are present in preparations of *Richardia* sp. and *Odontomera ferruginea* MACQUART studied by me.

(3) 7th left spiracle ($\male$) lying within 7th sternum.
(4) 6th tergum ($\male$) completely lost.
(5) Postgonites ($\male$) lost.
(6) Vein r_1 thickened where the subcosta ends, then forming short convex bow (HENDEL 1936).

Most Richardiidae are further characterized by the presence of double telomeres, but this condition may not have been present in the groundplan of the family since it is not shown by *Epiplatea*. The inner pair of telomeres can hardly represent the postgonites, as implied by STEYSKAL (1958c). They must either be neomorphous structures or have arisen through division of the telomeres.

D. 13. Family Piophilidae (sensu lato)
(including Thyreophoridae and Neottiophilidae)

Sources of information

HENNIG (1941a, 1943) gave information on the male postabdomen and genitalia of the Piophilidae. I studied preparations of the following: – *Neottiophilum praeustum* (MEIGEN); *Centrophlebomyia furcata* F.; *Actenoptera hilarella* (ZETTERSTEDT); *Piophila (Allopiophila) vulgaris* FALLÉN; *Piophila (Lasiopiophila) pilosa* STAEGER; *Piophila (Prochyliza) xanthostoma* (WALKER).

Description of male postabdomen and genitalia

6th and 7th abdominal spiracles absent.

Postabdomen asymmetrical. 6th tergum absent in most species, represented by small lateroventral sclerite on right side in *Actenoptera*. 6th sternum asymmetrical, expanded on left side where it is linked to or fused with the 7th sternum, in many species with expanded area also on right side. 7th sternum asymmetrically expanded on left side, where it is fused dorsally with the 8th sternum. 8th sternum asymmetrically developed on left side. 7th tergum absent or represented by small lateroventral sclerite on right side. 8th tergum absent.

Periandrium usually with discrete telomeres (except *Lasiopiophila*): interparameral sclerites variably developed, absent in some species. Cerci discrete or partly fused. Aedeagal apodeme and hypandrium extensively fused: in many species the hypandrial arms are fused to form a bridge above the base of the aedeagus. Aedeagus long, flexible (coiled when at rest), conspicuously pubescent in many groups but largely membranous, without evident pubescence, in

250

Allopiophila and *Actenoptera*, at most with weak differentiation of distal glans; phallophore consisting of sclerotized ring which can be swung through a wide arc against the aedeagal apodeme, in many species bearing an epiphallus. Postgonites well developed, contrastingly dark coloured in many species. Ejaculatory apodeme variably developed.

Characterization and discussion

The Piophilidae *(sensu lato)* are well characterized as a monophyletic group by the following groundplan conditions which are apomorphous with respect to the groundplan of the Tephritidae family-group.
(1) Face (prefrons) with membranous central area (HENNIG, in press).
(2) 6th and 7th abdominal spiracles (♂) lost on both sides.

This same condition is shown by members of the Tephritidae *s.l.* The females both of the Piophilidae and of the Tephritidae retain the normal complement of seven pairs of abdominal spiracles.

(3) 6th tergum (♂) much reduced, represented at most by small lateroventral vestige on right side.
(4) 8th tergum vestige (♂) lost.
(5) Aedeagal apodeme (♂) and hypandrium extensively fused.

The extensive fusion of the aedeagal apodeme and hypandrium shown by the Piophilidae is possibly a further modification of the type of condition shown by *Aenigmatomyia* (fig. 145), if the latter is closely related to the Piophilidae (see below). Among members of the Tephritidae family-group the aedeagal apodeme is free from the hypandrium and body wall only in some Richardiidae.

(6) Two elongate spermathecae (♀) (HENNIG 1958 and in press).
(7) 7th tergum and sternum (♀) attenuated posteriorly, where they become reduced to pairs of thin strips broadly separated by membrane (HENNIG 1958, figs. 185 and 186).

In my opinion this is probably an autapomorphous condition of the Piophilidae, not an earlier stage in the evolution of a completely sclerotized oviscape as implied by HENNIG's (1958) use of the phrase 'Vorstufe der Legrohrscheide'. It is interesting that in the South American *Aenigmatomyia* the 7th tergum and sternum are large, but separated by a membranous area in which the 7th spiracles lie. The condition shown by *Aenigmatomyia* seems a suitable antecedent, from which both the completely sclerotized oviscape shown by the Tephritidae *s.l.* and the flexible oviscape with reduced sclerotization shown by the Piophilidae could have been derived. The reduction of the sclerotization of the oviscape in Piophilidae was probably due to a change in the medium in which the eggs are deposited (oviposition in corpses).

It no longer seems necessary to argue the case for grouping the 'Thyreophoridae' and 'Neottiophilidae' with the Piophilidae in the narrow sense, since the monophyly of the Piophilidae *sensu lato* (Piophilariae in HENNIG's sense) now seems conclusively demonstrated by the considerations presented by HENNIG

251

(1958 and in press). In agreement with HENNIG I exclude the Australian genus *Tapeigaster* from this group (see under Rhinotoridae).

I examined preparations of the male postabdomen of three species belonging to southern hemisphere genera which have been classified in the 'Pallopteridae', namely *Heloparia bicolor* (WALKER), *Aenigmatomyia unipuncta* MALLOCH (figs. 144-145) and *Neomaorina bimacula* (MALLOCH). I find no grounds for supposing that these are monophyletic with the holarctic genus *Palloptera*, the type-genus of the Pallopteridae, which I refer to the Tephritidae *s.l.* The presence of only a single *ors* in all the genera which have been referred to the 'Pallopteridae' may possibly correspond to the groundplan condition for the Tephritidae family-group, since all Richardiidae and Eurygnathomyiidae show this condition, as well as some species of the other families. It is noteworthy that these southern hemisphere species all have well-developed dark postgonites, like those of the Piophilidae. The question of whether the possession of large dark postgonites is an apomorphous condition thus arises. The postgonites are small or absent in most other groups of Tephritoinea (except Mormotomyiidae and Cnemospath-idae), and to the best of my knowledge are completely absent in the Eurygna-thomyiidae, Richardiidae and Tephritidae *s.l.* (including *Palloptera*). Therefore we should bear in mind the possibility that the large postgonites of the Pio-philidae and 'southern Pallopteridae' are secondary structures (apomorphous), and that postgonites were absent in the groundplan of the Tephritidae family-group (which would require deletion of this condition from the characterization of particular families given in this work). This problem of interpretation is compounded by the apparently plesiomorphous condition of the female ovi-scape (7th segment) in *Aenigmatomyia*. Unfortunately I have no information on the form of the oviscape in other genera of 'southern Pallopteridae'. Another possible synapomorphy between the Piophilidae and 'southern Pallopteridae' is the reduction of the prosternal basisternum indicated by SPEIGHT's (1969) data. However the interpretation of this condition is problematical, as the Richardiidae also show a reduced prosternal basisternum.

In the present work I leave open the question of how the southern hemisphere genera of 'Pallopteridae' should be classified. There is clearly a need for more detailed morphological information on these genera before their relationships can be assessed. Perhaps they hold the key to an understanding of the relation-ships between the four groups here recognized as families. One noteworthy relatively plesiomorphous condition is shown by *Heloparia bicolor* (WALKER). This retains the 6th tergum as a narrow dorsal band. It is the only species of the Tephritidae family-group known to me in which the 6th tergum extends across the dorsum.

I do not think that HENNIG (1969b) was justified in postulating that the Baltic amber fossil *Proneottiophilum* is closely related to the Piophilidae. This insect does not show the modifications of the female postabdomen characteristic of the Piophilidae, nor even of the Tephritoinea as a whole, since according to HEN-NIG's description it has an unmodified 7th abdominal segment ($\female$). The condi-tions suggested as synapomorphous by HENNIG (presence of vibrissae; presence of a costal break; anal cross-vein recurved) are unconvincing, since they are of wide distribution among the Schizophora.

D. 14. Family Tephritidae (sensu lato) (= *Trypetidae*)
(including Platystomatidae, Pyrgotidae, Tachiniscidae, Otitidae, Ulidiidae, Pterocallidae and Palloptera)

I use the name Tephritidae rather than Trypetidae on grounds of priority, following the advice of C. W. SABROSKY.

Sources of information

Information on the male postabdomen and genitalia of the Tephritidae *s.l.* was given by HENNIG (1936a, 1936c, 1939b, 1940, 1941a), MUNRO (1947), RIVOSECCHI (1957), STEYSKAL (1961) and MORGE (1967). I studied preparations of the following: – *Homalocephala similis* (CRESSON); *Euxesta notata* (WIEDEMANN); *Physiphora demandata* (F.); *Chaetopsis massyla* (WALKER); *Seioptera vibrans* (L.); *Pterocalla strigula* LOEW; *Tetanops* sp.; *Otites stigma* (HENDEL); *Melieria ochricornis* (LOEW); *Platystoma seminationis* (F.); *Rivellia flavimana* LOEW; *Senopterina foxleei* SHEWELL; *Pyrgota undata* WIEDEMANN; *Orellia occidentalis* (SNOW); *Euaresta aequalis* (LOEW); *Paroxyna murina* (DOANE); *Epochra canadensis* (LOEW); *Palloptera trimaculata* (MEIGEN); *Palloptera superba* LOEW.

Description of male postabdomen and genitalia (fig. 146)

6th and 7th abdominal spiracles absent.

Postabdomen asymmetrical. 6th tergum absent or represented by small lateroventral sclerite on right side. 6th sternum asymmetrical, expanded on left side where it is linked to or fused with the 7th sternum. 7th sternum asymmetrically expanded on left side, where it is fused dorsally with the 8th sternum; in some groups the 7th sternum also extends around the right side of the venter as a narrow strip of sclerotization; usually the 7th sternum is clearly delimited from the 8th sternum, but these sclerites appear fully fused in *Platystoma*. 8th sternum asymmetrically developed on left side. 7th tergum absent or represented by small lateroventral sclerite on right side. 8th tergum absent.

Periandrium with discrete telomeres in *Physiphora*, but in most groups the latter are absent or fused with the periandrium: interparameral sclerites variably developed, absent in some species. Cerci discrete or fused. Aedeagal apodeme fused with body wall posteriorly, bearing a pair of lateroventral processes from its middle which in some groups are broad and confluent laterally with the hypandrium (e.g. *Palloptera, Tetanops*); in the groups customarily classified in the Platystomatidae, Pyrgotidae, Tachiniscidae and Tephritidae *(sensu stricto)* these processes consist of well-defined angulate rods. Aedeagus long, flexible (coiled when at rest), pubescent or smooth, with conspicuous spinules in some genera (e.g. *Melieria*), with variable differentiation of distal glans: phallophore consisting of sclerotized ring which can be swung through wide arc against aedeagal apodeme. Postgonites absent. Ejaculatory apodeme variably developed.

Characterization and discussion

The family name Tephritidae is here used in a wider sense than by previous authors, including several subordinate groups which have often been accorded family rank. The Tephritidae in this wide sense are characterized by the following groundplan conditions which are apomorphous with respect to the groundplan of the Tephritidae family-group.

(1)   6th and 7th abdominal spiracles (♂) lost on both sides.

This condition is possibly due to synapomorphy with the Piophilidae *s.l.* The females retain the normal complement of seven pairs of abdominal spiracles (as also in the Piophilidae).

(2)   6th tergum (♂) much reduced, represented at most by small lateroventral vestige on right side.
(3)   8th tergum vestige (♂) lost.
(4)   Aedeagal apodeme (♂) fused with body wall posteriorly, bearing pair of lateroventral processes from its middle.

The aedeagal apodeme with its processes forms the roof of a cavity within the hypandrium which protects the base of the aedeagus when at rest. Only in some genera of 'Otitidae' are sclerotized processes of the aedeagal apodeme not evident. This condition is most probably secondary (apomorphous), as it is associated with marked reduction in the size of the postabdomen. In the groups customarily classified as Tephritidae (*sensu stricto*), Tachiniscidae, Platystomatidae and Pyrgotidae the processes of the apodeme (called 'vanes' by MUNRO 1947) are very clearly defined, forming with the apodeme the structure commonly called the 'fultella' in descriptions (fig. 146). However this characteristic condition may not belong to the groundplan of the Tephritidae *s.l.* In *Palloptera* and *Tetanops* the processes of the apodeme are broader and confluent laterally with the hypandrial arms, similar to the condition of *Aenigmatomyia* (fig. 145). It is possible that the fultella condition is a further modification of such a condition, produced by the development of strongly sclerotized rods instead of a broader area of weak sclerotization.

(5)   Postgonites (♂) absent.

Whether this condition is apomorphous with respect to the groundplan of the Tephritidae family-group requires clarification, as discussed above under the Piophilidae.

(6)   7th tergum and sternum (♀) fused, forming completely sclerotized oviscape (basal cone of ovipositor); 7th spiracles lying within 7th sternum.

This condition is also shown by the Richardiidae. In many Tephritidae *s.l.* the suture between the 7th tergum and sternum is visible (as also in the Richardiidae).

254

The groups here referred to the Piophilidae *sensu lato* and Tephritidae *sensu lato* were classified by HENNIG (1958) in two superfamilies, the Pallopteroidea and Otitoidea, which he recognized as probably closely related. I am doubtful whether *Palloptera* was correctly placed in this dichotomy, although there are grounds for postulating that the southern hemisphere genera referred to the 'Pallopteridae' are more closely related to the Piophilidae than to the Tephritidae *s.l.* (see the discussion under Piophilidae). The presence of two elongate spermathecae in *Palloptera* perhaps supports HENNIG's classification, but this evidence is hardly conclusive since STURTEVANT (1925) reports that only two spermathecae (but pear-shaped) are present in many species of Tephritidae *sensu stricto*. The conditions of the male genitalia of *Palloptera*, especially the absence of telomeres and postgonites, and its possession of a completely sclerotized oviscape (♀) suggest to me that the genus more probably belongs to the Tephritidae *sensu lato* (part of HENNIG's Otitoidea). The traditional grouping of the 'Pallopteridae' and Lonchaeidae is surely untenable. The structure of the male postabdomen and genitalia of *Palloptera* leave no room for doubting that the genus belongs to the Tephritidae family-group. Therefore *Palloptera* cannot be more closely related to the Lonchaeidae than to other members of the Tephritidae family-group, irrespective of any doubts which can be raised about where the genus should be classified within this group.

Within the Tephritidae *s.l.* we can probably recognize (following STEYSKAL 1961) a major subordinate group containing the so-called families Platystomatidae, Pyrgotidae, Tachiniscidae and Tephritidae *(sensu stricto)*. These are characterized by reduction or loss of the aedeagal pubescence, and by the fultella condition of the aedeagal apodeme (see character 4 above). I have not seen any member of the 'Tachiniscidae', but include them in this group on the evidence of sketches of the male genitalia of species of *Anthophasia* and *Tachinisca* kindly sent me by STEYSKAL. These sketches indicate that the fultella condition is typically developed in both genera. Apart from this major subordinate group, the subdivision of the Tephritidae *sensu lato* is still in doubt (see the discussion by HENNIG 1958).

Most recent authors have classified the groups which I refer to the Tephritidae *s.l.* in a series of so-called families. My reasons for suspecting that such 'families' are not equivalent in age to most of the other families of Schizophora are twofold. First their relatively low position in the sequence of subordination here constructed (below the D level). Secondly the absence of specimens of these groups, so abundant in the Recent fauna, in Baltic amber collections (probably of Eocene age). The only Baltic amber fossil which belongs to the Tephritidae family-group is *Pallopterites*, assigned by HENNIG (1967) to the 'Pallopteridae'. However there are no clear grounds for referring this insect to any particular subordinate group of the Tephritidae *s.l.*, and HENNIG states that the species can have no descendants in the Recent fauna. Consequently the discovery of *Pallopterites* does not provide evidence for dating the radiation of the groups of Tephritidae *s.l.* in the Recent fauna. The suspicion is growing in my mind that this radiation did not occur before the late Tertiary, and I am therefore reluctant to accord family rank below the D level.

255

C. 52. Family Canacidae

The name of this family has usually been spelt 'Canaceidae', but this is grammatically incorrect. ENDERLEIN's (1936) correction to Canacidae should be upheld, in accordance with the advice on the formation of family names given in Appendix D of the International Code of Zoological Nomenclature (1961).

Sources of information

The male postabdomen and genitalia of the Canacidae have never been described in detail, although WIRTH (1951) gave some helpful figures. I studied preparations of the following: – *Canace snodgrassii* COQUILLETT; *Canaceoides* sp.; *Xanthocanace nigrifrons* MALLOCH.

Description of male postabdomen and genitalia (figs. 147-148)

6th abdominal spiracles in membrane or within dorsal pregenital sclerite: 7th spiracles within this sclerite.

Postabdomen fully symmetrical. Only one dorsal sclerite (6th tergum or syntergum 6 + 7) lies between the 5th tergum and the hypopygium. There are no clearly defined ventral sclerites between the 5th sternum and the hypopygium, but weak indications of sclerotization representing the 6th and possibly also the 7th sterna are sometimes visible.

Telomeres usually discrete from periandrium, but fused in *Xanthocanace*: interparameral sclerites absent in *Canace* and *Canaceoides*, but small discrete plates present in *Xanthocanace*. Cerci well separated. Aedeagal apodeme long and rod-like, free from hypandrium. Hypandrium short, not extending anteriorly beyond periandrium, its arms fused above aedeagus where they extend posteriorly to form a conspicuous spine between the anus and the aedeagus; in *Canaceoides* and *Xanthocanace* the hypandrial arms do not meet anteriorly below the aedeagal apodeme, but remain widely separated. Aedeagus short, with variable sclerotization, remaining posteriorly directed when at rest in *Canaceoides* and *Xanthocanace* (not able to be swung against aedeagal apodeme), but apparently able to be swung against apodeme to some extent in *Canace*. Postgonites present, joined above aedeagus by pubescent fold (aedeagal mantle). Ejaculatory apodeme present.

Characterization and discussion

The Canacidae are well characterized as a monophyletic group by the following conditions which are apomorphous with respect to the groundplan of the Schizophora.

(1) Only a single dorsal sclerite ('pregenital sclerite') between 5th tergum and hypopygium (fig. 147); no well-defined ventral sclerites between 5th sternum and hypopygium.

The homology of the dorsal pregenital sclerite of the Canacidae requires clarification. Probably the 6th tergum is the main element involved in its formation. HENNIG (1958) states that in *Xanthocanace ranula* (LOEW) this sclerite bears a distinct suture (situated on its posterior half), and in the species studied by me a corresponding change in the intensity of sclerotization is visible. It thus seems probable that the pregenital sclerite includes a posterior band of sclerotization which is of different segmental origin (representing either the 7th tergum or the inverted 8th sternum). If the pregenital sclerite consists of fused 6th and 7th terga, then this condition is closely similar to that shown by the Drosophilidae. However I agree with HENNIG (in press) that the Canacidae do not belong to the Drosophiloidea, because they lack the characteristic autapomorphous conditions of that group. Therefore, if the pregenital sclerite of the Canacidae has the same segmental origin as that of the Drosophilidae (fused 6th and 7th terga), this resemblance should be attributed to convergence or parallelism.

(2) 7th spiracles (♂) lying within pregenital sclerite.
(3) Hypandrial arms (♂) fused above aedeagus, where they extend posteriorly to form a conspicuous spine between the anus and the aedeagus (fig. 148).

The formation of this spine is one of the most characteristic conditions of the male genitalia of all the species of Canacidae studied by me. I did not find this condition in any other family.

(4) Postgonites (♂) joined above aedeagus by pubescent fold (aedeagal mantle) (fig. 148).
(5) Aedeagus (♂) shortened.

The aedeagus is very short in all Canacidae, and is clearly apomorphous in this respect. Its precise condition in the groundplan of the family requires clarification.

(6) Female postabdomen characterized by hook-shaped cerci, and inclusion of 7th spiracles within 7th tergum (HENNIG 1958).
(7) Two spermathecae (♀) (STURTEVANT 1926, HENNIG 1958).
(8) Fronto-orbitals *(ors)* all inclined outwards over eye margins (HENNIG 1958).
(9) Vibrissae present (HENNIG 1958).
(10) Peristomal opening large; face (prefrons) convexly arched (FREY 1921, HENNIG 1958).
(11) Costa broken at end of subcosta (HENNIG 1958).
(12) Anal vein shortened, represented only by short remmant (HENNIG 1958).

The Canacidae were formerly classified as a subordinate group of the Ephydridae (Drosophiloidea), but STURTEVANT (1926) already recognized that this was incorrect because the Canacidae do not show the modifications of the female reproductive system characteristic of the Ephydridae. I am satisfied on the basis of the new discussion on the Drosophiloidea presented by HENNIG (in press) that there are no grounds for classifying the Canacidae in the Drosophiloidea, let alone in one of the included families.

HENNIG (1958) suggested that the Canacidae are closely related to the Tethinidae or Milichiidae (Muscoidea, Tephritoinea). The Canacidae certainly do not belong to the Chloropidae family-group (in which I include Milichiidae), since they retain well developed spermathecae. I do not rule out the possibility that they are closely related to the Tethinidae. The autapomorphous conditions of the male postabdomen and genitalia of the Canacidae could have been derived from the conditions shown by the Tethinidae; on this interpretation the dorsal pregenital sclerite of the Canacidae would be considered to represent the fused 6th tergum and 8th sternum (which are fused in the Tethinidae). However the structure of the female postabdomen of the Canacidae (see WIRTH 1951 and HENNIG 1958, fig. 322) is difficult to reconcile with this interpretation, since there is no indication of the elongation of the 7th and 8th segments characteristic of the Tephritoinea. The interpretation that the Canacidae are closely related to the Tethinidae thus involves postulating extensive modifications in the structure of the female postabdomen.

It is interesting that among the so-called 'Heleomyzidae' described from South America by MALLOCH (1933) are a few genera (e.g. *Dihoplomyza*) whose females have hook-shaped cerci like those of the Canacidae. Unfortunately I was unable to obtain material of any of these genera for study. The possibility that they are closely related to the Canacidae clearly merits investigation.

The genus *Macrocanace* belongs to the Tethinidae (see the discussion under that family).

C. 53. Family Fergusoninidae

Sources of information

The male genitalia of the Fergusoninidae were treated by TONNOIR (1937). I studied preparations of *Fergusonina scutellata* MALLOCH.

Description of male postabdomen and genitalia (figs. 149-151)

6th and 7th abdominal spiracles symmetrically situated in membrane below the single protandrial sclerite.

Postabdomen fully symmetrical. The 5th tergum and 5th sternum are the last well-developed sclerites before the hypopygium, but a narrow dorsal protandrial sclerite ('x' on fig. 149) is also present in some species.

Telomeres poorly differentiated from periandrium, at most delimited by partial suture, joined by weakly sclerotized transverse interparameral sclerite. Cerci absent. Aedeagal apodeme short, linked to hypandrium by ventral process. Aedeagus sheathed dorsally by elongate aedeagal mantle, near the apex of which is situated the pair of small postgonites ('y' on fig. 151): aedeagus not able to be swung against aedeagal apodeme. Ejaculatory apodeme absent.

Characterization and discussion

All known species of this group are currently classified in the single genus *Fergusonina*, known only from Australia where the larvae form galls on various species of *Eucalyptus*. The genus was formerly referred to the Agromyzidae, presumably on account of the phytophagous habit of the larvae since there are no morphological grounds for such a classification. HENNIG (1958) rightly removed *Fergusonina* from the Agromyzidae, and accorded the group family rank. The Fergusoninidae are highly modified in many respects, and I have not reached any conclusion on their relationships with other families of Schizophora. But I think it likely that subsequent studies will show that they are referable to one of the groups treated as superfamilies in this paper, since their characteristic apomorphous conditions are not combined with any plesiomorphous conditions which suggest their early divergence from other Schizophora.

The Fergusoninidae show the following apomorphous conditions with respect to the groundplan of the Schizophora.

(1)   Vibrissae present (TONNOIR 1937).
(2)   Frontal lunule (= ptilinal suture) very large (TONNOIR 1937).
(3)   Costa only reaching apex of vein r_{4+5} (TONNOIR 1937).
(4)   Anal vein shortened, not reaching wing margin (HENNIG 1958).
(5)   Protandrial segments (6-8) ($\male$) much reduced, at most with single dorsal sclerite retained ('x' on fig. 149).

HENNIG (1958) stated that there are no postabdominal terga or sterna in the male between the 5th segment and the hypopygium. However this is not the groundplan condition, since in the species studied by me a small additional dorsal sclerite can be found infolded beneath the 5th tergum. The 6th and 7th abdominal spiracles lie in the infolded membrane below this sclerite. TONNOIR (1937) noted this sclerite (as 'eighth tergite') in *Fergusonina newmani* TONNOIR.

(6)   Aedeagus ($\male$) sheathed dorsally by elongate aedeagal mantle, near the apex of which is situated the pair of small postgonites ('y' on fig. 151).

The term 'aedeagal mantle' is here proposed for the dorsal semi-cylindrical structure which TONNOIR (1937) refers to as the 'hypophallus', since the term 'Penismantel' is used for an analogous structure developed in some *Drosophila* species (see NATER 1953). Near the apex of the mantle is situated the pair of lobes referred to as 'titillators' or 'paraphalli' by TONNOIR, who states that 'when the penis slides forward on the (mantle) the paraphalli are tipped sideways; their function is apparently to serve as anchorage during copulation'. These conditions are highly apomorphous and pose problems of interpreting the homologies of the observed structures. On present information I think the most probable interpretation is that the true aedeagus consists solely of the uniformly sclerotized ventral tube (called penis by TONNOIR) and that the mantle is a neomorphous structure, not part of the aedeagus itself. TONNOIR's description implies that the mantle does not enter the female genital opening during copulation. The distal 'titillators' possibly represent the postgonites. On this interpretation TONNOIR's use of the terms 'paraphalli' and 'hypophallus' is inappropriate, since these terms properly refer to sclerites in the walls of the aedeagus itself. However I emphasize that this interpretation is provisional, because I have not had access to fresh specimens to check the relevant musculature. If the so-called titillators are in fact postgonites, I expect them to possess extrinsic musculature originating on the aedeagal apodeme.

(7)   Aedeagal apodeme ($\male$) short, linked to hypandrium by ventral process (fig. 150).

259

Probably this is a general condition of the Fergusoninidae, since TONNOIR (1937) states that the aedeagal apodeme 'is attached by a transverse sclerotized bridge' to the hypandrium.

(8) Telomeres (♂) poorly differentated from periandrium, at most delimited by partial suture.

TONNOIR (1937: 128) has discussed the variation in the form of the telomeres, which he called 'coxites'.

(9) Cerci (♂) absent.

(10) Ejaculatory apodeme (♂) absent.

(11) 6th abdominal spiracles (♀) absent: 7th spiracles lying within 7th tergum.

This condition of the postabdominal spiracles was illustrated by HENNIG (1958) for *Fergusonina tillyardi* TONNOIR. I have confirmed that the same condition is shown by *F. scutellata* MALLOCH.

(12) 6th tergum and sternum (♀) fused to form sclerotized ring; 7th segment fusiform, with large tergum and sternum only narrowly separated by membrane (HENNIG 1958, fig. 291); 8th segment extremely long and slender, retracted within 6th and 7th segments when at rest.

This strengthening of the 6th, as well as of the 7th, segments is a highly characteristic autapomorphous condition of the Fergusoninidae. I know of no other group in which the 6th segment has become involved in the formation of an ovipositor sheath in this way. The condition in a preparation of *Fergusonina scutellata* MALLOCH studied by me is similar to that illustrated by HENNIG for *F. tillyardi* TONNOIR.

(13) Spermathecae (♀) reduced.

I list this character provisionally, as I can find no sclerotized spermathecae in the preparation of *Fergusonina scutellata* MALLOCH before me. The precise nature of the condition of the spermathecae can only be established from fresh material.

(14) Puparium with comb-like dorsal plate on boundary between 1st and 2nd abdominal segments (HENNIG 1958, fig. 294).

The immature stages of Fergusoninidae have not yet been described on a comparative basis. Future studies will probably reveal additional apomorphous conditions, correlated with the unusual biology of the larvae.

C. 54. Notomyzidae familia nova (type-genus Notomyza MALLOCH 1933)

Sources of information

I studied a preparation of the male postabdomen and genitalia of *Notomyza edwardsi* MALLOCH. Some limited information on the male postabdomen and genitalia of other species was given in MALLOCH's (1933) treatment, but this does not cover all the characters considered here.

260

Description of male postabdomen and genitalia (figs. 152-154)

6th abdominal spiracles symmetrically situated in membrane below 6th tergum: 7th spiracles absent.

Postabdomen fully symmetrical. 6th tergum large, almost as long as the 5th tergum: there are no further dorsal sclerites between the 6th tergum and the hypopygium. The 5th sternum is the last ventral sclerite before the hypopygium.

Periandrium large, bearing discrete telomeres: sclerotization of inner side of telomeres extending inwards to form horseshoe-shaped interparameral sclerite. Cerci fused below anus. Hypandrium with slender ventral processes anteriorly. Aedeagal apodeme rod-like, linked anteriorly to apex of hypandrium by ventral process. Aedeagus of simple tubular structure, uniformly sclerotized, shielded laterally by large postgonites: from the base of the aedeagus an epiphallus-like structure projects posteriorly: aedeagus remaining posteroventrally directed when at rest, not able to be swung through wide arc against aedeagal apodeme. Ejaculatory apodeme small and slender.

Characterization and discussion

MALLOCH'S (1933) classification of the South American genus *Notomyza* in the Heleomyzidae seems to me untenable. In the preparation which I studied I saw no trace of the asymmetrical arrangement of the postabdominal sclerites characteristic of the Heleomyzidae and most other Muscoidea, and the aedeagus is a simple tubular structure which remains posteroventrally directed when at rest (without swinging mechanism). The appearance of the male postabdomen and genitalia of *Notomyza* is similar to that of some Drosophiloidea, but *Notomyza* can hardly belong to that superfamily because the apomorphous conditions of frontal chaetotaxy and antennal structure characteristic of the Drosophiloidea are lacking. As far as I can judge, *Notomyza* cannot be referred to any family yet proposed, and I therefore propose the new family Notomyzidae based on this genus. As here proposed the family is monobasic.

Apomorphous conditions of *Notomyza* with respect to the groundplan of the Schizophora are as follows.

(1) Sclerites of 7th and 8th abdominal segments completely lost (♂); the last dorsal sclerite before the hypopygium is the 6th tergum (fig. 152).
(2) 6th sternum lost (♂): the last ventral sclerite before the hypopygium is the 5th sternum (fig. 152).
(3) 7th abdominal spiracles (♂) lost.
(4) Aedeagal apodeme (♂) linked to apex of hypandrium by ventral process (fig. 154).
(5) Hypandrium (♂) with pair of slender ventral processes anteriorly (fig. 154).
(6) Epiphallus-like structure (♂) projecting posteriorly from base of aedeagus (fig. 154).
(7) Cerci (♂) fused below anus (fig. 153).
(8) Postverticals convergent.

261

(9) Vibrissae present.

(10) Costa broken near end of vein r_1.

(11) Subcosta fading distally, not reaching wing margin.

(12) Anal vein shortened, not reaching wing margin.

(13) Two spermathecae ($\mathcal{Q}$) (HENNIG 1958).

C. 55. Family Braulidae

Sources of information

Information on the male postabdomen and genitalia of *Braula* (the sole genus) was given by HENNIG (1938c, 1938d). I studied preparations of a *Braula* species from Sicily.

Description of male postabdomen and genitalia (figs. 31-32)

6th abdominal spiracles symmetrically situated within 6th pair of pleural sclerites. 7th spiracles absent.

Postabdomen fully symmetrical. 6th segment with large tergum (longer than preceding terga), pair of small pleural sclerites (last and smallest of series of six pleural sclerites) and small sternum (smaller than preceding sterna). Between the 6th segment and the hypopygium lies only one further sclerite, a weak ill-defined sclerite in dorsal position (probably the 7th tergum).

Periandrium bearing discrete telomeres in apical position: the small cerci lie between the bases of the telomeres. A pair of broad interparameral sclerites is present which are linked to the bases of the telomeres. Hypandrium long, becoming narrow anteriorly where it extends as far as the apex of the aedeagal apodeme. Aedeagal apodeme rod-like, free from hypandrium. Aedeagus of simple tubular structure, uniformly sclerotized, remaining posteriorly directed when at rest (not able to be swung through wide arc against aedeagal apodeme). Postgonites well developed, variable in shape. Ejaculatory apodeme absent.

Characterization and discussion

It seems scarcely necessary to discuss some of the well-known features of *Braula*, such as loss of the wings and halteres, reduction of the eyes, concentration of the thorax and modification of the claws. These are autapomorphous conditions which demonstrate the monophyly of the Braulidae, but have not proved helpful in clarifying the relationship of the family to other Schizophora. I direct my present discussion mainly to characters of the male abdomen.

My interpretation of the segmentation of the male abdomen is the same as that given in HENNIG's first treatment of *Braula* (HENNIG 1938c). I do not agree

262

with the change of segment numbering proposed in his subsequent treatment in 'Die Fliegen der paläarktischen Region' (HENNIG 1938d). This change was due to his attempt to homologize the postabdominal sclerites of *Braula* with those of the Sphaeroceridae. The possibility of different interpretations of the segmentation of the male abdomen of *Braula* arose because there are only five spiracle-bearing segments with large terga and sterna. However there are clear indications that the first spiracle-bearing segment is the 2nd segment. The sternum of this segment is the largest abdominal sternum, which suggests that it is the 2nd sternum because the sternum of the first abdominal segment is relatively small in all Schizophora (see YOUNG 1921). The apparent first tergum is large and discrete, without any indication of the 'adventitious suture' which is found on the true 1st tergum in Schizophora (see YOUNG 1921). For this reason I think that it represents the 2nd tergum (the tergum of the second abdominal segment). And finally, there is an additional pair of pleural sclerites before the first spiracle-bearing pair, as noted by HENNIG (1938c), a condition which is better compatible with the interpretation that the first pair of spiracles belongs to the 2nd segment. Consistent with this interpretation I number the five pairs of spiracles in male *Braula* as spiracles 2-6, and I consider the last large dorsal sclerite to represent an unreduced 6th tergum. Between this and the periandrium lies a weak additional sclerite, which probably represents the 7th tergum.

In the light of the preceding discussion I consider that the Braulidae show the following conditions of the male abdomen and genitalia which are apomorphous with respect to the groundplan of the Schizophora.

(1)   1st abdominal tergum lost.

This reduction may be considered as correlated with the very marked reduction of the thorax in *Braula*, which was well discussed by HENNIG (1938d).

(2)   Only five pairs of spiracles retained (belonging to segments 2-6).

This condition (like the preceding) is probably also shown by the female, since HENNIG (1938d) did not indicate any spiracles on the 7th segment in his figure of the female abdomen (Textfig. 7). In the male the spiracles lie within the pleural sclerites, but HENNIG's figure indicates that in the female they lie within the margins of the terga.

(3)   Abdomen (♂) with series of six pleural sclerites lying between terga and sterna; all except first pair enclosing spiracles (fig. 31).

The presence of these pleural sclerites is a unique condition of male *Braula*, not known in any other group.

(4)   Only single ill-defined dorsal sclerite between 6th segment and hypopygium (♂); no ventral sclerites between 6th sternum and hypopygium (fig. 31).
(5)   Hypandrium (♂) long, becoming narrow anteriorly where it extends as far as the apex of the aedeagal apodeme (fig. 32).
(6)   Ejaculatory apodeme (♂) lost.

The relationships of *Braula* were discussed by HENNIG (1938c, 1938d, 1952,

1958) and IMMS (1942). This question is still not clarified, but I think that the structure of the male abdomen and genitalia allow me to narrow down the range of possibilities. According to my interpretation of the segmentation the 6th segment of male *Braula* is relatively large and of similar structure to the preceding segments. Therefore *Braula* cannot be included in any group which shows a reduced 6th segment in its groundplan. The structure of the male genitalia seems surprisingly plesiomorphous in some respects, particularly in respect of the rod-like aedeagal apodeme (free from the hypandrium) and the posteriorly directed simple tubular aedeagus (without swinging mechanism) (fig. 32). I think that the structure of the aedeagus rules out any possibility that the Braulidae belong to the Muscoidea or Nothyboidea. I accordingly reject the view formerly held by HENNIG that the Braulidae are closely related to the Sphaeroceridae (Muscoidea, Anthomyzoinea). They also show none of the characteristic autapomorphous conditions of the Drosophiloidea. Thus the field in which close relatives of the Braulidae may be sought can be narrowed to the superfamilies Lonchaeoidea and Lauxanioidea. If the Braulidae do not belong to either of these groups, then they can only be referred to a separate monobasic superfamily. Such an isolated position for the Braulidae was implied by ROHDENDORF's (1964) proposal of a monobasic infraorder Braulomorpha, although this rank is too high for a phylogenetic system because the group should be subordinated to the Schizophora. IMMS (1942) concluded that the Braulidae are closely related to the Chamaemyiidae (Lauxanioidea, Chamaemyioinea) because of similarities in larval structure. However HENNIG (1952a : 325) reviewed the evidence presented by IMMS and found it unconvincing. I do not rule out the possibility that the Braulidae belong to the Chamaemyioinea, but agree with HENNIG that no convincing evidence for this has yet been produced. Because of these unresolved questions I prefer to leave open for the present the classification of the Braulidae at the superfamily level.

6.3. Key to the families of Schizophora

The key below is not intended for quick sorting of material, since some of the characters used can only be seen in preparations. It is intended primarily to provide systematists with a first approximation of where doubtful species should be classified, to be supplemented of course by reference to the preceding discussion in section 6.2. The key can be used for males alone, or for males and females in association; but not for females alone.

The keys hitherto available include early divisions on the basis of such characters as the presence or absence of the costal break and vibrissae, and the direction of the postvertical bristles. Since these characters are subject to variation within some groups, such divisions in the key do not reflect the relationships between the families; and radical misclassification can occur because the distinctions drawn are equivocal in some cases. In the past many genera and species, particularly from the tropics, have been described in the wrong families due to the inadequacies of the available keys and an absence of comprehensive structural information of the kind which I have tried to present in this work. For this reason I think there may be some value in my presenting an alternative key which makes extensive use of characters of the male postabdomen.

The best previous key to the families of Schizophora with worldwide coverage is that given by HENDEL (1936).

1. Adults parasitic or phoresic, with enlarged or pectinate claws. Often wingless or brachypterous; if with functional wings (some Hippoboscidae), three radial branches crowded on fore-margin of wing and costa not extending beyond r_{4+5} 2
 – Adults not parasitic or phoresic, except for a few Milichiidae and Carnidae (in which the wing venation is not of the above type); if blood-sucking, not remaining on host. Usually with functional wings; if brachypterous (e.g. *Mormotomyia*, some Sphaeroceridae), claws not modified for clinging 3

2. Small (1-1.5 mm.) phoresic commensals of the honeybee (*Apis*). Mouthparts not piercing. Wings, halteres and scutellum lost. Claws pectinate. Male abdomen with series of six pleural sclerites between terga and sterna (fig. 31) . . . . **Braulidae** (C.55)
 – Larger, blood-sucking parasites of birds and mammals, with piercing mouthparts. Scutellum present. Claws enlarged, hook-shaped. Male abdomen without pleural sclerites, often with much reduced sclerotization. Apterous, brachypterous or fully winged . . . .
 . . . . **Hippoboscidae** (*sensu lato*) (D.2)
 (Compare the closely related Glossinidae, couplet 26)

Note. Individuals of *Carnus* (Carnidae) which have shed their wings may also be taken to this couplet. The mouthparts of *Carnus* are short, not adapted for piercing.

3. Uniformly dark flies with dark halteres. Frons densely clothed with setulae, at most with single *ors*. Female abdomen with normal terga and sterna only as far as 6th segment. Male aedeagus remaining posteroventrally directed when at rest, rigid and uniformly sclerotized, often upcurved (figs. 21 and 23) . . . . (Lonchaeoidea) 4
 – Halteres usually pale; if dark (some Milichiidae and Agromyzidae), more than one *ors* present 5

4. Arista present; 2nd antennal article with suture; *ors* present; 7th abdominal segment (♀) extensively sclerotized, forming oviscape . .
 . . . . **Lonchaeidae** (B.1)
 – Arista lost; 2nd antennal article without suture; *ors* scarcely differentiated; female postabdomen membranous after 6th segment . . . .
 . . . . **Cryptochetidae** (B.2)

5. Frons usually with one or two reclinate *ors* and single proclinate *ors* (except in some Ephydridae); postverticals convergent (except in some

Ephydridae); 2nd antennal article with suture. Male postabdomen fully symmetrical, with 8th sternum reduced to pair of dorsolateral vestiges or completely lost; aedeagus not able to be swung through wide arc against aedeagal apodeme. (Groups of Ephydridae in which the chaetotaxy of the head is not as stated above may be recognized *inter alia* by loss of the anal cell and lower cross-vein, and by the sclerotized ventral receptacle ($\female$); see couplet 6) . (Drosophiloidea)   6
- Frontal chaetotaxy not as stated above, except in some Milichiidae. Other conditions not shown in combination 10
6. Spermathecae ($\female$) rudimentary; ventral receptacle heavily sclerotized. 6th and 7th terga ($\male$) very narrow (*Diastata*, fig. 41) or lost (other genera). 7th abdominal spiracles lost in both sexes. Anal vein and anal cross-vein usually absent (except in *Diastata*) . . . . **Ephydridae** (C.7)
- Two spermathecae ($\female$) present; ventral receptacle not sclerotized. Anal cell closed except in Camillidae 7
7. 5th and 6th terga ($\male$) reduced (fig. 35); 4th segment the largest abdominal segment. Female postabdomen largely membranous. Anal vein and anal cross-vein lost. 7th abdominal spiracles absent in both sexes . . . . **Camillidae** (C.4)
- 5th tergum ($\male$) large. 6th and 7th segments ($\female$) with well developed terga and sterna 8
8. Aedeagus ($\male$) grossly enlarged (fig. 40). Subcosta distinct to wing margin . . . . **Curtonotidae** (C.5)
- Aedeagus ($\male$) not as above. Subcosta faded or fused with r_1 distally   9
9. 6th and 7th terga ($\male$) discrete (fig. 37). Proclinate *ors* lying between reclinate *ors* and eye margin . . . . **Campichoetidae** (C.6)
- 6th and 7th terga ($\male$) fused, forming large syntergum (fig. 33). Proclinate *ors* lying before or to inner side of reclinate *ors* . . . . .
. . . . **Drosophilidae** (C.3)
10. Aedeagus ($\male$) sheathed dorsally by elongate aedeagal mantle, near the apex of which is situated the pair of small postgonites: protandrial segments much reduced, at most with single dorsal sclerite between 5th segment and hypopygium (figs. 149-151). Female postabdomen modified for boring, with 6th tergum and sternum fused to form sclerotized ring and 7th segment fusiform (with large tergum and sternum only narrowly separated by membrane) . **Fergusoninidae** (C.53)
- Aedeagus ($\male$) not as stated above. If female with boring ovipositor, 6th tergum and sternum discrete or reduced (some Tephritidae) . . 11
11. Aedeagus ($\male$) short; postgonites joined above aedeagus by pubescent fold (aedeagal mantle); postabdomen symmetrical, with only single dorsal sclerite between 5th segment and hypopygium (figs. 147-148). Female with hook-shaped cerci. *Ors* all inclined outwards over eye margin . . . . **Canacidae** (C.52)
- Male genitalia not as above. Female cerci not hook-shaped 12

Note. Certain South American genera with hook-shaped female cerci (e.g. *Dihoplomyza*) cannot be taken beyond this couplet. Their family position is unclarified.

12. Hypandrium (♂) with pair of slender ventral processes; cerci fused below anus; aedeagus posteroventrally directed when at rest (not swung through wide arc against aedeagal apodeme): postabdomen symmetrical, with all sclerites of 7th and 8th segments lost (6th tergum the last dorsal sclerite before the hypopygium) (figs. 152-154) . . . .
. . . . **Notomyzidae** (C.54)
- Hypandrium (♂) without such ventral processes; other conditions stated above not shown in combination 13
13. Aedeagus (♂), when well developed, of simple tubular structure, posteroventrally directed when at rest (not swung through wide arc against aedeagal apodeme); 7th tergum usually developed as dorsal sclerite (lost only in some Chamaemyiidae); sclerites of 8th segment lost (figs. 24-30). Costa unbroken; anal vein not reaching wing margin. Postverticals, when well developed, convergent. . . (Lauxanioidea) 14
- Aedeagus (♂) in most groups able to be swung through wide arc against aedeagal apodeme to anteriorly directed rest position (see figs. 52-53). If not, other conditions stated above not shown in combination . . . 16
14. 6th tergum (♂) much shorter than preceding terga: periandrium without discrete telomeres (fig. 30). 4 spermathecae (♀). Subcosta contiguous with r_1 before its apex . . . . **Chamaemyiidae** (C.2)
- 6th tergum (♂) about as long as preceding terga: periandrium bearing discrete telomeres (figs. 24 and 26). Subcosta free throughout . . . 15
15. Head transverse, without macrochaetae. 4 spermathecae (♀) . . . .
. . . . **Eurychoromyiidae** (C.1)
- Head usually with macrochaetae (except in *Celyphus* and its relatives, which are readily recognizable by their enlarged scutellum). 3 spermathecae (♀) . . . . **Lauxaniidae** (B.3)
16. Male postabdomen (figs. 43, 45, 48 and 51) fully symmetrical, except in *Somatia* (Periscelididae); 6th tergum usually as long as preceding terga (except Nothybidae); symmetrical 6th sternum always retained; usually a single dorsal sclerite between 6th segment and hypopygium (lost only in some Psilidae). Postverticals, when well developed, divergent . . . . (Nothyboidea) 17
- Male postabdomen usually asymmetrical, with 6th and 7th sterna more strongly developed on left side where the latter is linked to the large 8th sternum (inverted); only in Heteromyzidae and Micropezoinea (figs. 56 and 71) is a discrete symmetrical 6th sternum present; in groups in which this asymmetrical arrangement of sterna is no longer apparent due to loss or fusion of sclerites, the last symmetrical sternum in ventral position is the 5th sternum. 6th tergum (♂) variably developed (in some groups reduced or lost); 7th tergum reduced to lateroventral vestige on right side or lost . . . . . (Muscoidea) 20
17. Ejaculatory apodeme (♂) reduced or lost. Spermathecae (♀) not sclerotized, represented by branched tubes. Female postabdomen elongate, forming retractile ovipositor. Subcosta cut off distally, connected with wing margin by hyaline stria . . . . **Psilidae** (C.11)

267

–   Ejaculatory apodeme ($\male$) and spermathecae ($\female$) well developed. Female postabdomen not elongate   18

18. 7th abdominal spiracles ($\male$) within pregenital sclerite (fig.46); aedeagus slender and ribbon-like throughout, without complex distal section (fig. 47). 7th abdominal tergum and sternum fused in female. Anal vein short, cut off apically. Postverticals usually well developed (except *Somatia*) . . . .   **Periscelididae** (C.10)
–   7th abdominal spiracles ($\male$) in membrane (Nothybidae) (fig. 43) or lost (Teratomyzidae) (fig. 51); if aedeagus partly ribbon-like (Nothybidae), it bears a complex distal section (fig. 44). 7th abdominal tergum and sternum ($\female$) discrete or fused. Anal vein long, close to wing margin. Postverticals reduced or lost   19

19. Subcosta faded or fused with r₁ distally; costa broken; posterior cross-vein (m-m) displaced towards wing base. 1 *ors*. 7th abdominal spiracles lost in both sexes . . . .   **Teratomyzidae** (C.9)
–   Subcosta distinct to wing margin; costa unbroken; posterior cross-vein remote from wing base. 2 *ors*. 7th abdominal spiracles present in both sexes . . . .   **Nothybidae** (C.8)

20. Abdomen with petiolate appearance, with basal segments elongate but distal segments short and wide (fig. 77). Postverticals absent; only one vertical bristle and at most one *ors* present. 6th sternum ($\male$) well developed only in *Centrioncus* (Diopsidae, fig. 77); in other groups reduced to pair of fragments . . . .   (Diopsioinea)   21
–   Abdomen not petiolate; if markedly elongate (e.g. Micropezoinea, Megamerinidae), the elongation is more uniform, affecting both basal and distal segments   22

21. Front femora thickened and armed with ventral tubercles. Lower cross-vein lost. Scutellum with pair of spines . . . .   **Diopsidae** (C.22)
–   Hind femora thickened. Lower cross-vein usually present. Scutellum without spines . . . .   **Syringogastridae** (C.23)

22. Periandrium ($\male$) elongate; aedeagus borne on long basal cone; 6th tergum as long as preceding terga; 6th sternum subtriangular, more or less symmetrical (enlarged in some Micropezidae) (figs. 71-73). 7th tergum and sternum ($\female$) lengthened to form more or less closed ovipositor sheath . . . .   (Micropezoinea)   23
–   Male postabdomen and genitalia not as above   25

23. Lower cross-vein lost; subcosta not distinct to wing margin; costa broken. Ocellar bristles well developed . . . .   **Cypselosomatidae** (C.20)
–   Lower cross-vein usually present; subcosta distinct to wing margin; costa broken or unbroken. Ocellar bristles reduced or lost   24

24. 2nd antennal article with fingerlike process on inner side; arista apical or subapical. Front legs long . . . .   **Neriidae** (D.3)
–   2nd antennal article without fingerlike process; arista medial to sub-basal. Front legs relatively short and weak . . . .   **Micropezidae** (D.4)

25. 2nd antennal article with suture; *ori* usually present. Preabdominal spiracles within terga (except in Glossinidae). 7th left spiracle ($\male$)

within 7th sternum or synsternum (7 + 8) (except in *Haematobia*, Muscidae). Inner mouth-opening with hyoid (except in Stomoxyinae, Muscidae) . . . . (Calyptratae, part) 26

– 2nd antennal article usually without suture (except in *Mormotomyia*, some Megamerinidae and some Tephritidae). Preabdominal spiracles usually in membrane. Hyoid absent 31

26. The three radial branches and m_{1+2} crowded on fore-margin of wing. Mouthparts piercing (adapted for blood sucking). 6th to 8th sterna (♂) fused to form composite pregenital sclerite (fig. 67); 6th tergum as long as 5th tergum . . . . **Glossinidae** (D.1)

– Veins usually not crowded on fore-margin of wing; or, if somewhat crowded (*Oestrus, Gasterophilus* and relatives), the mouthparts are reduced. 6th tergum (♂) shorter than 5th tergum, in some groups much reduced or lost; 6th sternum usually discrete (except Fanniidae) . . 27

27. Hypopleuron with a row of outstanding bristles, except in *Gasterophilus* which may be recognized by its reduced mouthparts . . . .
. . . . **Tachinidae** (*sensu lato*) (C.18)

– Hypopleuron without outstanding bristles, at most with fine pubescence (*Eginia*) 28

28. Lower fold of squama ('thoracic squama') consisting of narrow strip. Frons not sexually dimorphic (eyes widely separated in male). Aedeagus (♂) swung through wide arc against aedeagal apodeme to anteriorly directed rest position, with well-developed epiphallus; articulated pregonites present (fig. 60) . . . . **Scatophagidae** (C.14)
(limits of family not clarified)

– Lower fold of squama expanded (except Fucelliinae, in which the aedeagus is posteriorly directed when at rest and the epiphallus lost). Frons sexually dimorphic (narrower in male) in many groups 29

29. 6th to 8th sterna (♂) fused to form symmetrical pregenital sclerite, within which both the 6th and 7th pairs of spiracles are situated (fig. 63); aedeagus small (often membranous); hypandrium without articulated pregonites . . . . **Fanniidae** (C.16)

– Male postabdomen not as above; 6th sternum retained as discrete asymmetrical sclerite; articulated pregonites usually present (except in some Muscidae) 30

30. 6th and 7th abdominal spiracles usually lost in female (at most with 6th pair retained in *Acanthiptera*). Processus longi (♂) absent. Anal vein not distinct to wing margin . . . . **Muscidae** (C.17)

– Anal vein distinct to wing margin . . . . **Anthomyiidae** (C.15)
(limits of family not clarified)

31. 7th sternum (♂) and 7th tergum vestige fused to form complete ring of sclerotization; 6th tergum as long as 5th tergum; periandrium divided into two halves (figs. 74-76) . . . . **Australimyzidae** (B.8)

– 7th sternum and tergum (♂) not thus fused, except in *Helcomyza*

(Dryomyzidae) in which the 6th tergum is reduced; periandrium entire, except in *Aphaniosoma* (Chyromyidae) 32

32. 6th sternum (♂) large and discrete, symmetrical or somewhat asymmetrical (if asymmetrical, the 6th tergum is as long as the 5th tergum); epiphallus well developed (figs. 54-58). 7th abdominal spiracles lost in both sexes. Frons sexually dimorphic (narrower in male) in many species . . . . (Tanypezoinea) 33

– 6th sternum (♂) strongly asymmetrical (more strongly developed on left side), except when reduced or fused with succeeding sterna to form composite pregenital sclerite. Frons not sexually dimorphic, except in *Neomaorina* (Tephritidae family-group) 34

33. 6th sternum (♂) asymmetrical (extending towards 7th sternum on left side); telomeres fused with periandrium; aedeagal apodeme linked to hypandrium by ventral process or pair of ventral processes (figs. 54-55). Postverticals divergent . . . . **Tanypezidae** (C.12)

– 6th sternum (♂) symmetrical; telomeres discrete; aedeagal apodeme free from hypandrium (figs. 56-58). Postverticals convergent . . . .
. . . . **Heteromyzidae** (C.13)

34. Aedeagus (♂) flexible, coiled when at rest, extended by pressure of body fluid (often very long and distinctly pubescent); but reduced in Milichiidae, Chloropidae and some Conopidae (without distinct sclerites beyond the basal phallophore) (figs. 122-146). Female usually with retractile ovipositor formed by elongation of 7th and 8th abdominal segments (except in some genera of Chloropidae and Tethinidae) .
. . . . (Tephritoinea) 35

– Aedeagus more or less rigid, not coiled when at rest, with its walls usually supported by well differentiated sclerites (exceptionally with coiled distal section in some Agromyzidae or coiled basal section in some Clusiidae); without extensive pubescence except in Dryomyzidae 48

35. Wings vestigial; whole body densely clothed with ochraceous pubescence; eyes very small; ocelli lost. 7th abdominal spiracles lost in both sexes. 6th tergum (♂) large; hypandrium modified to form pubescent trough-like structure (fig. 126) . . . . **Mormotomyiidae** (C.45)

– Not as above 36

36. 2nd antennal article elongate; haustellum elongate. 7th and 8th sterna (♂) fully fused (fig. 139), forming more or less symmetrical pregenital sclerite with which the 6th sternum is joined on both sides; both 6th and 7th pairs of abdominal spiracles symmetrically situated within this pregenital sclerite (except in the highly modified *Stylogaster* in which there are no clearly defined sclerites between the 5th segment and hypopygium). Ovipositor (♀) (segments 7 and following) usually directed obliquely in relation to longitudinal axis of abdomen (except in *Stylogaster*) . . . . **Conopidae** (C.50)

– 2nd antennal article and haustellum not both elongate in combination (one or other is elongate in a few groups, such as Milichiidae and some Tephritidae); postabdominal structure not as stated above 37

37. 7th abdominal segment (♀) extremely long (at least twice as long as 6th segment); 8th segment extremely long, retractile within 7th segment; female cerci more or less fused. Spermathecae (♀) well developed. 8th sternum (♂) asymmetrically developed on left side of abdomen: aedeagus usually extremely long and coiled (but membranous in some groups of Piophilidae, and somewhat shortened in *Heloparia*) (figs. 140-146) . . . . (Tephritidae family-group) 38
– Elongation of 7th and 8th abdominal segments (♀) less extreme; female cerci more or less discrete except in Carnidae (in which the spermathecae are reduced). 8th sternum (♂) not strongly asymmetrical, in some groups fused with preceding sclerites or reduced (Milichiidae) 41
38. 6th tergum (♂) represented by two fragments, one of which lies above the 6th left spiracle on the left side (fig. 140). 7th abdominal spiracles lost in both sexes . . . . **Eurygnathomyiidae** (D.11)
– 6th tergum (♂) represented at most by narrow band (*Heloparia*) or lateroventral fragment on right side. 6th and 7th abdominal spiracles retained in all females, but lost in most males (except Richardiidae) 39
39. 6th and 7th abdominal spiracles (♂) retained on the left or on both sides (fig. 142). 7th tergum and sternum (♀) fused, forming completely sclerotized oviscape. Vein r₁ thickened where the subcosta ends, then forming short convex bow . . . . **Richardiidae** (D.12)
– 6th and 7th abdominal spiracles (♂) lost on both sides 40
40. 7th tergum and sternum (♀) fused, forming completely sclerotized oviscape. Postgonites absent (♀) . . . . **Tephritidae** (*sensu lato*) (D.14)
– 7th tergum and sternum (♂) discrete, becoming attenuated to pairs of thin strips posteriorly. Large postgonites present (♂) . . . .
. . . . **Piophilidae** (*sensu lato*) (D.13)

Note. Certain southern hemisphere genera previously classified in the 'Pallopteridae' (e.g. *Heloparia*, *Aenigmatomyia* and *Neomaorina*) are not included in the above dichotomy. See the discussion under Piophilidae (section 6.2).

41. Aedeagus (♂) reduced (figs. 135 and 136), virtually membranous beyond the basal phallophore (at most with weak traces of pubescence). Spermathecae (♀) rudimentary 42
– Aedeagus (♂) elongate; or, if rather short (as in Odiniidae), still conspicuously pubescent 43
42. Male postabdomen symmetrical (fig. 133), with only single narrow dorsal sclerite (in *Desmometopa* reduced to pair of lateral strips) between 5th segment and hypopygium. Anal cell and lower cross-vein retained. Ocellar plate not enlarged . . . . **Milichiidae** (D.9)
– Male postabdomen variable in structure, with asymmetrical 6th and 7th sterna clearly delimited in some genera (e.g. *Lasiopleura*, fig. 137), but lost in others (e.g. *Chlorops*, *Meromyza*) in which the postabdomen appears more or less symmetrical. Anal cell and lower cross-vein lost. Ocellar plate enlarged . . . . **Chloropidae** (D.10)
43. Male postabdomen fully symmetrical (fig. 129); 6th tergum fused with

8th sternum; 6th and 7th sterna lost. Hypandrium (♂) much shorter than aedeagal apodeme. Postverticals convergent. Spermathecae (♀) well developed . . . . **Tethinidae** (C.48)

–   Male postabdomen usually strongly asymmetrical (with 6th and 7th sterna clearly delimited); if almost symmetrical due to fusion of sterna (Acartophthalmidae), 6th tergum lost but 6th sternum retained . . .   44

44. 7th and 8th sterna (♂) fused, forming almost symmetrical pregenital sclerite with which the 6th sternum is joined on both sides (fig. 132). Sclerotized spermathecae absent (♀). Costal break situated shortly beyond humeral cross-vein. Postverticals divergent . . . .
. . . . **Acartophthalmidae** (D.7)

–   Male postabdomen not as above. Costal break at end of subcosta or r₁   45

45. 7th tergum vestige (♂) on centre-line of venter (fig. 134). Sclerotized spermathecae absent (♀); female cerci fused to form stylus-like structure . . . . **Carnidae** (D.8)

–   7th tergum vestige (♂) situated on right side of venter (Chiropteromyzidae, fig. 122) or lost (Cnemospathidae, Odiniidae). Sclerotized spermathecae present (♀)   46

46. 7th abdominal spiracles lost in both sexes. Periandrium (♂) with pair of lateral notches on anterior margin; telomeres not discrete; aedeagal apodeme short (fig. 127). Postverticals convergent   **Cnemospathidae** (C.46)

–   7th abdominal spiracles retained. Periandrium (♂) without lateral notches, with or without discrete telomeres; aedeagal apodeme long and rod-like   47

47. 7th tergum vestige (♂) lost; hypandrium long and narrow anteriorly; aedeagus rather short (fig. 128); ejaculatory apodeme lost. Postverticals divergent; 3 *ors* present . . . . **Odiniidae** (C.47)

–   7th tergum vestige (♂) present; hypandrium shorter; aedeagus very long (figs. 122-123); ejaculatory apodeme present. Postverticals convergent; 2 *ors* present . . . . **Chiropteromyzidae** (C.44)

48. Male postabdomen fully symmetrical (fig. 120); 5th sternum the last ventral sclerite before the hypopygium; 6th tergum the last large dorsal sclerite before the hypopygium (with the inverted 8th sternum represented at most by a narrow band). Female with piercing ovipositor; 7th abdominal segment forming conical ovipositor sheath; 8th segment bearing numerous anteriorly directed denticles; proctiger elongate. One or more pairs of inwardly directed *ori* present; divergent postverticals usually present (except in *Penetagromyza*)   **Agromyzidae** (C.43)

–   Male and female postabdomen not as above   49

49. 2nd antennal article with angular projection on its outer side. 6th tergum (♂) well developed (at least half as long as 5th tergum). At least three pairs of orbital bristles present, usually including a pair of inwardly directed *ori*; postverticals, when present, divergent . . . . . .
. . . . **Clusiidae** (C.42)

–   2nd antennal article without such angular projection   50

50. Metatarsus of hind legs short and thick. 6th tergum ($\male$) lost; 6th and 7th left spiracles both within 6th sternum; 7th right spiracle lost (fig. 113) . . . . **Sphaeroceridae** (C.39)
– Metatarsus of hind legs slender (similar to metatarsi of front two pairs of legs) 51
51. Costa unbroken 52
– Costa broken near end of subcosta or r_1 . . . (Anthomyzoinea, part) 62
52. Anal cell and lower cross-vein lost. 7th abdominal spiracles lost in both sexes. Postverticals reduced . . . . **Asteiidae** (C.37)
– Anal cell and lower cross-vein present . . . . . . (Sciomyzoinea) 53
53. Ocellar and postvertical bristles lost. Hind femora thickened and armed with spines. Abdominal segments elongate, except reduced male 6th and 7th segments (fig. 97) . . . . **Megamerinidae** (C.30)
– Ocellar bristles present 54
54. Metastigmatal bristles present (posterior to metathoracic spiracles) 55
– Metastigmatal bristles absent; metathoracic spiracles with fine pubescence only 56
55. Hypandrial arms ($\male$) fused above base of aedeagus, where they bear a conspicuous upcurved process (fig. 94); aedeagus with long membranous distal area; telomeres discrete . . . . . . **Ropalomeridae** (D.5)
– Hypandrial arms ($\male$) not bearing upcurved process; aedeagus with distal sclerotization; telomeres fused with periandrium . **Sepsidae** (D.6)
56. Subcostal cell infuscated. Postverticals reduced . . **Cremifaniidae** (C.31)
– Subcostal cell not infuscated. Postverticals well developed 57
57. Male with thumbnail-like apical ventral processes on at least front and hind metatarsi; aedeagus extensively pubescent (fig. 88) **Dryomyzidae** (C.26)
– Metatarsi without such processes; aedeagus without extensive pubescence 58
58. 6th tergum ($\male$) well developed, symmetrical, about half as long as 5th tergum (fig. 82) . . . . *Heterocheila* (? Coelopidae)
– 6th tergum ($\male$) further reduced or lost, in some groups asymmetrically developed on right side 59
59. 7th and 8th tergum vestiges present ($\male$); 6th tergum asymmetrically reduced, developed mainly on right side (figs. 81 and 85) 60
– 7th and 8th tergum vestiges lost ($\male$); 6th tergum reduced to symmetrical narrow band or lost 61
60. Postverticals convergent. 6th and 7th left spiracles ($\male$) in membrane . . . . **Coelopidae** (C.24)
– Postverticals divergent. 6th and 7th left spiracles ($\male$) within the 6th and 7th sterna (fig. 86) . . . . **Phaeomyiidae** (C.25)
61. Postverticals parallel or convergent. Aedeagus ($\male$) bilobed distally (fig. 93). 3 spermathecae ($\female$) . . . . **Helosciomyzidae** (C.28)
– Postverticals divergent. Aedeagus ($\male$) variously modified. 2 spermathecae ($\female$) . . . . **Sciomyzidae** (C.27)
62. Postverticals divergent; only one pair of vertical bristles present. Aedeagus ($\male$) slender . . . . *Stenomicra* (? Anthomyzidae)

- Postverticals, if well developed, convergent; two pairs of vertical bristles present 63
63. Ocellar bristles reduced; postverticals lost. 2 *ors*, of which the anterior pair are directed inwards. Aedeagus (♂) short, with very complex sclerotization (fig. 116) . . . . **Aulacigastridae** (C.41)
- Ocellar bristles well developed 64
64. Male postabdomen symmetrical, sometimes with much reduced sclerotization (*Aphaniosoma*); 6th tergum with lateroventral extensions, entire in *Chyromya* (fig. 115) but divided in *Aphaniosoma*; 6th and 7th sterna lost; aedeagus with expanded complex distal area (fig. 117) . .
. . . . **Chyromyidae** (C.40)
- Male postabdomen usually obviously asymmetrical (except in some Rhinotoridae); 6th tergum, when present, without lateroventral extensions 65
65. Postverticals reduced or lost; only 1 *ors* present. Aedeagus (♂) with very complex sclerotization (fig. 109); telomeres not differentiated from periandrium; 6th tergum reduced to narrow band or lost. 7th abdominal spiracles lost in both sexes . . . . . . **Opomyzidae** (C.38)
- Not as above. Postverticals usually present (except in some Rhinotoridae and Anthomyzidae) 66
66. Aedeagus (♂) with expanded medial area from which arises a slender distal section (fig. 104); ejaculatory apodeme reduced **Anthomyzidae** (C.34)
- Aedeagus (♂) not as above; ejaculatory apodeme well developed . . 67
67. Aedeagus (♂) not swung through wide arc against aedeagal apodeme; 5th sternum (♂) usually divided (fig. 106); male cerci small. Femora (especially of front legs) somewhat thickened, usually with spines on their ventral surface . . . . **Rhinotoridae** (C.33)
- Aedeagus (♂) swung through wide arc against aedeagal apodeme to anteriorly directed rest position; 5th sternum (♂) entire 68
68. Slender epiphallus present (♂); 7th tergum vestige present (♂); aedeagal apodeme free from hypandrium (figs. 110-111). 7th abdominal spiracles retained in both sexes . . . . **Borboropsidae** (C.35)
- Epiphallus (♂) short or absent; 7th tergum vestige usually absent (except in *Anorostoma* and *Orbellia*, Heleomyzidae) 69
69. 7th abdominal spiracles lost in both sexes. Aedeagal apodeme (♂) free from hypandrium (fig. 107) . . . . **Trixoscelididae** (C.36)
(limits of family not clarified)

- 7th abdominal spiracles usually retained (except in *Allophylopsis*). Aedeagal apodeme (♂) linked to hypandrium by single or paired ventral processes . . . . **Heleomyzidae** (C.32)
(limits of family not clarified)

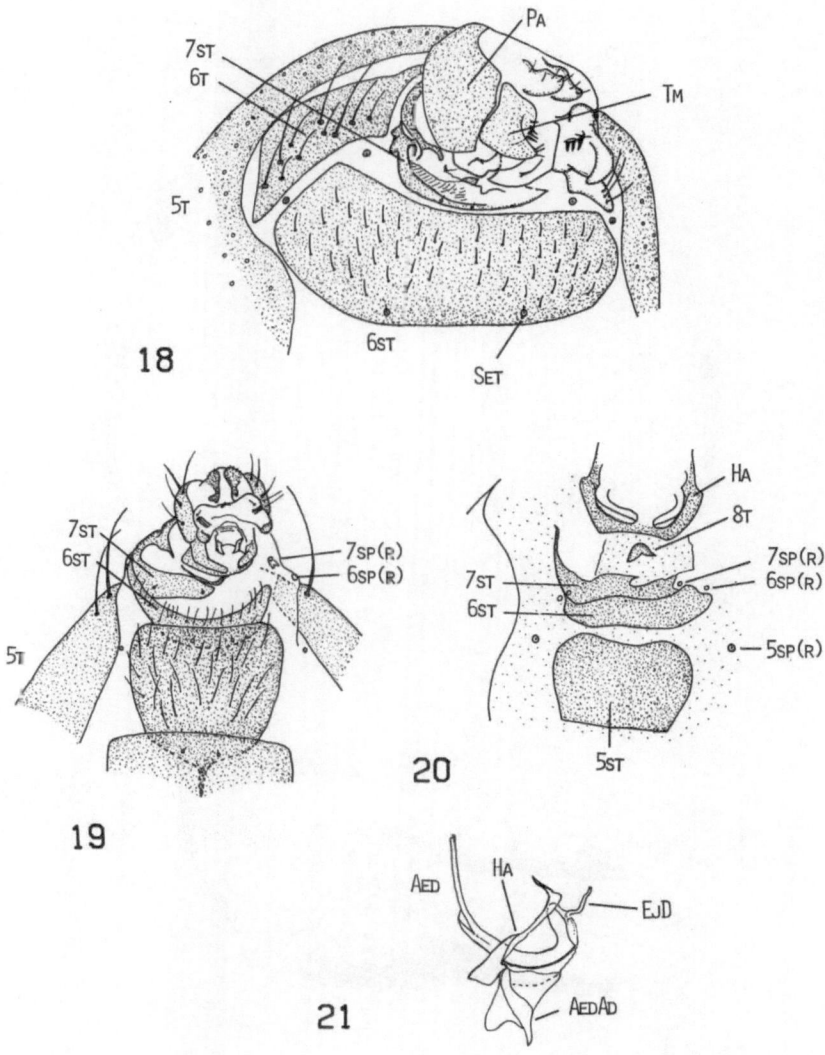

Figs. 18 - 21. 18. *Dasiops relicta* McAlpine (Lonchaeidae), postabdomen (♂) in ventral view (after J. F. McAlpine 1962b). 19. *Protearomyia obscura* (Walker) (Lonchaeidae), postabdomen (♂) in ventral view (after J. F. McAlpine 1962b). 20. *Lonchaea subpolita* Malloch (Lonchaeidae), ventral postabdominal sclerites (♂) (after J. F. McAlpine 1962b). 21. *Lamprolonchaea aurea* (Macquart) (Lonchaeidae), aedeagus, hypandrium and associated structures (♂) in lateral view (after J. F. McAlpine 1962b).

275

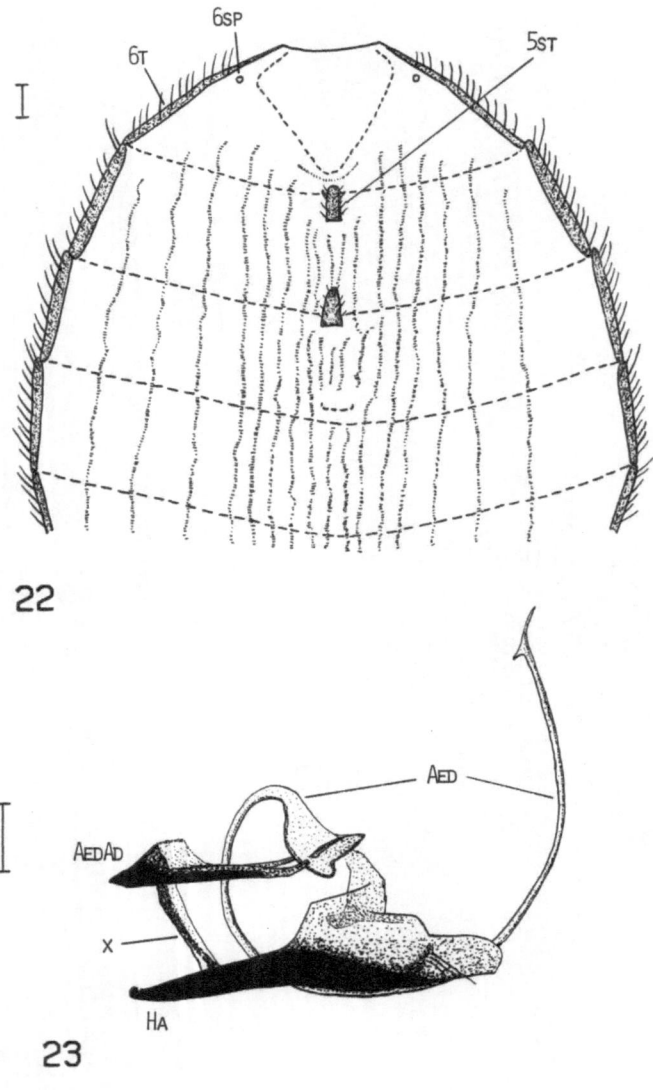

Figs. 22 - 23. 22. *Cryptochetum nipponense* Tokunaga (Cryptochetidae), abdomen (♂) (part) in ventral view (with hypopygium removed). 23. *Cryptochetum nipponense* To-kunaga, aedeagus, hypandrium and associated structures (♂) in lateral view.

276

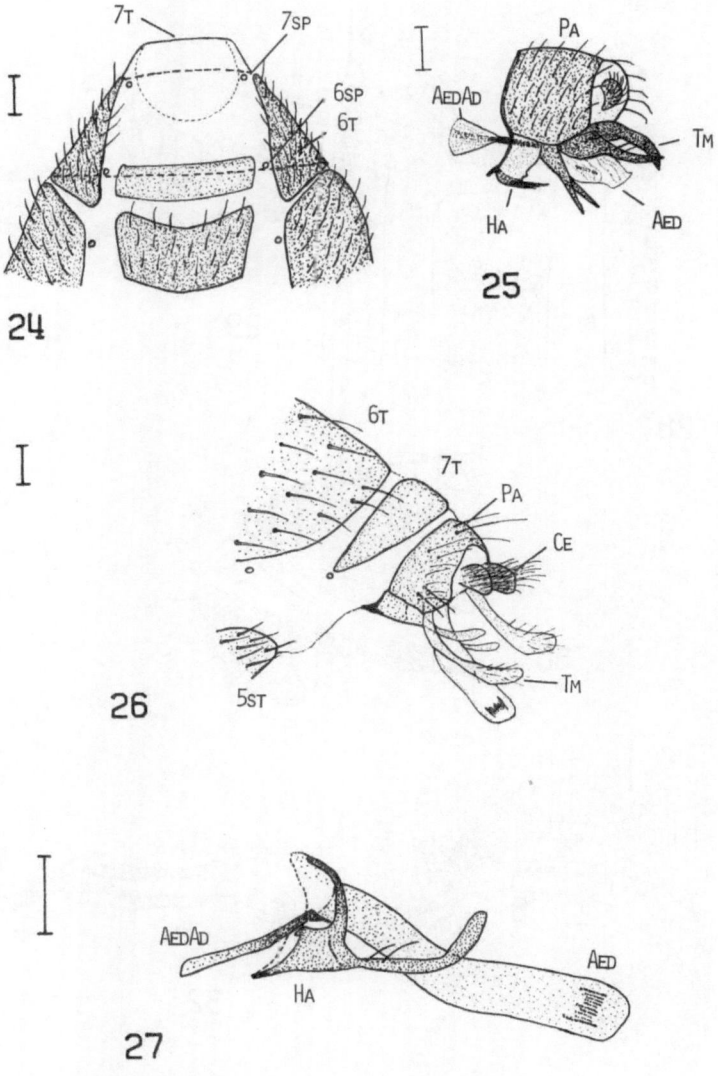

Figs. 24 - 27. 24. *Lauxania cylindricornis* (F.) (Lauxaniidae), postabdomen (♂) in ventral view (with hypopygium removed). 25. *Lauxania cylindricornis* (F.), hypopygium (♂) in lateral view. 26. *Camptoprosopella borealis* SHEWELL (Lauxaniidae), postabdomen (♂) in lateral view. 27. *Camptoprosopella borealis* SHEWELL, aedeagus, hypandrium and associated structures (♂) in lateral view.

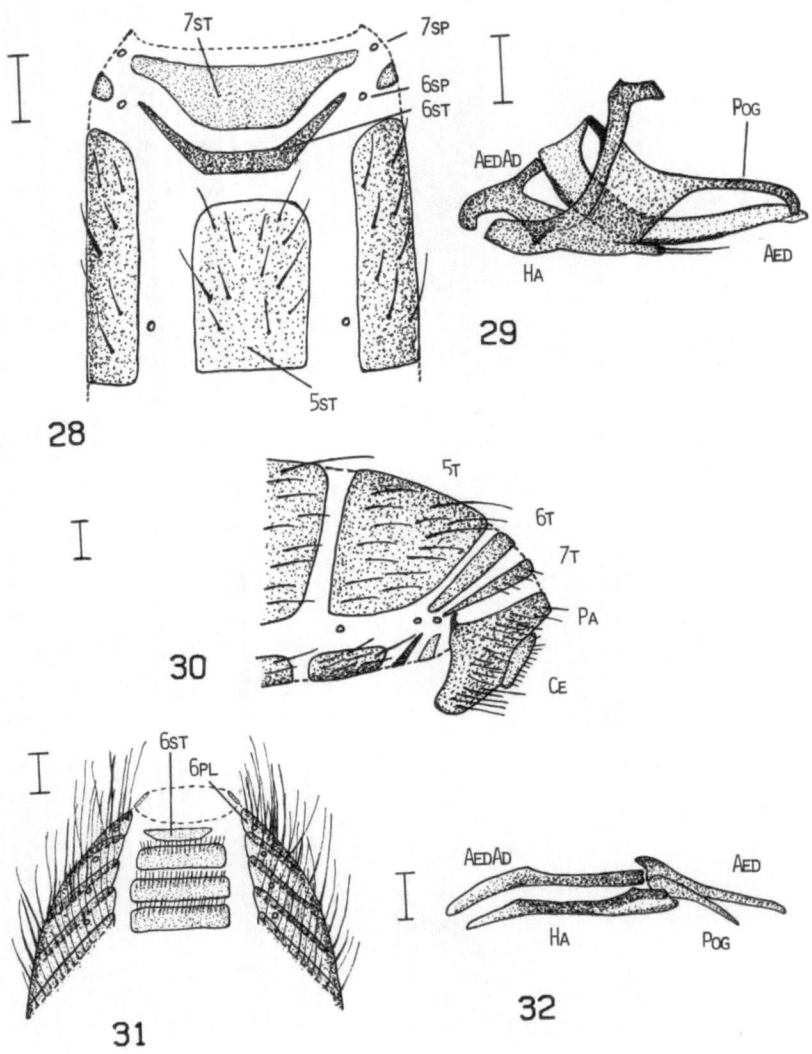

Figs. 28 - 32. 28. *Chamaemyia juncorum* (FALLÉN) (Chamaemyiidae), postabdomen (♂) in ventral view (with hypopygium removed). 29. *Chamaemyia juncorum* (FALLÉN), aedeagus, hypandrium and associated structures (♂) in lateral view. 30. *Chamaemyia juncorum* (FALLÉN), postabdomen (♂) in lateral view. 31. *Braula* sp. (Braulidae), abdomen (part) (♂) in ventral view (with hypopygium removed). 32. *Braula* sp., aedeagus, hypandrium and associated structures (♂) in lateral view.

278

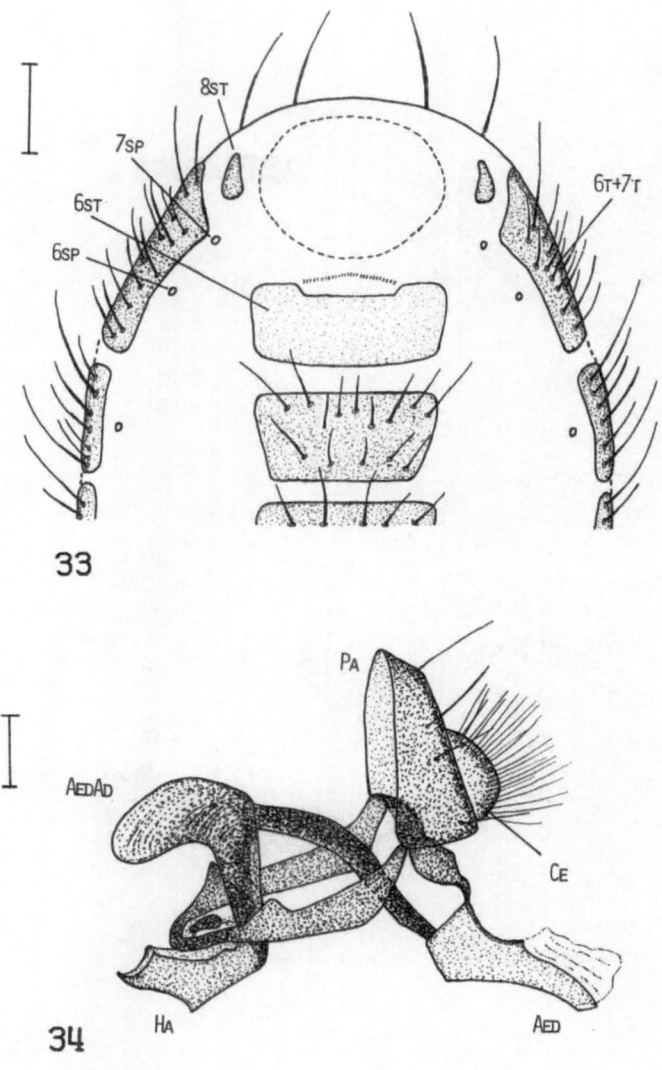

Figs. 33 - 34. 33. *Drosophila* sp. cf. *melanogaster* MEIGEN (Drosophilidae), postabdomen ($\male$) in ventral view (with hypopygium removed). 34. *Amiota picta* (COQUILLETT) (Drosophilidae), hypopygium ($\male$) in lateral view.

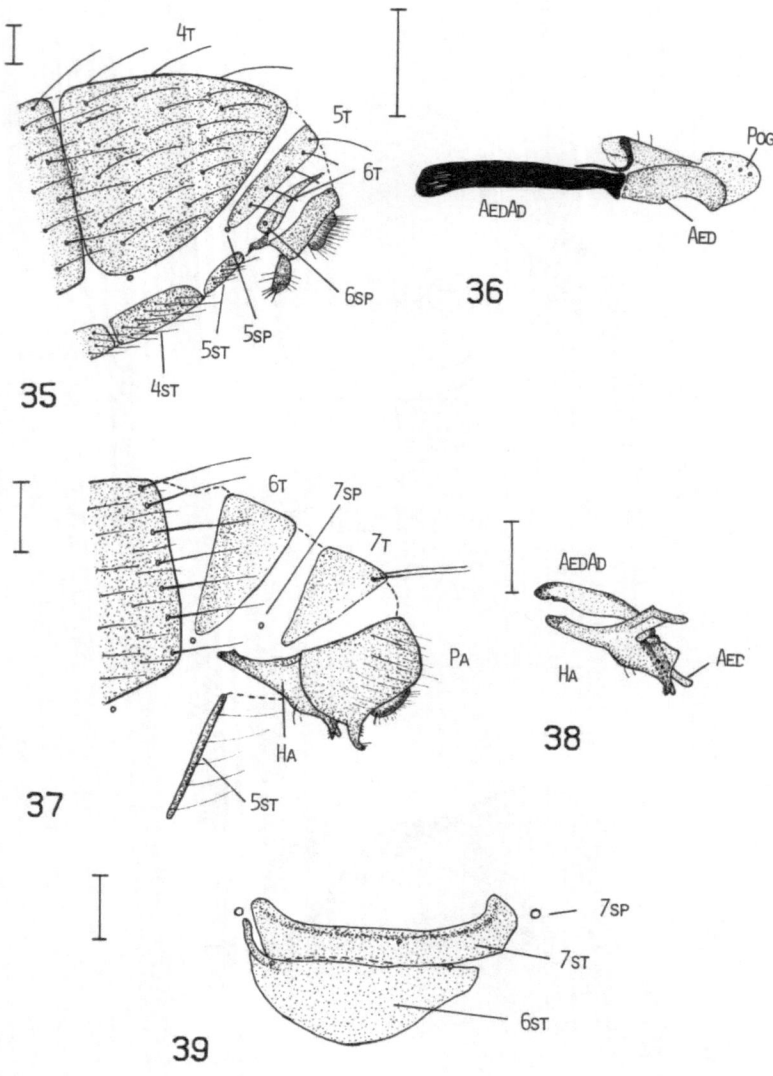

Figs. 35 - 39. 35. *Camilla atripes* Duda (Camillidae), abdomen (part) (♂) in lateral view. 36. *Camilla atripes* Duda, aedeagus and associated structures (♂) in lateral view (with the left postgonite removed). 37. *Campichoeta griseola* (Zetterstedt) (Campichoetidae), postabdomen (♂) in lateral view. 38. *Campichoeta griseola* (Zetterstedt), aedeagus, hypandrium and associated structures (♂) in lateral view. 39. *Curtonotum helvum* (Loew) (Curtonotidae), 6th and 7th sterna (♂) in posteroventral view.

280

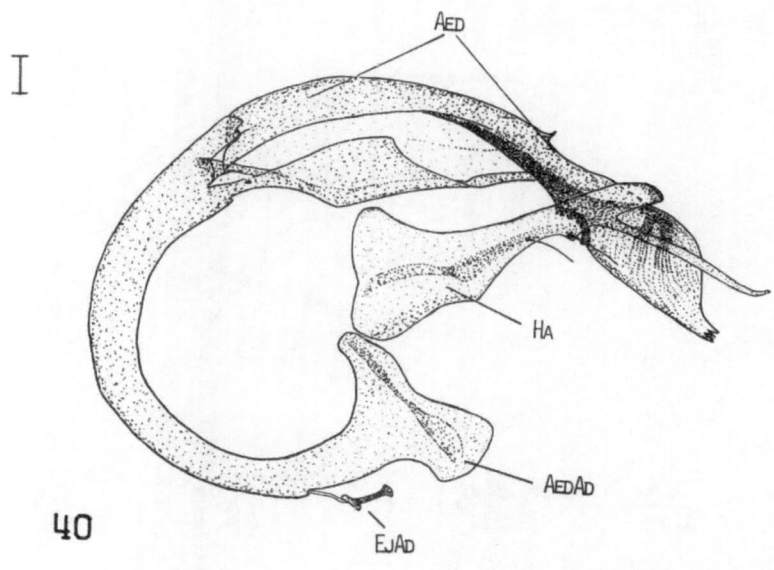

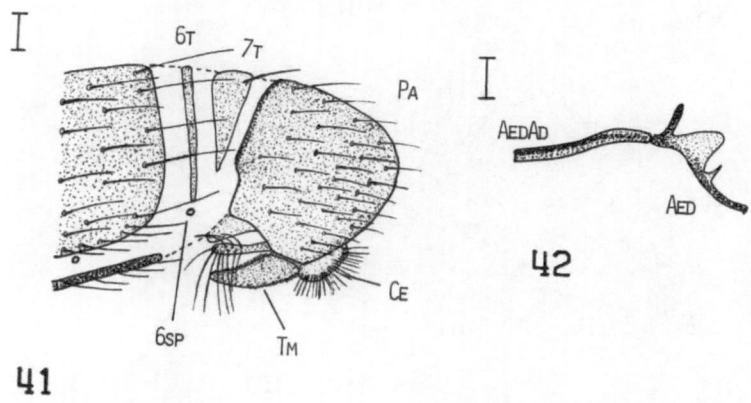

Figs. 40 - 42. 40. *Curtonotum helvum* (LOEW) (Curtonotidae), aedeagus, hypandrium and associated structures (♂) in lateral view. 41. *Diastata vagans* LOEW (Ephydridae), postabdomen (♂) in lateral view. 42. *Diastata vagans* LOEW, aedeagus and aedeagal apodeme (♂) in lateral view.

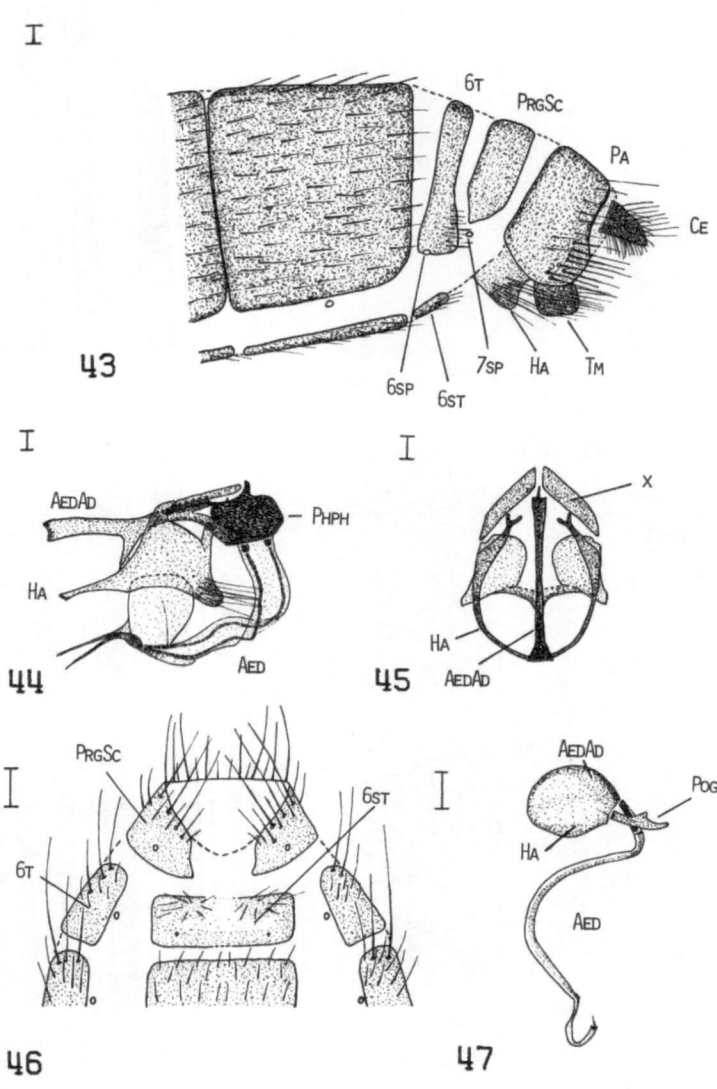

Figs. 43 - 47. 43. *Nothybus longithorax* Rondani (Nothybidae), postabdomen (♂) in lateral view. 44. *Nothybus longithorax* Rondani, aedeagus, hypandrium and associated structures (♂) in lateral view. 45. *Nothybus longithorax* Rondani, hypandrium and aedeagal apodeme (♂) in dorsal view. 46. *Scutops maculipennis* Malloch (Periscelididae), postabdomen (♂) in ventral view (with hypopygium removed). 47. *Scutops maculipennis* Malloch, aedeagus, hypandrium and associated structures (♂) in lateral view.

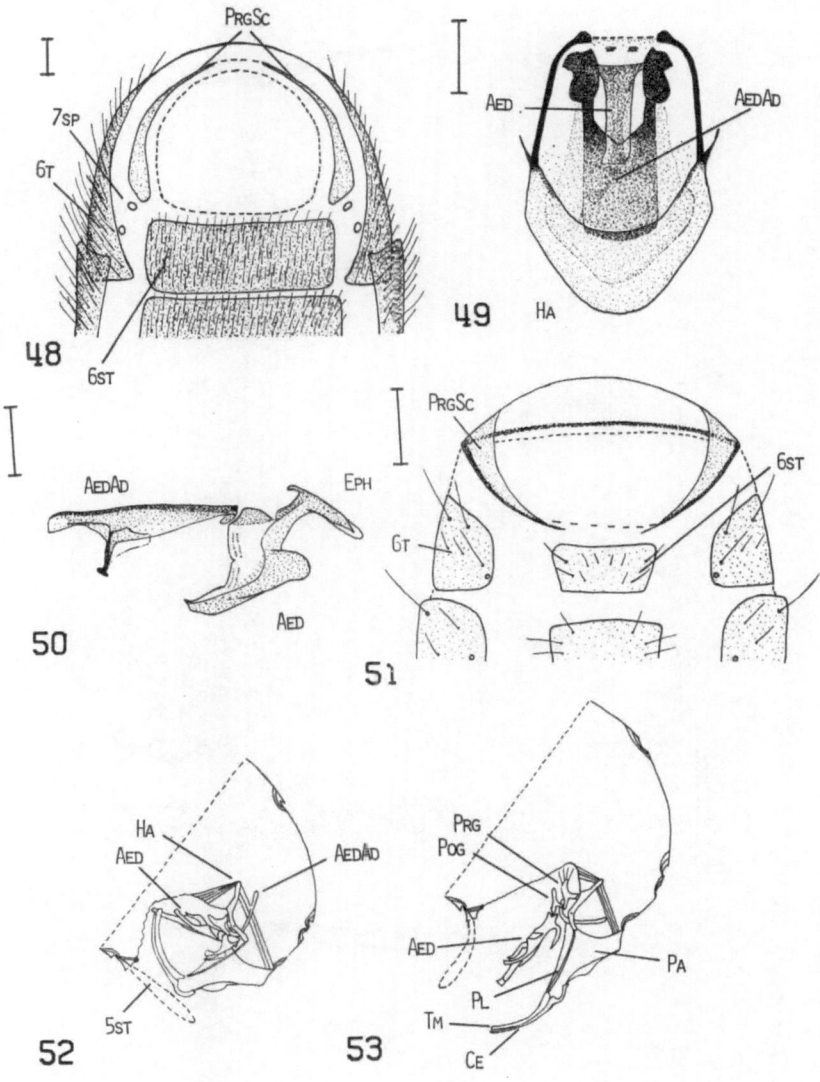

Figs. 48 - 53. 48. *Loxocera cylindrica* SAY (Psilidae), postabdomen (♂) in ventral view (with hypopygium removed). 49. *Psila rosae* (F.) (Psilidae), aedeagus, hypandrium and associated structures (♂) in ventral view. 50. *Teratomyza* sp. (Australia) (Teratomyzidae), aedeagus and associated structures (♂) in lateral view. 51. *Teratomyza* sp. (Australia), postabdomen (♂) in ventral view (with hypopygium removed). 52. *Calliphora erythrocephala* MEIGEN (Tachinidae *s.l.*), schematic representation of the male postabdomen in rest position (lateral view) (after SALZER 1968). 53. As the preceding, but in copulatory position (after SALZER 1968).

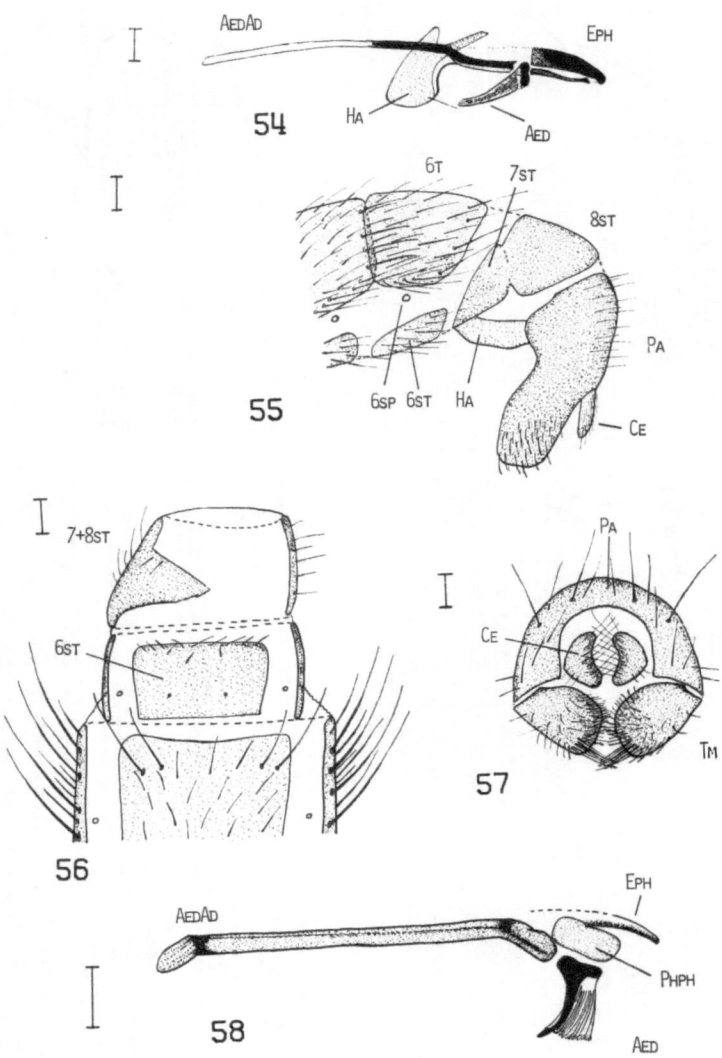

Fig. 54 - 58. 54. *Neotanypeza elegans* (WIEDEMANN) (Tanypezidae), aedeagus, hypandrium and associated structures (♂) in lateral view. 55. *Tanypeza luteipennis* KNAB & SHANNON (Tanypezidae), postabdomen (♂) in lateral view. 56. *Heteromyza atricornis* MEIGEN (Heteromyzidae), postabdomen (♂) in ventral view (with hypopygium removed). 57. *Heteromyza atricornis* MEIGEN, hypopygium (♂) in posterior view. 58. *Heteromyza atricornis* MEIGEN, aedeagus and associated structures (♂) in lateral view.

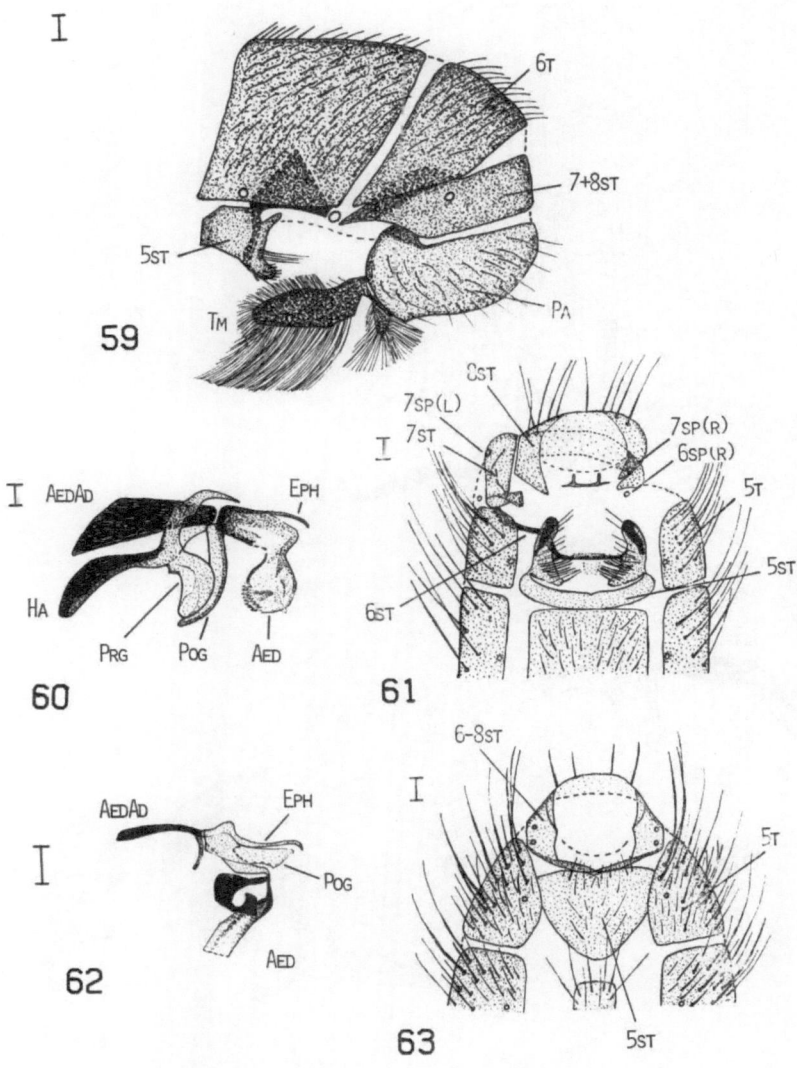

Figs. 59 - 63. 59. *Scatophaga aldrichi* (MALLOCH) (Scatophagidae), postabdomen (♂) in lateral view. 60. *Scatophaga aldrichi* (MALLOCH), aedeagus, hypandrium and associated structures (♂) in lateral view. 61. *Pegomya versicolor* (MEIGEN) (Anthomyiidae), postabdomen (♂) in ventral view (with hypopygium removed). 62. *Pegomya versicolor* (MEIGEN), aedeagus and associated structures (♂) in lateral view. 63. *Fannia canicularis* (L.) (Fanniidae), postabdomen (♂) in ventral view (with hypopygium removed).

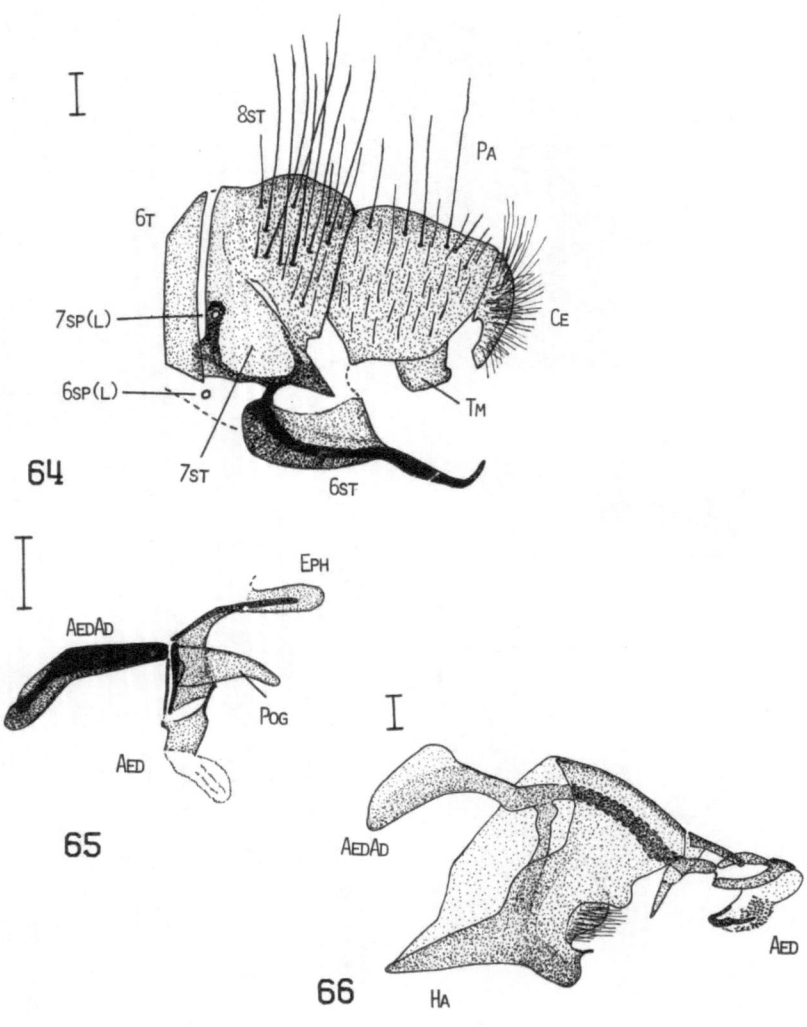

Figs. 64 - 66. 64. *Muscina assimilis* (FALLÉN) (Muscidae), postabdomen (♂) in lateral view. 65. *Muscina assimilis* (FALLÉN), aedeagus and associated structures (♂) in lateral view. 66. *Glossina morsitans* WESTWOOD (Glossinidae), aedeagus, hypandrium and associated structures (♂) in lateral view.

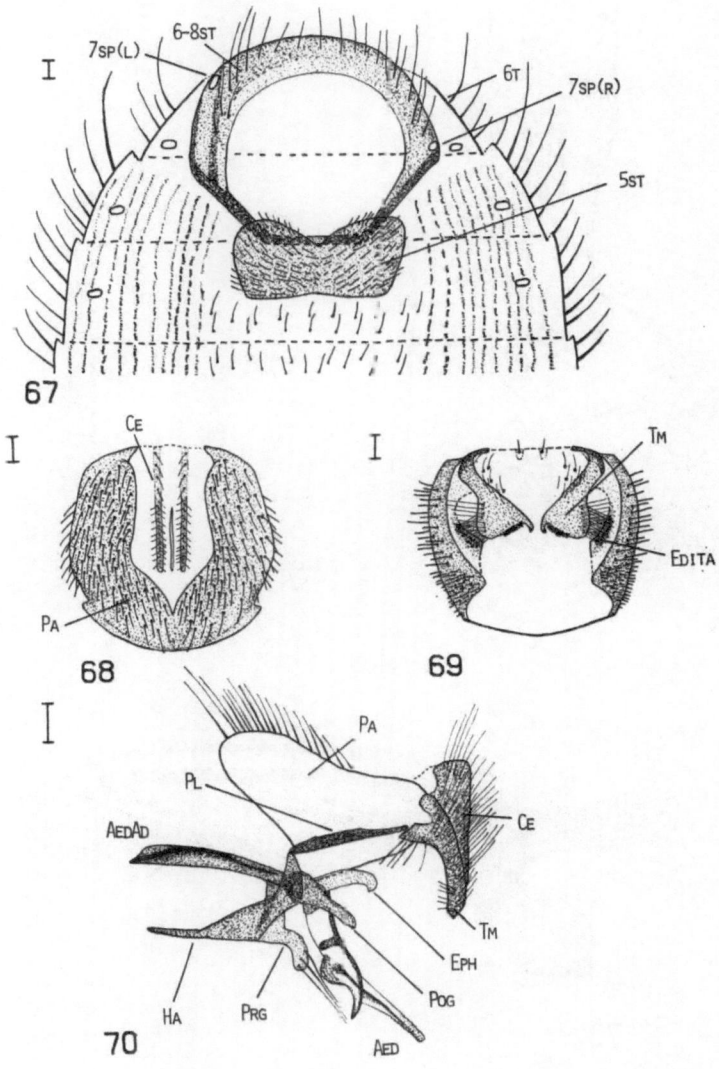

Fig. 67 - 70. 67. *Glossina morsitans* WESTWOOD (Glossinidae), abdomen (part) (♂) in ventral view (with hypopygium removed). 68. *Glossina morsitans* WESTWOOD, hypopygium (♂) in dorsal view. 69. *Glossina morsitans* WESTWOOD, periandrium, telomeres and associated structures (♂) in ventral view. 70. *Eucalliphora lilaea* (WALKER) (Tachinidae *s.l.*), hypopygium (♂) in lateral view (with left side of periandrium removed).

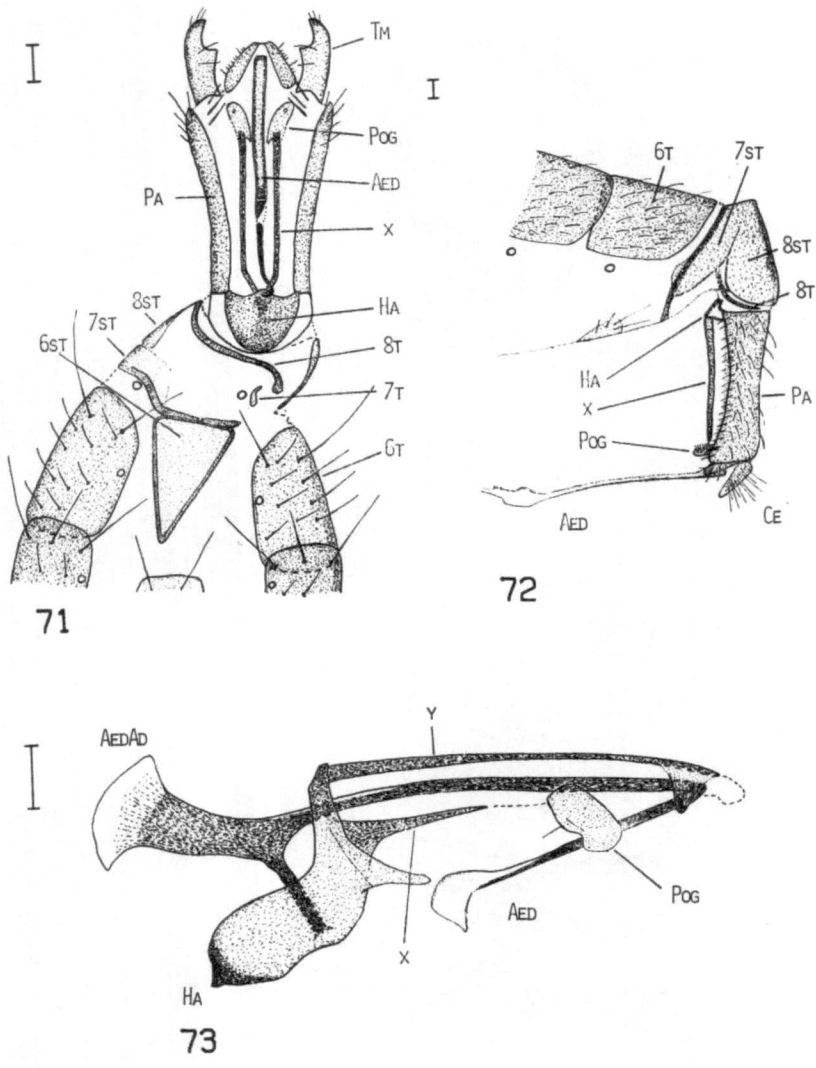

Figs. 71 - 73. 71. *Heloclusia imperfecta* MALLOCH (Cypselosomatidae), postabdomen (♂) in ventral view (with hypopygium turned back). 72. *Gymnonerius fuscus ceylanicus* HENNIG (Neriidae), postabdomen (♂) in lateral view. 73. *Micropeza lineata* VAN DUZEE (Micropezidae), aedeagus, hypandrium and associated structures (♂) in lateral view.

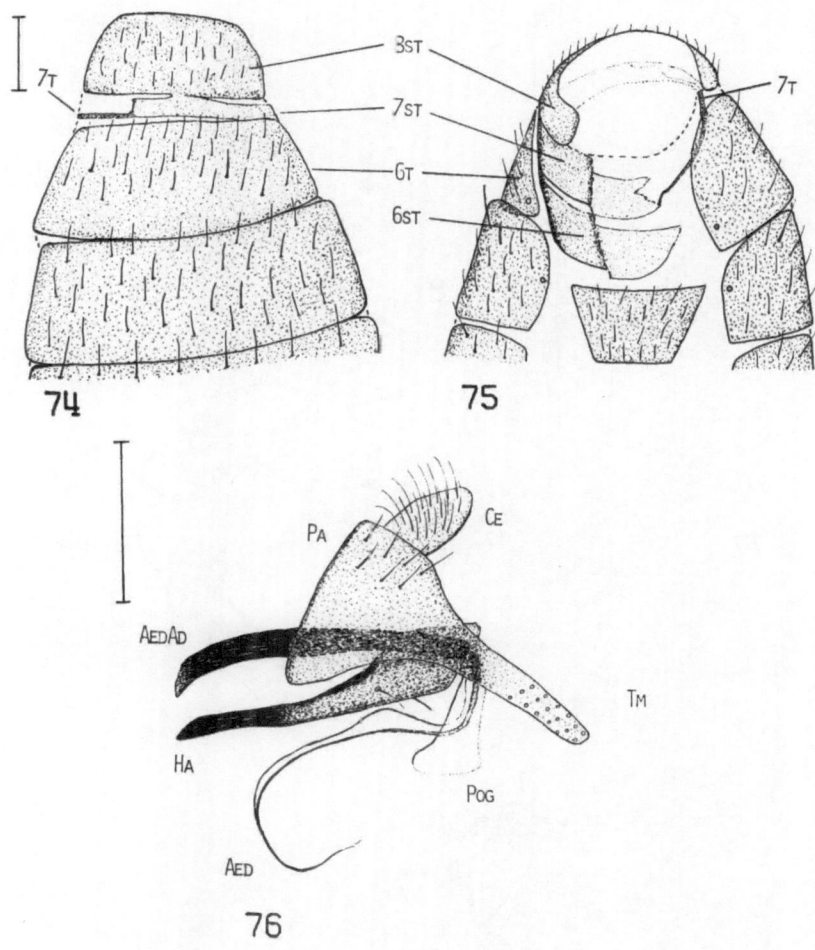

Figs. 74 - 76. 74. *Australimyza anisotomae* HARRISON (Australimyzidae), postabdomen
(♂) in dorsal view (with hypopygium removed). 75. *Australimyza anisotomae* HAR-
RISON, postabdomen (♂) in ventral view (with hypopygium removed). 76. *Australimyza
anisotomae* HARRISON, hypopygium (♂) in lateral view.

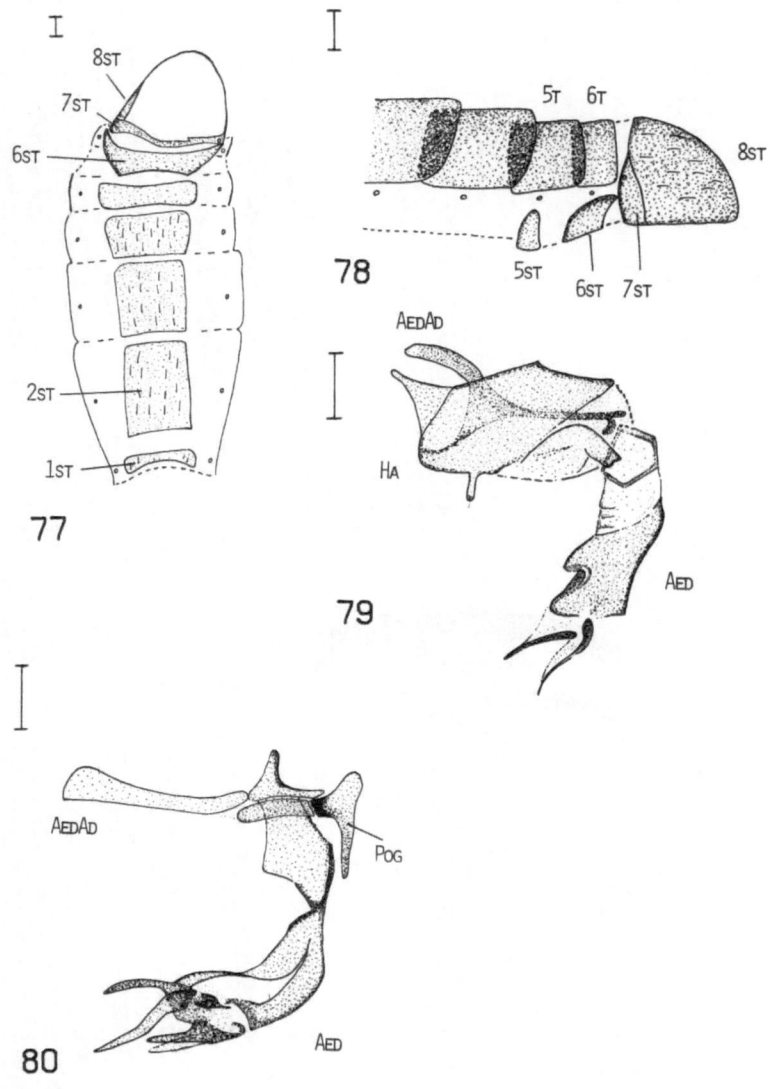

Figs. 77 - 80. 77. *Centrioncus prodiopsis* SPEISER (Diopsidae), abdomen (♂) in ventral view (with hypopygium removed). 78. *Centrioncus prodiopsis* SPEISER, abdomen (part) (♂) in lateral view (with hypopygium removed). 79. *Centrioncus prodiopsis* SPEISER, aedeagus, hypandrium and associated structures (♂) in lateral view. 80. *Syringogaster rufa* CRESSON (Syringogastridae), aedeagus and associated structures (♂) in lateral view.

290

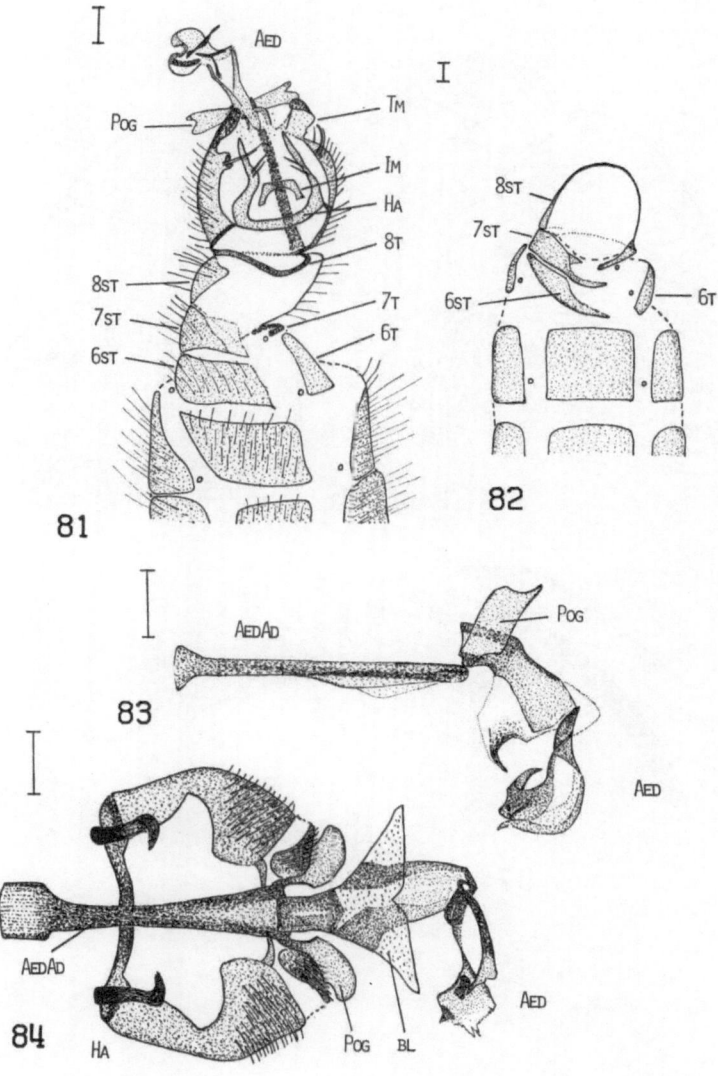

Fig. 81 - 84. 81. *Malacomyia sciomyzina* (HALIDAY) (Coelopidae), postabdomen (♂) in ventral view (with hypopygium turned back). 82. *Heterocheila hannai* (COLE) (Coelopidae), postabdomen (♂) in ventral view (with hypopygium removed, and with the dense pubescence omitted). 83. *Malacomyia sciomyzina* (HALIDAY), aedeagus and associated structures (♂) in lateral view. 84. *Heterocheila hannai* (COLE), aedeagus, hypandrium and associated structures (♂) in dorsal view (with aedeagus in copulatory position).

291

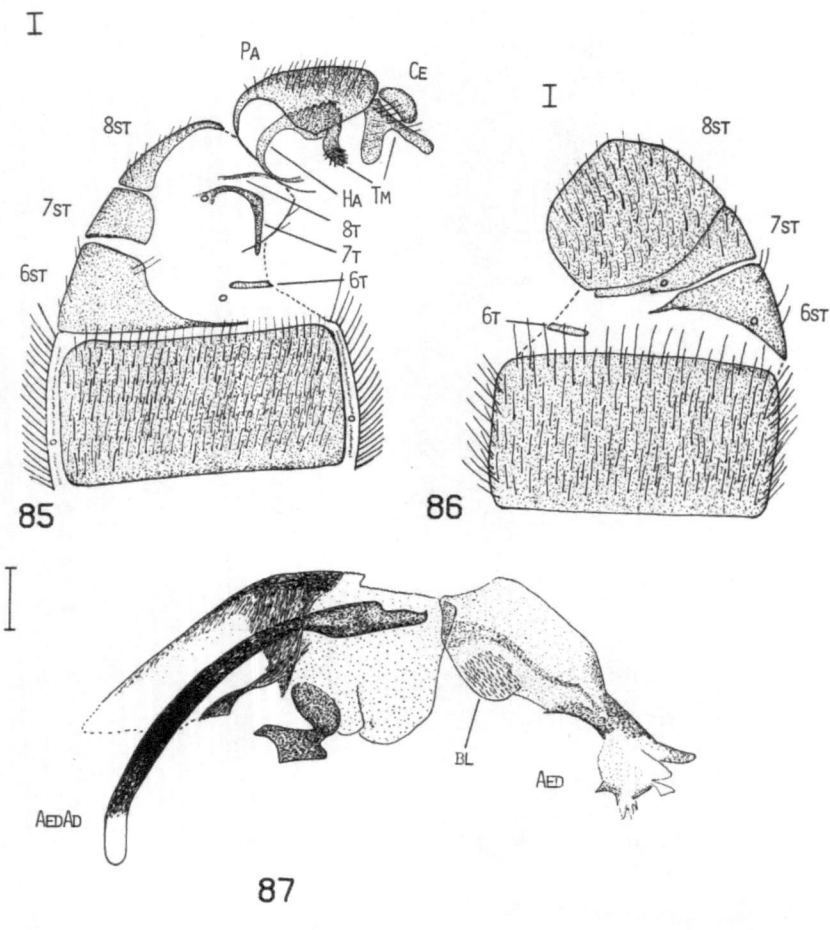

Figs 85 - 87. 85. *Pelidnoptera nigripennis* (F.) (Phaeomyiidae), postabdomen (♂) in ventral view, with the left side of the hypopygium (turned laterally). 86. *Pelidnoptera nigripennis* (F.), postabdomen (♂) in dorsal view (with hypopygium removed). 87. *Pelidnoptera nigripennis* (F.), aedeagus and associated structures (♂) in lateral view.

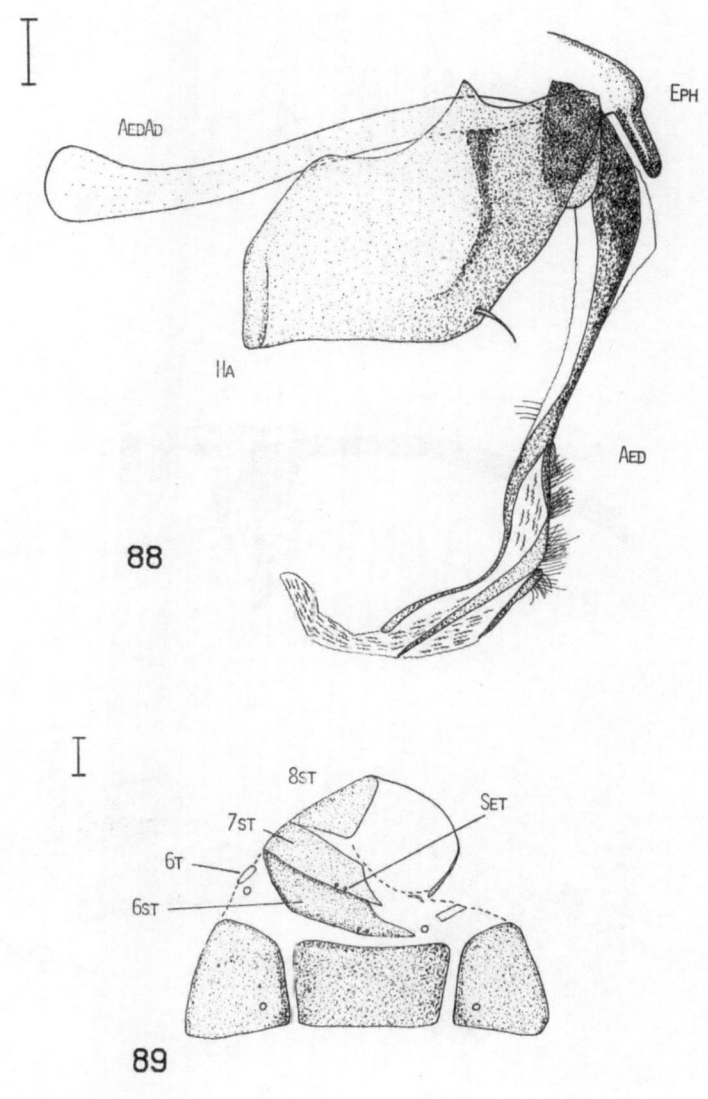

Figs. 88 - 89. 88. *Maorimyia bipunctata* (Hutton) (Dryomyzidae), aedeagus, hypandrium and associated structures (♂) in lateral view. 89. *Dryomyza anilis* (Fallén) (Dryomyzidae), postabdomen (♂) in ventral view (with hypopygium removed, and with the dense pubescence omitted).

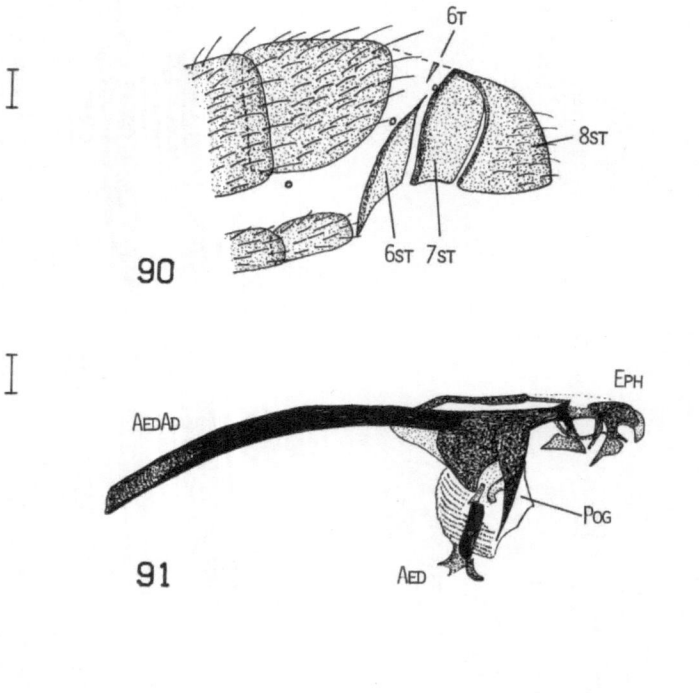

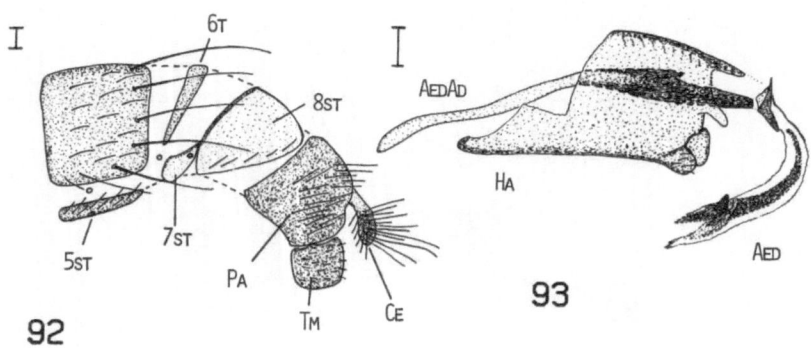

Figs. 90 - 93. 90. *Pteromicra apicata* (LOEW) (Sciomyzidae), postabdomen (♂) in lateral view (with hypopygium removed). 91. *Pherbellia albocostata* (FALLÉN) (Sciomyzidae), aedeagus and associated structures (♂) in lateral view. 92. *Helosciomyza subspinicosta* TONNOIR & MALLOCH (Helosciomyzidae), postabdomen (♂) in lateral view. 93. *Helosciomyza subspinicosta* TONNOIR & MALLOCH, aedeagus, hypandrium and associated structures (♂) in lateral view.

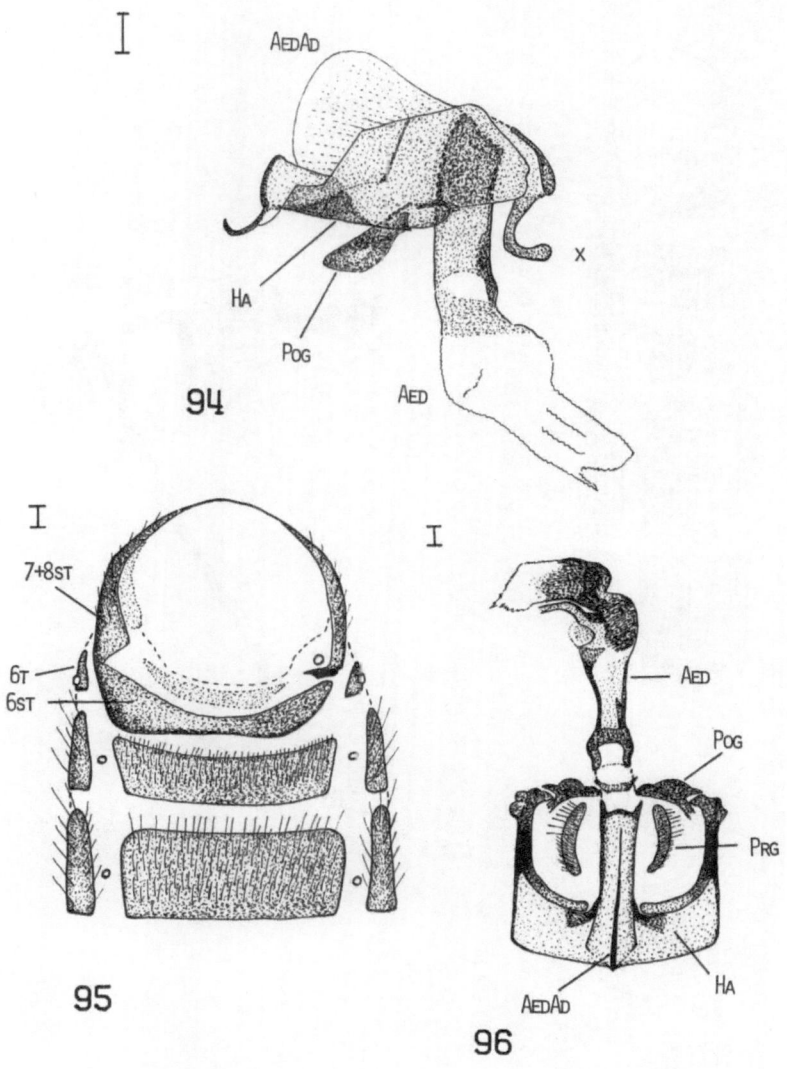

Figs. 94 - 96. 94. *Rhytidops floridensis* (ALDRICH) (Ropalomeridae), aedeagus, hypandrium and associated structures (♂) in lateral view. 95. *Orygma luctuosum* MEIGEN (Sepsidae), postabdomen (♂) in ventral view (with hypopygium removed). 96. *Orygma luctuosum* MEIGEN, aedeagus, hypandrium and associated structures (♂) in ventral view (with aedeagus in copulatory position).

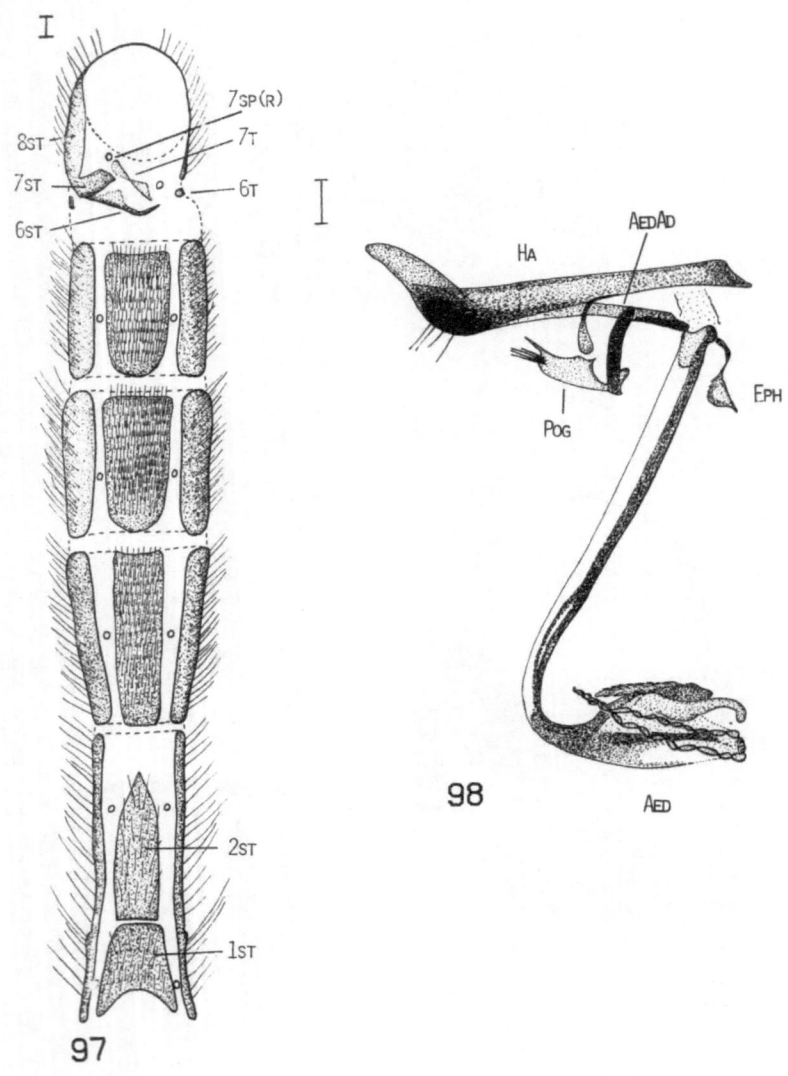

Figs. 97 - 98. 97. *Megamerina dolium* (F.) (Megamerinidae), abdomen (♂) in ventral view (with hypopygium removed). 98. *Megamerina dolium* (F.), aedeagus, hypandrium and associated structures (♂) in lateral view.

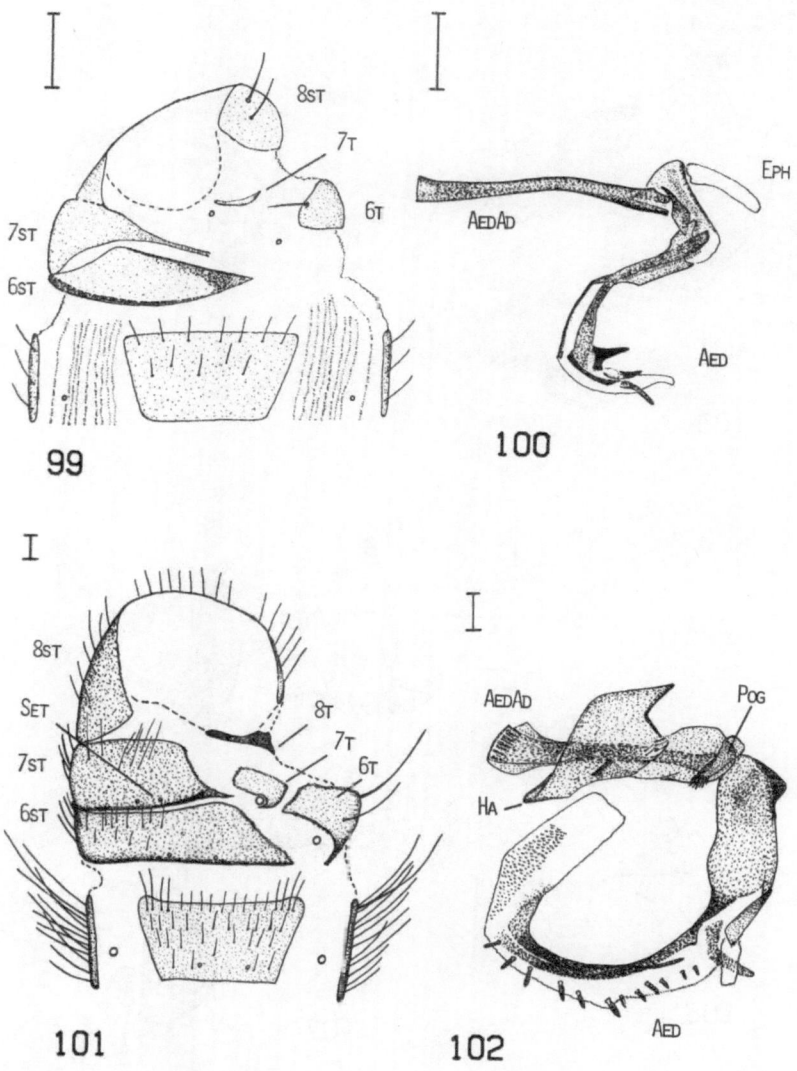

Figs. 99 - 102. 99. *Cremifania nigrocellulata* CzERNY (Cremifaniidae), postabdomen (♂) in ventral view (with hypopygium removed). 100. *Cremifania nigrocellulata* CzER-NY, aedeagus and associated structures (♂) in lateral view. 101. *Orbellia tokyoensis* CzERNY (Heleomyzidae), postabdomen (♂) in ventral view (with hypopygium removed). 102. *Orbellia tokyoensis* CzERNY, aedeagus, hypandrium and associated structures (♂) in lateral view.

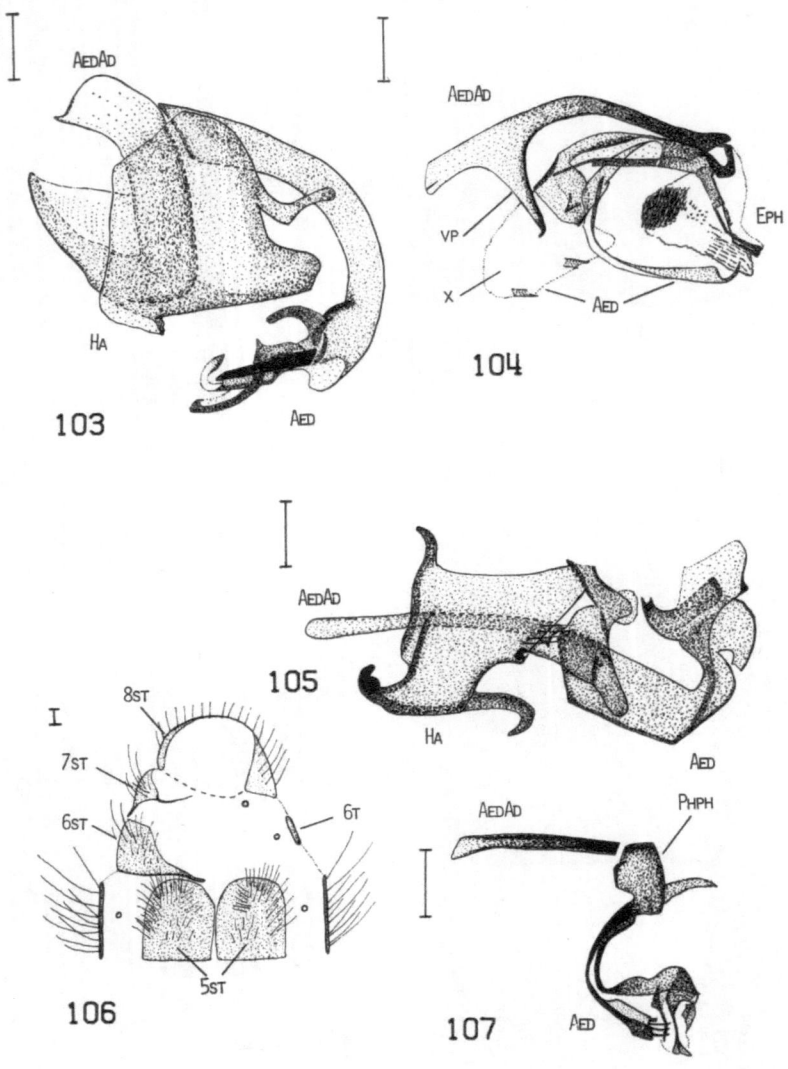

Figs. 103 - 107. 103. *Allophylopsis scutellata* (Hutton) (Heleomyzidae), aedeagus, hypandrium and associated structures (♂) in lateral view. 104. *Anthomyza* sp. (Anthomyzidae), aedeagus and associated structures (♂) in lateral view. 105. *Tapeigaster marginifrons* Bezzi (Rhinotoridae), aedeagus, hypandrium and associated structures (♂) in lateral view. 106. *Tapeigaster marginifrons* Bezzi, postabdomen (♂) in ventral view (with hypopygium removed). 107. *Trixoscelis* sp. (Trixoscelididae), aedeagus and associated structures (♂) in lateral view.

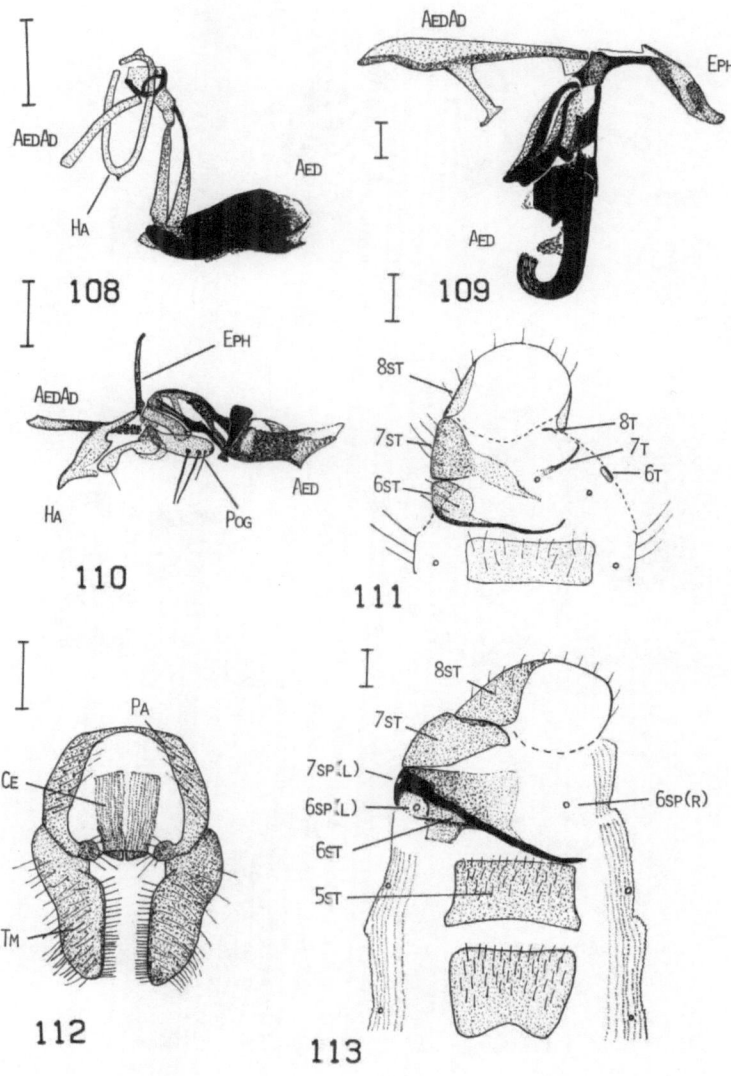

Figs. 108 - 113. 108. *Asteia amoena* MEIGEN (Asteiidae), aedeagus, hypandrium and associated structures (♂) in anterolateral view. 109. *Opomyza germinationis* (L.) (Opomyzidae), aedeagus and associated structures (♂) in lateral view. 110. *Borboropsis puberula* (ZETTERSTEDT) (Borboropsidae), aedeagus, hypandrium and associated structures (♂) in lateral view (with aedeagus in copulatory position). 111. *Borboropsis puberula* (ZETTERSTEDT), postabdomen (♂) in ventral view (with hypopygium removed). 112. *Sphaerocera curvipes* LATREILLE (Sphaeroceridae), hypopygium (♂) in posterior view. 113. *Sphaerocera curvipes* LATREILLE, abdomen (part) (♂) in ventral view (with hypopygium removed).

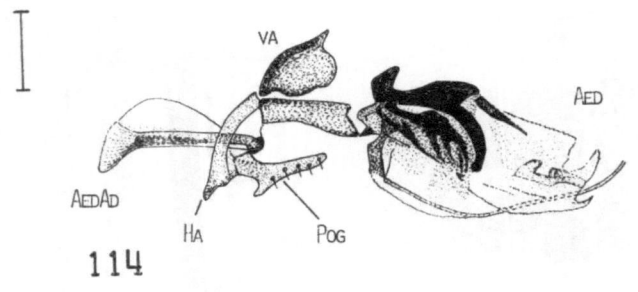

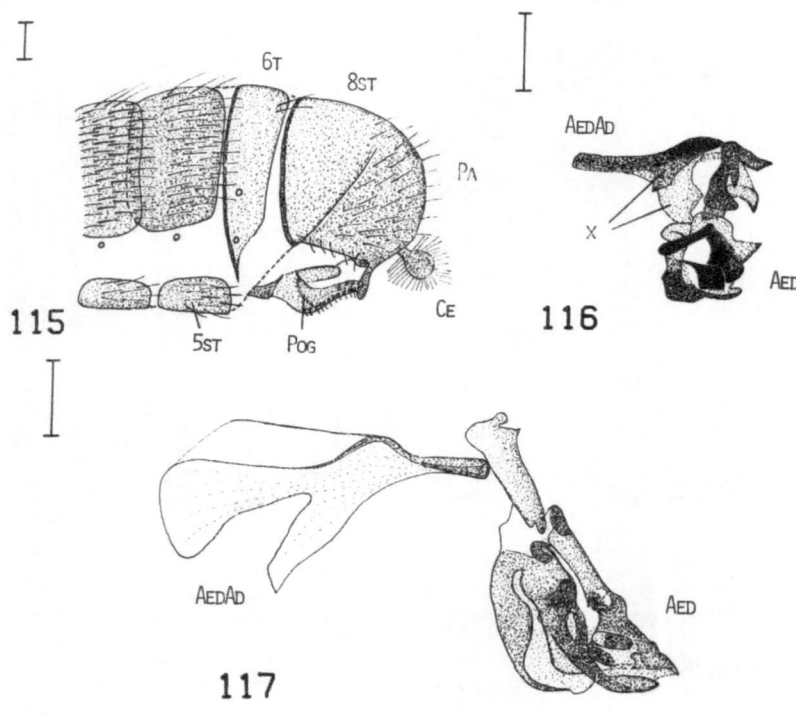

Figs. 114 - 117. 114. *Sphaerocera curvipes* LATREILLE (Sphaeroceridae), aedeagus, hypandrium and associated structures (♂) in lateral view (with aedeagus in copulatory position). 115. *Chyromya flava* (L.) (Chyromyidae), postabdomen (♂) in lateral view. 116. *Cyamops nebulosus* MELANDER (Aulacigastridae), aedeagus and associated structures (♂) in lateral view. 117. *Chyromya flava* (L.), aedeagus and associated structures (♂) in lateral view.

300

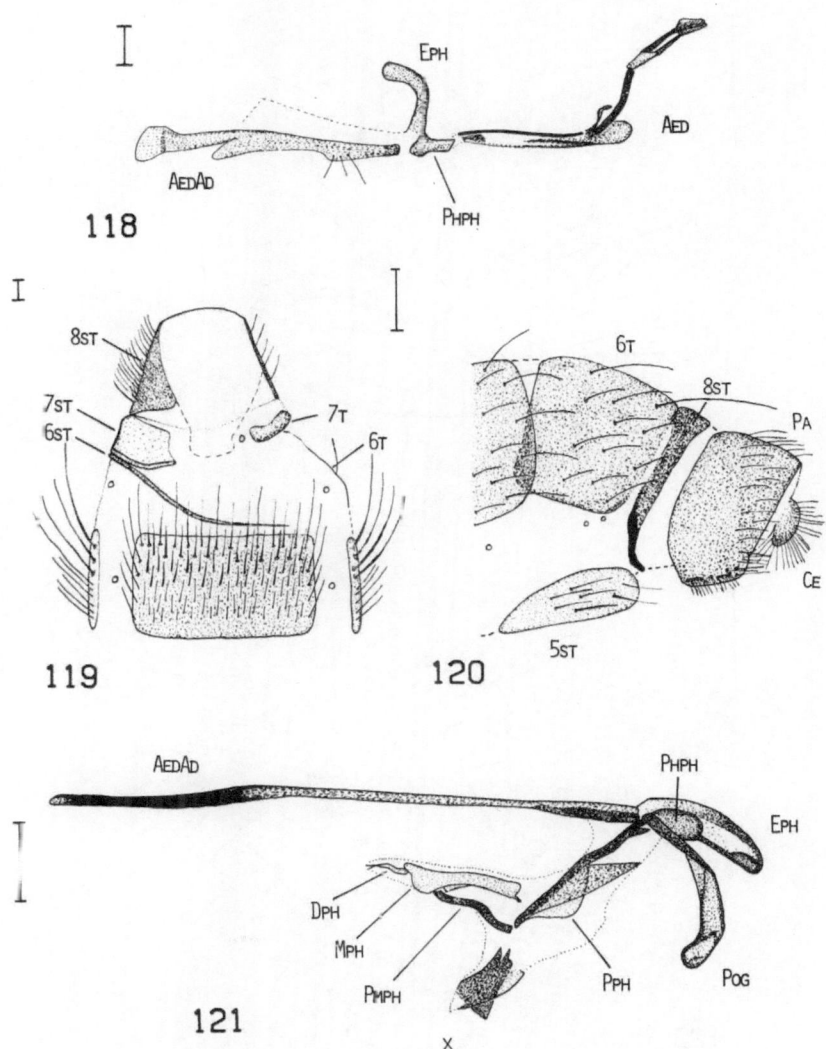

Fig. 118 - 121. 118. *Clusia lateralis* (WALKER) (Clusiidae), aedeagus and associated structures (♂) in lateral view (with aedeagus in copulatory position). 119. *Clusia lateralis* (WALKER), postabdomen (♂) in ventral view (with hypopygium removed). 120. *Phytomyza cineracea* HENDEL (Agromyzidae), postabdomen (♂) in lateral view. 121. *Phytomyza symphyti* HENDEL, aedeagus and associated structures (♂) in lateral view.

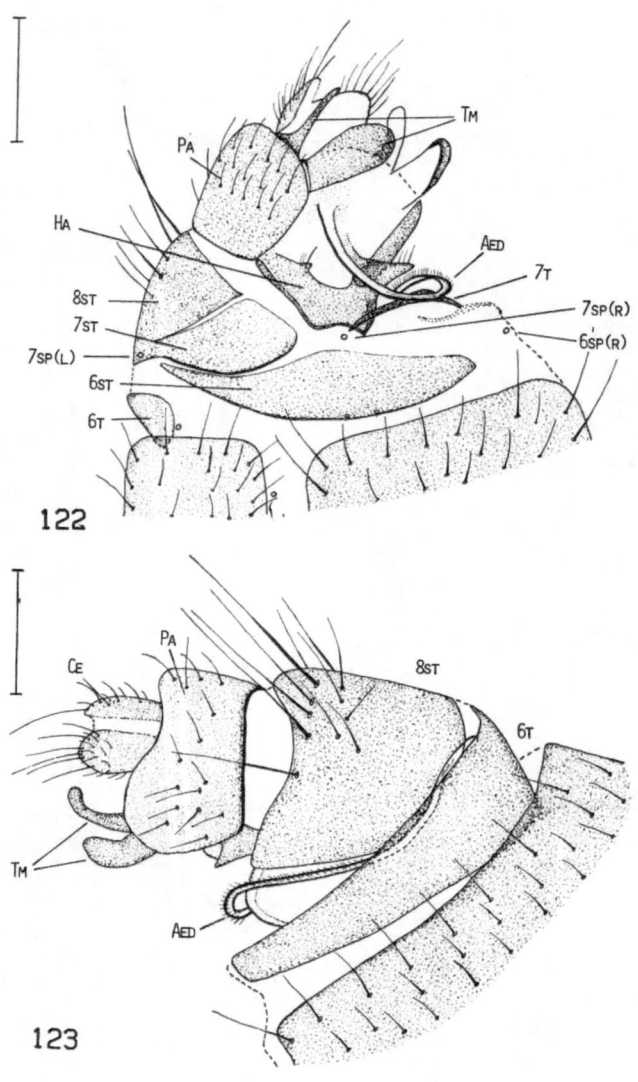

Figs. 122 - 123. 122. *Chiropteromyza wegelii* FREY (Chiropteromyzidae), postabdomen (♂) in ventral view. 123. *Chiropteromyza wegelii* FREY, postabdomen (♂) in dorsal view.

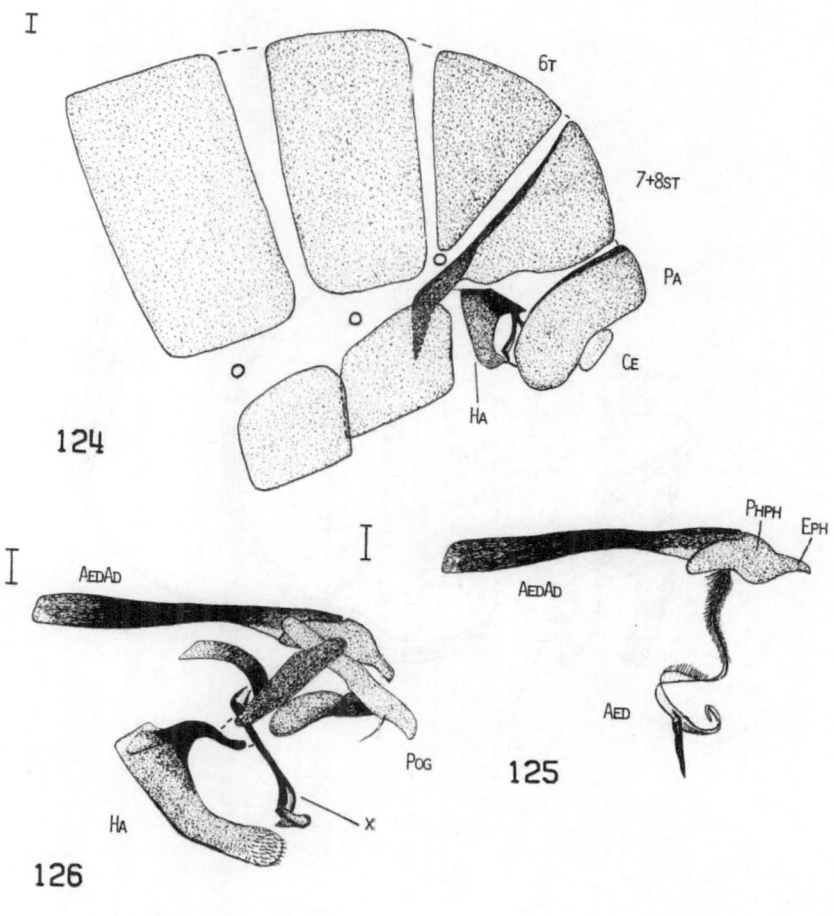

Figs. 124 - 126. 124. *Mormotomyia hirsuta* AUSTEN (Mormotomyiidae), abdomen (part) (♂) in lateral view (with the dense pubescence omitted). 125. *Mormotomyia hirsuta* AUSTEN, aedeagus and aedeagal apodeme (♂) in lateral view. 126. *Mormotomyia hirsuta* AUSTEN, hypandrium, aedeagal apodeme and associated structures (♂) in lateral view.

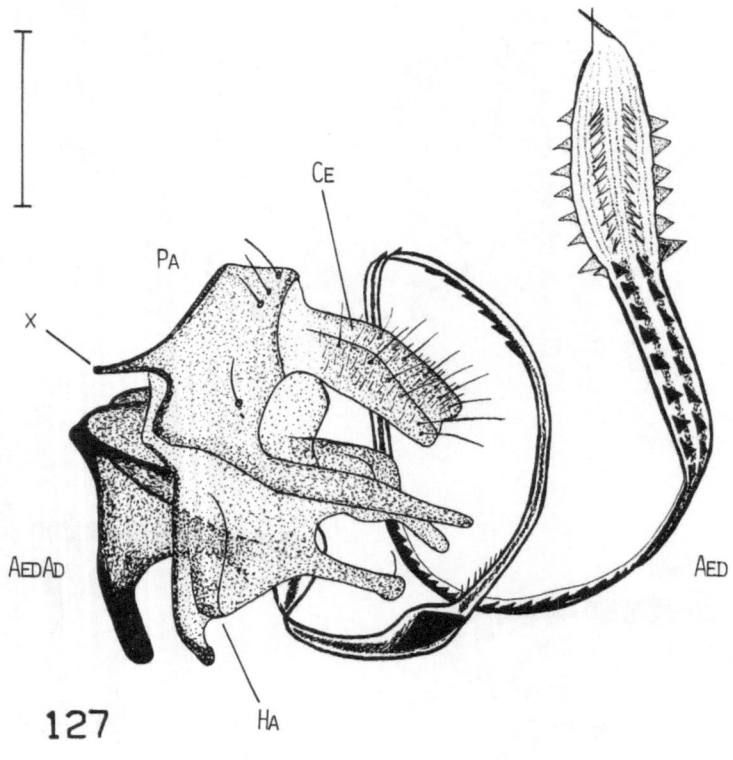

127

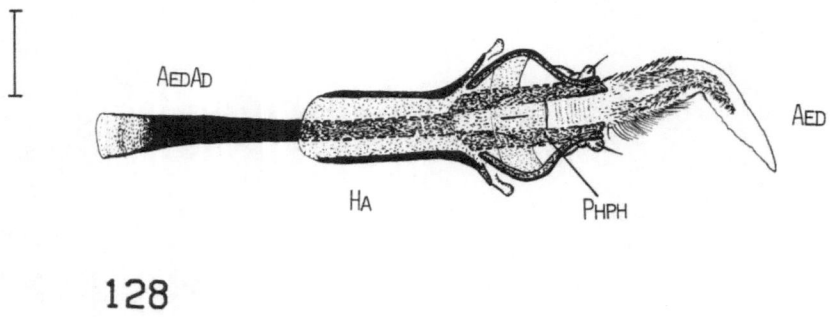

128

Figs. 127 - 128. 127. *Prosopantrum inerme* MALLOCH (Cnemospathidae), hypopygium (♂) in lateral view. 128. *Odinia boletina* (ZETTERSTEDT) (Odiniidae), aedeagus, hypandrium and associated structures (♂) in ventral view (with aedeagus in copulatory position).

304

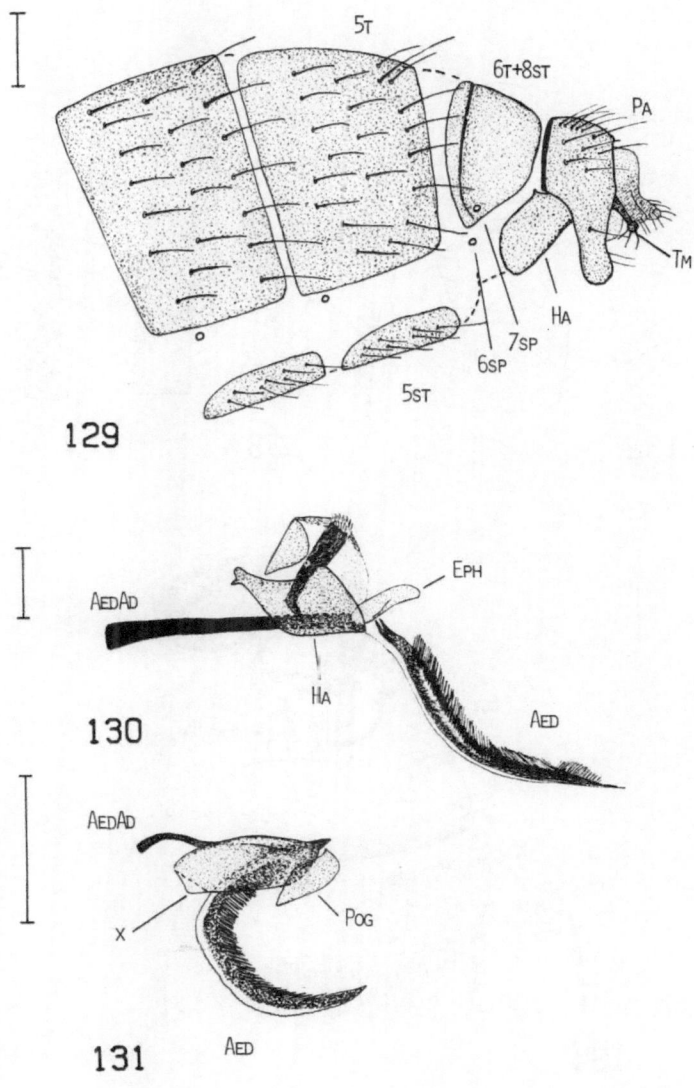

Figs. 129 - 131. 129. *Pelomyiella melanderi* (STURTEVANT) (Tethinidae), abdomen (part) (♂) in lateral view. 130. *Pelomyiella melanderi* (STURTEVANT), aedeagus, hypandrium and associated structures (♂) in lateral view. 131. *Acartophthalmus nigrinus* (ZETTER-STEDT) (Acartophthalmidae), aedeagus and associated structures (♂) in lateral view.

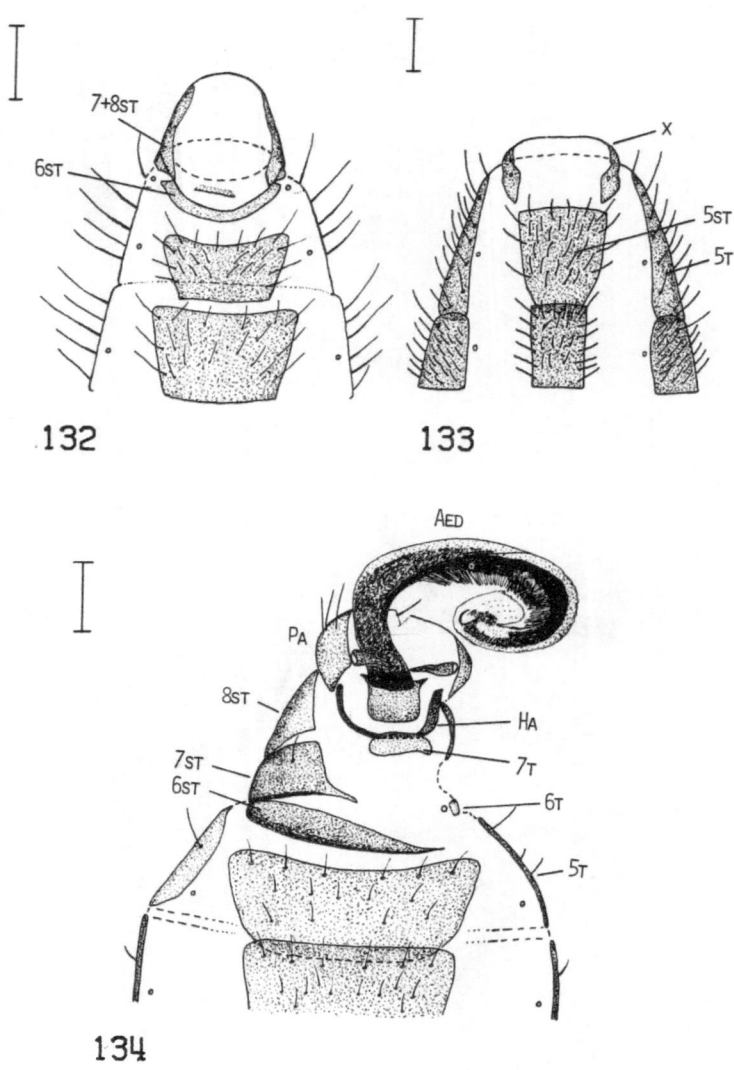

Fig. 132 - 134. 132. *Acartophthalmus nigrinus* (ZETTERSTEDT) (Acartophthalmidae), abdomen (part) (♂) in ventral view (with hypopygium removed). 133. *Leptometopa latipes* (MEIGEN) (Milichiidae), abdomen (part) (♂) in ventral view (with hypopygium removed). 134. *Hemeromyia* sp. (Carnidae), abdomen (part) (♂) in ventral view.

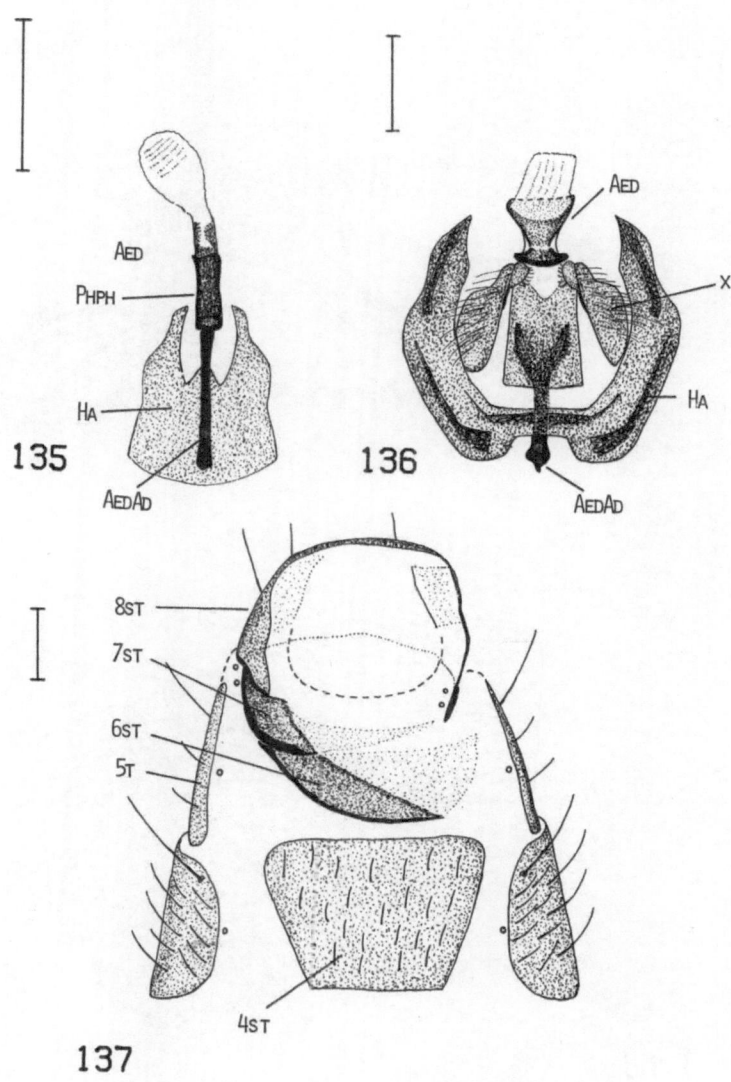

Fig. 135 - 137. 135. *Leptometopa latipes* (MEIGEN) (Milichiidae), aedeagus, hypandrium and associated structures (♂) in dorsal view. 136. *Lipara lucens* MEIGEN (Chloropidae), aedeagus, hypandrium and associated structures (♂) in ventral view. 137. *Lasiopleura shewelli* SABROSKY (Chloropidae), abdomen (part) (♂) in ventral view (with hypopygium removed).

307

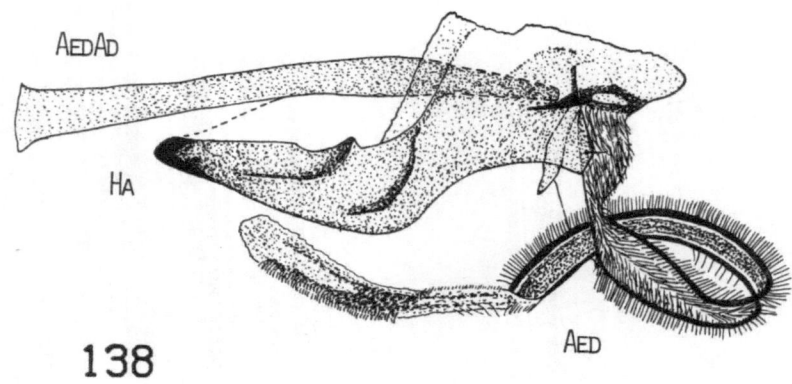

138

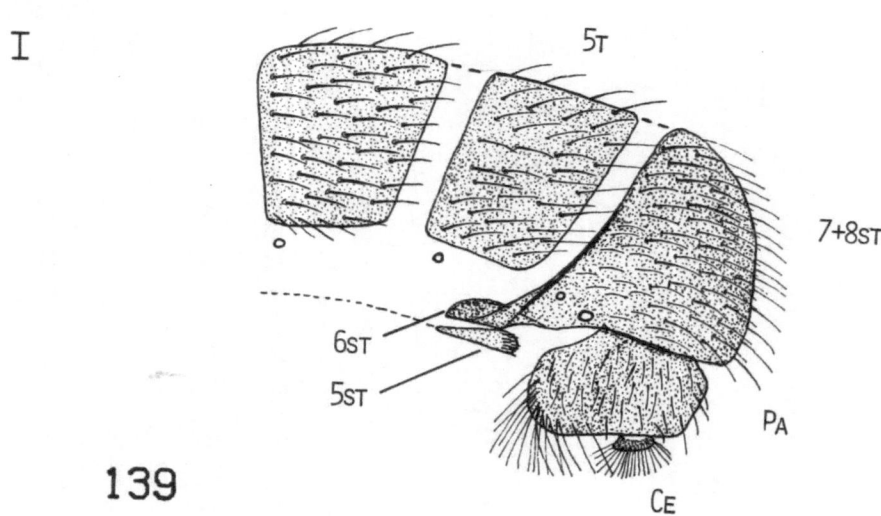

139

Figs. 138 - 139. 138. *Dalmannia nigriceps* Loew (Conopidae), aedeagus, hypandrium and associated structures (♂) in lateral view. 139. *Zodion fulvifrons* Say (Conopidae), abdomen (part) (♂) in lateral view.

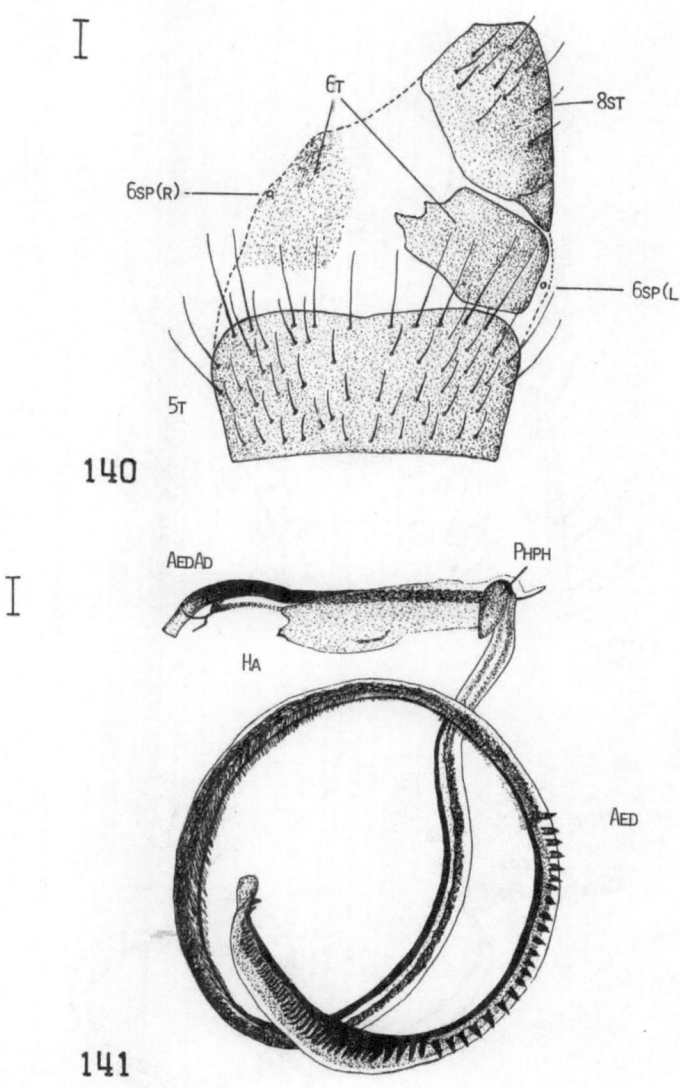

Figs. 140 - 141. 140. *Eurygnathomyia bicolor* (ZETTERSTEDT) (Eurygnathomyiidae), postabdomen (♂) in dorsal view (with hypopygium removed). 141. *Eurygnathomyia bicolor* (ZETTERSTEDT), aedeagus, hypandrium and associated structures (♂) in lateral view.

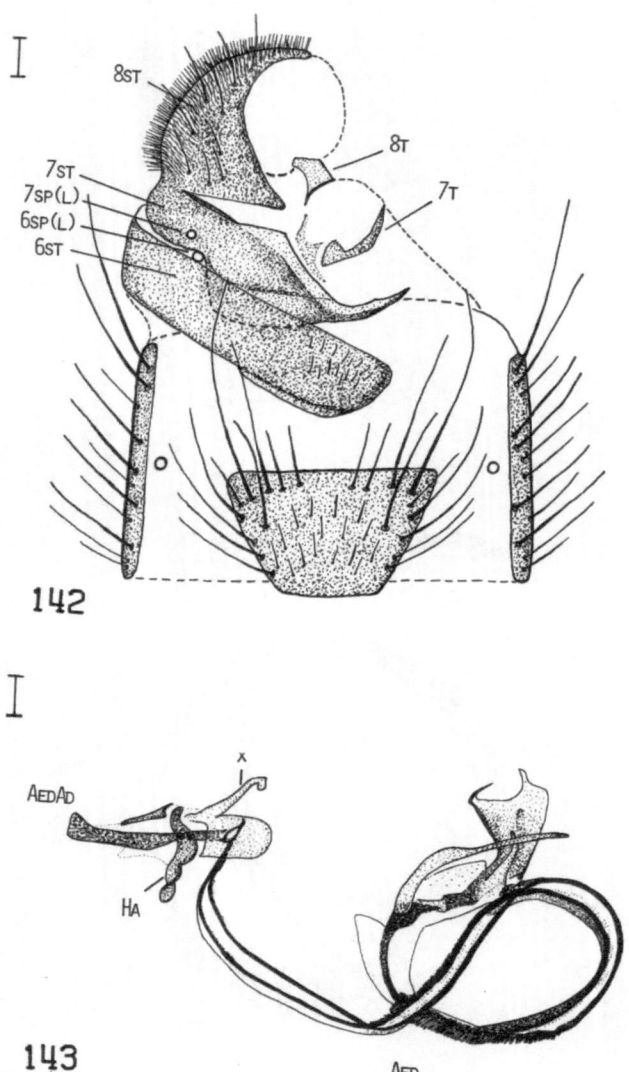

Figs. 142 - 143. 142. *Richardia* sp. (Richardiidae), postabdomen (♂) in ventral view (with hypopygium removed). 143. *Richardia* sp., aedeagus, hypandrium and associated structures (♂) in lateral view.

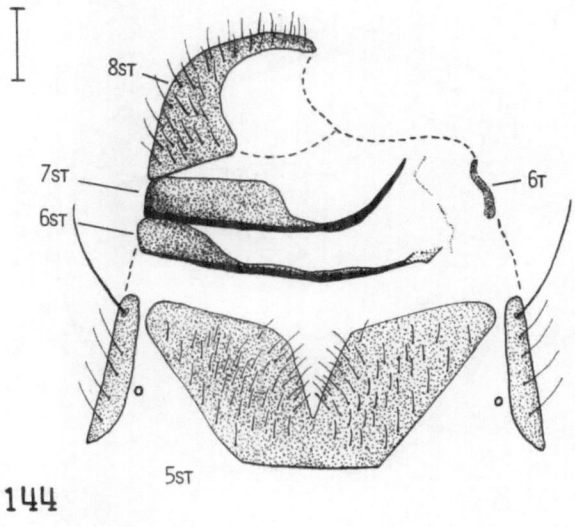

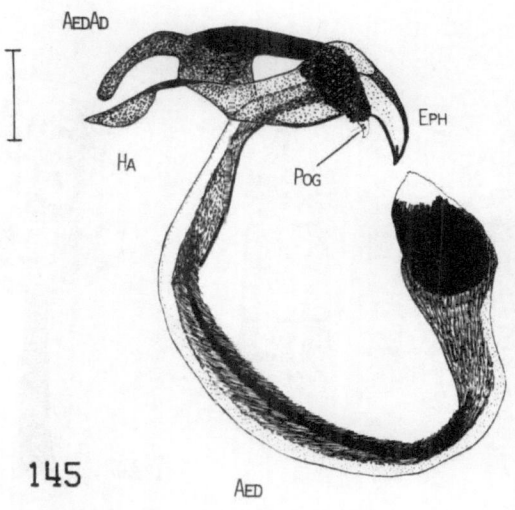

Figs. 144 - 145. 144. *Aenigmatomyia unipuncta* MALLOCH (see under Piophilidae *s.l.*), postabdomen (♂) in ventral view (with hypopygium removed). 145. *Aenigmatomyia unipuncta* MALLOCH, aedeagus, hypandrium and associated structures (♂) in lateral view.

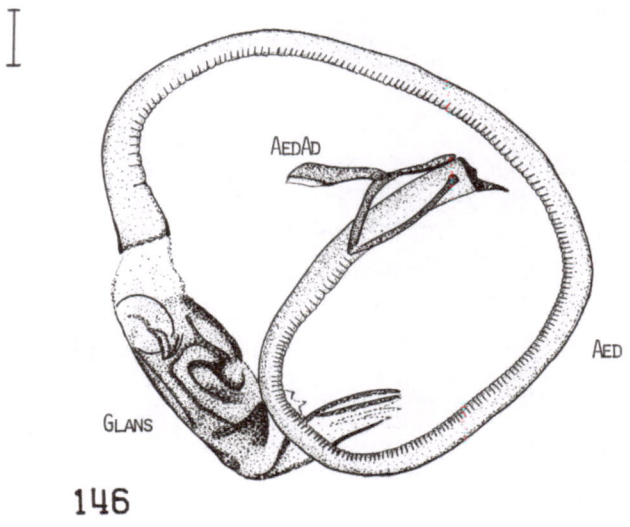

146

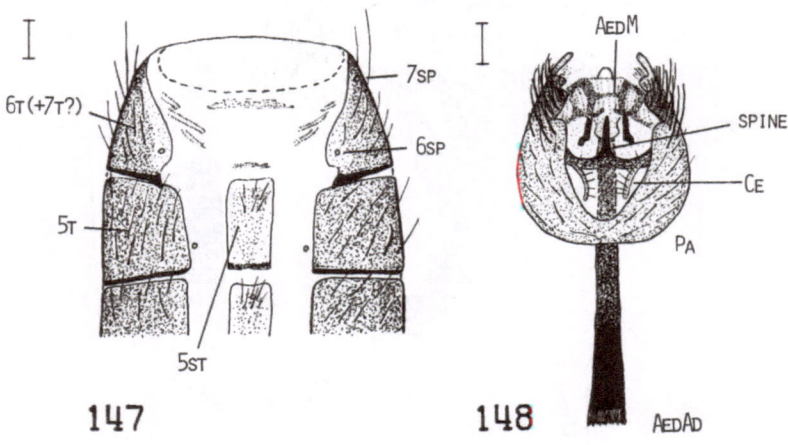

147   148

Figs. 146 - 148. 146. *Orellia occidentalis* (Snow) (Tephritidae *s.l.*), aedeagus and aedeagal apodeme ('fultella') (♂) in lateral view. 147. *Canace snodgrassii* Coquillett (Canacidae), postabdomen (♂) in ventral view (with hypopygium removed). 148. *Canace snodgrassii* Coquillett, hypopygium (♂) in dorsal view.

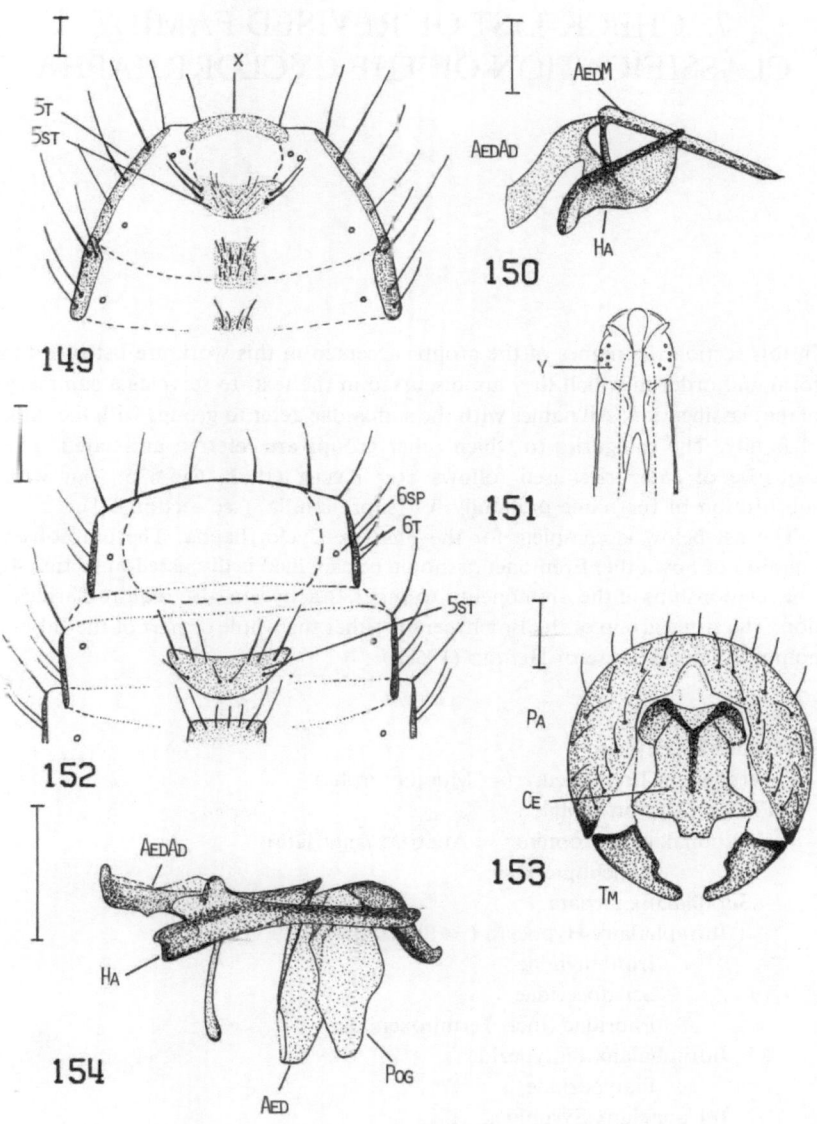

Figs. 149 - 154. 149. *Fergusonina scutellata* MALLOCH (Fergusoninidae), abdomen (part) (♂) in ventral view (with hypopygium removed). 150. *Fergusonina scutellata* MALLOCH, aedeagus, hypandrium and associated structures (♂) in lateral view. 151. *Fergusonina evansi* TONNOIR, 'tip of intromittent organ' (♂) in ventral view (after TONNOIR 1937). 152. *Notomyza edwardsi* MALLOCH (Notomyzidae), postabdomen (♂) in ventral view (with hypopygium removed). 153. *Notomyza edwardsi* MALLOCH, hypopygium (♂) in posterior view. 154. *Notomyza edwardsi* MALLOCH, aedeagus, hypandrium and associated structures (♂) in lateral view.

7. CHECK-LIST OF REVISED FAMILY CLASSIFICATION OF THE CYCLORRHAPHA

In this section the names of the groups accepted in this work are listed in the form and order in which they are discussed in the text, to serve as a summary of the classification. All names with the suffix -idae refer to groups with the rank of family. The categories to which other groups are referred are stated. The sequence of categories used follows VON KÉLER (1963: 635-636), but with substitution of the name prefamily for 'suprafamilia' (see section 6.1).

The list below is complete for the phalanx Cyclorrhapha. The unresolved question of how other Eremoneura should be classified is discussed in section 4. The relationships of the Eremoneura to other Brachycera also require clarification. The sister-group of the Brachycera is either the whole or part of the Bibionomorpha in the sense of HENNIG (1968b).

Infraorder Brachycera
  Superphalanx Eremoneura (= Muscomorpha)
    Phalanx Cyclorrhapha
      Subphalanx Acroptera (= Anatria, Anatriata)
             Lonchopteridae
      Subphalanx Atriata
        Infraphalanx Hypocera (= Phoridea)
             Ironomyiidae
             Sciadoceridae
             Phoridae (incl. Termitoxeniidae)
        Infraphalanx Platypezidea
          Platypezidae
        Infraphalanx Syrphidea
          Pipunculidae
          Syrphidae
        Infraphalanx Schizophora
          A.1. Superfamily Lonchaeoidea
            B.1. Lonchaeidae
            B.2. Cryptochetidae
          A.2. Superfamily Lauxanioidea
            B.3. Prefamily Lauxanioinea, Lauxaniidae (incl. Celyphidae)
            B.4. Prefamily Chamaemyioinea

C.1. Eurychoromyiidae
C.2. Chamaemyiidae
A.3. Superfamily Drosophiloidea
C.3. Drosophilidae
C.4. Camillidae
C.5. Curtonotidae
C.6. Campichoetidae **familia nova**
C.7. Ephydridae (incl. *Diastata*)
A.4. Superfamily Nothyboidea
C.8. Nothybidae
C.9. Teratomyzidae
C.10. Periscelididae (incl. Somatiidae)
C.11. Psilidae (= Loxoceridae)
A.5. Superfamily Muscoidea
B.5. Prefamily Tanypezoinea
C.12. Tanypezidae (incl. Strongylophthalmyiidae)
C.13. Heteromyzidae
B.6. Prefamily Calyptratae (= Thecostomata)
C.14. Scatophagidae (= Cordiluridae)
C.15. Anthomyiidae
C.16. Fanniidae
C.17. Muscidae
C.18. Tachinidae (incl. Oestridae, Hypodermatidae, Cutere-
bridae, Gasterophilidae, Calliphoridae and
Sarcophagidae)
C.19. Hippoboscidae family-group
D.1. Glossinidae
D.2. Hippoboscidae (incl. Nycteribiidae and Streblidae)
B.7. Prefamily Micropezoinea
C.20. Cypselosomatidae (incl. Pseudopomyzidae)
C.21. Micropezidae family-group
D.3. Neriidae
D.4. Micropezidae (incl. Calobatidae and Taeniapteridae)
B.8. Prefamily Australimyzoinea, Australimyzidae **familia nova**
B.9. Prefamily Diopsioinea
C.22. Diopsidae
C.23. Syringogastridae
B.10. Prefamily Sciomyzoinea
C.24. Coelopidae
C.25. Phaeomyiidae **status novus**
C.26. Dryomyzidae (incl. Helcomyzidae)
C.27. Sciomyzidae
C.28. Helosciomyzidae **status novus**
C.29. Sepsidae family-group
D.5. Ropalomeridae
D.6. Sepsidae

C.30. Megamerinidae

C.31. Cremifaniidae **status novus**

B.11. Prefamily Anthomyzoinea

C.32. Heleomyzidae

C.33. Rhinotoridae

C.34. Anthomyzidae

C.35. Borboropsidae **familia nova**

C.36. Trixoscelididae

C.37. Asteiidae

C.38. Opomyzidae

C.39. Sphaeroceridae (= Borboridae)

C.40. Chyromyidae

C.41. Aulacigastridae

B.12. Prefamily Agromyzoinea

C.42. Clusiidae

C.43. Agromyzidae

B.13. Prefamily Tephritoinea

C.44. Chiropteromyzidae

C.45. Mormotomyiidae

C.46. Cnemospathidae

C.47. Odiniidae

C.48. Tethinidae

C.49. Chloropidae family-group

D.7. Acartophthalmidae

D.8. Carnidae

D.9. Milichiidae

D.10. Chloropidae (incl. Mindidae)

C.50. Conopidae (incl. Stylogastridae)

C.51. Tephritidae family-group

D.11. Eurygnathomyiidae **status novus**

D.12. Richardiidae

D.13. Piophilidae (incl. Thyreophoridae and Neottiophilidae)

D.14. Tephritidae (= Trypetidae; incl. Platystomatidae, Pyrgotidae, Tachiniscidae, Otitidae, Ulidiidae, Pterocallidae and *Palloptera*)

Schizophora incertae sedis

C.52. Canacidae

C.53. Fergusoninidae

C.54. Notomyzidae **familia nova**

C.55. Braulidae

8. CONCLUDING REMARKS

In conclusion, I wish to emphasize that phylogenetic systematics is a dynamic subject and that no classification system should ever be regarded as final. It is my hope that publication of this work will stimulate increased interest in this field and indicate where new research efforts might be directed. I also hope that my work will prove helpful to workers in other fields of biology, since one of the main aims of systematics is to provide classifications which are widely useful in many fields of research.

A phylogenetic classification consists of a hierarchical system of hypotheses about the limits of monophyletic groups. Some of these hypotheses may be highly warranted, others poorly so. If a revised classification is better warranted than those which preceded it, progress has been achieved. The attempt may be judged to have failed, if it can be shown that some of the new hypotheses presented are less warranted than conflicting hypotheses contained in previous classifications. The classification of the Diptera Cyclorrhapha here presented is based on far more extensive data than was considered by authors of previous classifications. The need for substantial revision thus should not seem surprising. Further changes yet may be expected as a result of future studies. I have now worked on this revision for three-and-a-half years. It is time to make it publicly available.

9. ACKNOWLEDGEMENTS

I was fortunate at an early stage of this study to secure the cooperation of Dr. E. L. Kessel (California Academy of Sciences, San Francisco), who provided me with information on Platypezidae and material from his cultures. It is largely due to Dr. Kessel's help that I have been able to make progress with the interpretation of hypopygial rotation in Cyclorrhapha. Dr. W. Hennig (Staatliches Museum für Naturkunde, Ludwigsburg) gave valuable advice on an early draft of my manuscript, and was kind enough to provide a manuscript copy of one of his own works in advance of publication. Other persons who made substantive comments on all or part of the manuscript are: Dr. D. L. Hull (University of Wisconsin – Milwaukee), Dr. G. E. Ball, Dr. B. Hocking, Dr. D. A. McB. Craig and Dr. B. S. Heming (the last four at the University of Alberta, Edmonton). Dr. G. E. Ball's advice on presentational matters was particularly valuable.

Important information was provided in correspondence (in addition to persons already mentioned) by Dr. J. F. McAlpine and Dr. J. R. Vockeroth (Entomology Research Institute, Ottawa), and by Mr. G. Steyskal and Mr. C. W. Sabrosky (U.S. Department of Agriculture, Washington).

In addition to the collections available at the University of Alberta (assembled mainly by the late Dr. E. H. Strickland), I received material on loan for study through the good offices of the following: Dr. J. F. McAlpine (Entomology Research Institute, Ottawa), Mr. G. Steyskal (U. S. Department of Agriculture, Washington), Dr. W. Hackman (University Zoological Museum, Helsinki), Dr. P. M. Johns (University of Canterbury, New Zealand) and Mr. B. H. Cogan of the British Museum (Natural History).

I am grateful to Dr. B. Hocking (Chairman, Department of Entomology) and Dr. G. E. Ball for securing the necessary support for this project during my stay at Edmonton, from University funds and from National Research Council grant A-1399.

French and German translations of the summary were provided by Dr. G. Laugé (University of Paris, Orsay) and Mr. J. Rickert (Edmonton).

10. SUMMARY

This work presents a review of the classification of the Diptera Cyclorrhapha at the family level and above, based both on information available in the literature and on the results of my own analysis of characters of the male postabdomen and genitalia. The classification proposed is intended to be a phylogenetic classification consisting of probably monophyletic groups. A review of the principles and procedures of biological systematics is included.

The morphology of the male postabdomen and genitalia of the Diptera Cyclorrhapha is discussed. Recent observations on newly emerged Platypezidae provide conclusive evidence that the 8th abdominal segment in Cyclorrhapha is rotated through half the angle of hypopygial rotation, thus becoming inverted (rotated through 180°) when the hypopygium has become circumverted (rotated through 360°). In Platypezidae hypopygial rotation is not completed until after emergence from the puparium and is partly reversed during copulation, which takes place with the hypopygium in the inverse position. I conclude on the basis of comparative morphology that the so-called 'epandrium' of the Cyclorrhapha and some Empididae is formed by fusion of the basimeres across the dorsum, and is therefore not homologous with the epandrium of other Diptera (which has been completely lost in all Cyclorrhapha according to this interpretation). The new term periandrium is therefore proposed for the cyclorrhaphous 'epandrium'.

A review of the classification of the Schizophora leads to a proposed division of the group into five superfamilies: Lonchaeoidea, Lauxanioidea, Drosophiloidea, Nothyboidea and Muscoidea. The prefamily category is introduced between the superfamily and family categories. The Calyptratae are classified as one of the prefamilies of Muscoidea, which is the largest superfamily of Schizophora. The customary division of the Schizophora into the Calyptratae and 'Acalyptratae' is unacceptable in the phylogenetic system, since the latter group is not monophyletic. The new classification proposed is supported by the available ontogenetic evidence, which indicates that the morphological differences between the male postabdomen of Muscoidea and Drosophiloidea rest on differences in the number and development of the imaginal discs. A complete summary of the classification proposed is given in section 7.

RÉSUMÉ

Le présent ouvrage commente la classification systématique des Diptera Cyclorrhapha au niveau de la famille et au-dessus, d'après les données bibliographiques actuelles et les résultats d'une analyse personnelle des caractéristiques du postabdomen et des genitalia mâles. La classification proposée se présente comme une classification phylogénique de groupes vraisemblablement monophylétiques. Les principes et les méthodes utilisés en systématique biologique sont examinés.

La morphologie du postabdomen et des genitalia mâles des Diptères Cyclorrhaphes sont discutés. Des récentes observations, sur des Platypezidae juste après l'émergence, montrent de façon indubitable que le 8e segment abdominal a subi une rotation dont la valeur est égale à la moitié de celle de l'hypopygium. Le 8e segment est ainsi inversé (rotation de 180°), alors que l'hypopygium est circumversé (rotation de 360°). Chez les Platypezidae, la rotation de l'hypopygium n'est pas complète avant la sortie du puparium; pendant l'accouplement, la rotation subit un renversement partiel, et l'hypopygium est alors en position inversée. Une étude de morphologie comparative me conduit à conclure que l'"épandrium" des Cyclorrhaphes et de quelques Empididae est formé de la fusion des basimères dans la région dorsale; il n'est donc pas l'homologue de l'épandrium des autres Diptères qui, selon cette hypothèse, a totalement disparu chez les Cyclorrhaphes. Un terme nouveau: periandrium, est proposé pour désigner l'"épandrium" des Cyclorrhaphes.

Un examen de la classification des Schizophora permet de diviser le groupe en cinq superfamilles: Lonchaeoidea, Lauxanioidea, Drosophiloidea, Nothyboidea, Muscoidea. Une catégorie: la préfamille, est introduite entre la superfamille et la famille. Les Calyptratae correspondent à une prefamille des Muscoidea, ceux-ci représentant la superfamille la plus grande des Schizophora. La distinction entre Calyptratae et 'Acalyptratae' à l'intérieur des Schizophora est inacceptable dans le système phylogénique, puisque les Acalyptratae ne constituent pas un groupe monophylétique. La nouvelle classification proposé est corroborée par l'évidence ontogénique actuelle: les différences morphologiques entre le postabdomen des mâles de Muscoidea et de Drosophiloidea reposent sur des différences dans le nombre et le développement des disques imaginaux. Un résumé complet de la classification proposée est donné dans la section 7.

ZUSAMMENFASSUNG

Diese Arbeit enthält eine Übersicht der Klassifikation der cyclorrhaphen Fliegen auf der Familienebene und darüber, aufgebaut sowohl auf Information der erhältlichen Literatur, als auch auf meiner eigenen Analyse der Charaktere des Postabdomens und der Genitalien des Männchens. Die vorgeschlagene Klassifikation beabsicht eine phylogenetische zu sein, die aus wahrscheinlich monophyletischen Gruppen besteht. Eine Übersicht der Prinzipien und Prozeduren der biologischen Systematik ist mit einbegriffen.

Die Morphologie des Postabdomens und der Genitalien des Männchens der cyclorrhaphen Fliegen wird behandelt. Kürzlich gemachte Beobachtungen an gerade ausgeschlüpften Platypezidae liefern einen klaren Beweis, dass das achte abdominale Segment der Cyclorrhaphen durch den halben Winkel der hypopygialen Rotation gedreht ist und so eine entgegengesetzte Stellung einnimt (180° Rotation), während das Hypopygium eine volle Umdrehung macht (360°). Die hypopygiale Rotation der Platypezidae ist erst nach dem Ausschlüpfen aus dem Puparium beendet und ist während der Kopulation teilweise zurückgedreht. Diese findet statt, wenn das Hypopygium sich in der voll verdrehten Stellung befindet. Unter Beziehungnahme auf vergleichende Morphologie, schliesse ich, dass das sogenannte 'Epandrium' der Cyclorrhaphen und einiger Empididen durch ein Zusammenwachsen der Basimeren quer durch das Dorsum gebildet wird. Aus diesem Grunde kann es nicht als homolog mit dem Epandrium anderer Fliegen angesehen werden (welches nach dieser Ausdeutung bei allen Cyclorrhaphen verloren gegangen ist). Die neue Bezeichnung Periandrium, wird deshalb für das sogennante 'Epandrium' der Cyclorrhaphen vorgeschlagen.

Das Studium der Klassifizierung der Schizophora führt zu der vorgeschlagenen Unterteilung der Gruppe in fünf Überfamilien: Lonchaeoidea, Lauxanioidea, Drosophiloidea, Nothyboidea und Muscoidea. Die Kategorie der Praefamilie ist zwischen den Kategorien der Familie und der Überfamilie eingesetzt. Die Calyptraten sind als eine der Praefamilien der Muscoidea eingeteilt, welche die grösste Überfamilie der Schizophoren ist. Die althergebrachte Teilung der Schizophoren in Calyptratae und 'Acalyptratae' ist unannehmbar im phylogenetischen System, da die letztere Gruppe nicht monophyletisch ist. Die vorgeschlagene neue Klassifizierung ist durch das verfügbare ontogenetische Beweismaterial unterstützt, das andeutet, dass die morphologischen Unterschiede des Postabdomens des muscoiden und drosophiloiden Männchens auf Unterschieden in der Anzahl und Entwicklung der Imaginalscheiben beruht. Eine vollständige Zusammenfassung der vorgeschlagenen Klassifizierung ist in der 7. Sektion behandelt.

11. REFERENCES

Russian entries translated into English

ABUL NASR, S. E. 1950. Structure and development of the reproductive system of some species of Nematocera (order Diptera: suborder Nematocera). *Phil. Trans. R. Soc. (B)* 234: 339-395.

ACZÉL, M. L. 1948. Grundlagen einer Monographie der Dorilaiden (Diptera). Dorilaiden-Studien VI. *Acta zool. lilloana* 6: 5-168

ACZÉL, M. L. 1951. Morfologia externa y division sistematica de las Tanypezidiformes con sinopsis de las especies argentinas de Tylidae (Micropezidae) y Neriidae (Dipt.). *Acta zool. lilloana 11*: 483-589.

ACZÉL, M. L. 1954. Orthopyga and Campylopyga, new divisions of Diptera. *Ann. ent. Soc. Am.* 47: 75-80.

ACZÉL, M. L. 1955. Nothybidae, a new family of Diptera. *Treubia* 23: 1-18.

ANDERSON, D. T. 1966. The comparative embryology of the Diptera. *Ann. Rev. Ent.* 11: 23-46.

AUSTEN, E. E. 1936. A remarkable semi-apterous fly (Diptera) found in a cave in East Africa, and representing a new family, genus, and species. *Proc. zool. Soc. Lond.* 1936: 425-431.

BÄHRMANN, R. 1960. Vergleichend-morphologische Untersuchungen der männlichen Kopulationsorgane bei Empididen (Diptera). *Beitr. Ent.* 10: 485-540.

BÄHRMANN, R. 1966. Das Hypopygium von *Dolichopus* Latreille unter besonderer Berücksichtigung der Muskulatur und der Torsion (Dipera: Dolichopodidae). *Beitr. Ent.* 16: 61-72.

BECHER, E. 1882. Zur Kenntnis der Kopfbildung der Dipteren. *Wien. ent. Z.* 1: 49-54.

BLACK, V. H. 1966. The pupal development of the male genital disc of *Eucalliphora lilaea*. Sacramento State College (unpublished M. A. thesis). 40 pp.

BOLWIG, N. 1940. The reproductive organs of *Scatophila unicornis* Czerny (Diptera). *Proc. R. ent. Soc. Lond. (A)* 15: 97-102.

BRAUER, F. 1880. Die Zweiflügler des kaiserlichen Museums zu Wien. I. 2. Bemerkungen zur Systematik der Dipteren. *Denkschr. Akad. Wiss., Wien* 42: 108-118.

BRAUER, F. 1883. Die Zweiflügler des kaiserlichen Museums zu Wien. III. Systematische Studien auf Grundlage der Dipteren-Larven nebst einer Zusammenstellung von Beispielen aus der Literatur über dieselben und Beschreibung neuer Formen. *Denkschr. Akad. Wiss., Wien* 47: 1-100.

BRAUER, F. 1890. Ueber die Verbindungsglieder zwischen den orthorrhaphen und cyclorrhaphen Dipteren und solche zwischen Syrphiden und Muscarien. *Verh. zool.-bot. Ges. Wien* 40: 273-275.

BRÜEL, L. 1897. Anatomie und Entwicklungsgeschichte der Geschlechtsausführwege sammt Annexen von *Calliphora erythrocephala*. *Zool. Jb.* 10: 511-618.

BRUNDIN, L. 1966. Transantarctic relationships and their significance, as evidenced by chironomid midges, with a monograph of the subfamilies Podonominae and Aphroteniinae and the austral Heptagyiae. *K. svenska VetenskAkad. Handl.* 11, no. 1. 472 pp.

BUCK, R. C. & HULL. D. L. 1966. The logical structure of the Linnaean hierarchy. *Syst. Zool.* 15: 97-111.

CAIN, A. J. & HARRISON, G. A. 1960. Phyletic weighting. *Proc. zool. Soc. Lond.* 135: 1-31

CAMIN, J. H. & SOKAL, R. R. 1965. A method for deducing branching sequences in phylogeny. *Evolution* 19: 311-326.

CARMICHAEL, J. W. & SNEATH, P. H. A. 1969. Taxometric maps. *Syst. Zool.* 18: 402-415.

CHILLCOTT, J. G. T. 1958. The comparative morphology of the male genitalia of muscoid Diptera. *Proc. Xth int. Congr. Ent. Montreal* (1956) 1: 587-592.

CHILLCOTT, J. G. T. 1960. A revision of the Nearctic species of Fanniinae (Diptera: Muscidae). *Can. Ent.*, supplement 14. 295 pp.

CLAUSEN, P. J. 1965. A comparative study of the copulatory apparatus of selected Nearctic species of Parydrinae (Diptera: Ephydridae). University of Minnesota (unpublished M. Sc. thesis). 98 pp.

COLLESS, D. H. & McALPINE, D. K. 1970. Diptera. *In* The insects of Australia. Melbourne University Press, Carlton, Victoria. pp. 656-740.

COQUILLETT, D. W. 1901. A systematic arrangement of the families of the Diptera. *Proc. U. S. natn. Mus.* 23: 653-658.

CRAMPTON, G. C. 1941. The terminal abdominal structures of male Diptera. *Psyche* 48: 79-94.

CRAMPTON, G. C. 1942. The external morphology of the Diptera. In CRAMPTON, G. C., CURRAN, C. H. & ALEXANDER, C. P. Guide to the Insects of Connecticut. Part VI. The Diptera or true flies of Connecticut. First Fascicle. Bull. 64, St. geol. nat. Hist. Surv. Connecticut: 10-165.

CRAMPTON, G. C. 1944a. A comparative morphological study of the terminalia of male calypterate cyclorrhaphous Diptera and their acalypterate relatives. *Bull. Brooklyn ent. Soc.* 39: 1-31.

CRAMPTON, G. C. 1944b. Suggestions for grouping the families of acalypterate cyclorrhaphous Diptera on the basis of the male terminalia. *Proc. ent. Soc. Wash.* 46: 152-154.

CROWSON, R. A. 1970. Classification and biology. Heinemann Educational Books, London. 350 pp.

CZERNY, L. 1927. Heleomyzidae, Trichoscelidae und Chiromyidae. Fliegen palaearkt. Reg. 5, Teil 53. 56 pp.

DAHL, R. G. 1959. Studies on Scandinavian Ephydridae (Diptera Brachycera). *Opusc. ent.*, supplementum 15. 224 pp.

DARLINGTON, P. J. 1970. A practical criticism of Hennig-Brundin 'Phylogenetic Systematics' and Antarctic Biogeography. *Syst. Zool.* 19: 1-18.

DELUCCHI, V & PSCHORN-WALCHER, H. 1954. *Cremifania nigrocellulata* Czerny (Diptera, ? Chamaemyiidae), ein Räuber an *Dreyfusia* (*Adelges*) *piceae* Ratz. (Hemiptera, Adelgidae). *Z. angew. Ent.* 36: 84-107.

DEMEREC, M. 1950. Biology of *Drosophila*. Wiley, New York & London. 632 pp.

DOBZHANSKY, T. & BRIDGES, C. B. 1928. The reproductive system of triploid intersexes in *Drosophila melanogaster*. *Am. Nat.* 62: 425-434.

DÜBENDORFER, A. 1970. Entwicklungsleistungen transplantierter Genital- und Analanlagen von *Musca domestica* und *Phormia regina*. *Experientia* 26: 1158-1160.

DUFOUR, L. 1851. Recherches anatomiques et physiologiques sur les Diptères, accompagnées de considérations relatives à l'histoire naturelle de ces insectes. *Mém. présentés par divers savants à l'Akad. Sci. Inst. natn. Fr.* 11: 171-360.

EDWARDS, A. W. F. & CAVALLI-SFORZA, L. L. 1964. Reconstruction of evolutionary trees. *In* HEYWOOD, V. H. & McNEILL, J. (eds.). Phenetic and phylogenetic classification. The Systematics Association, London. pp. 67-76.

EGGLISHAW, H. J. 1960. Studies on the family Coelopidae (Diptera). *Trans. R. ent. Soc. Lond.* 112: 109-140.

EHRLICH, P. R. & EHRLICH, A. H. 1967. The phenetic relationships of the butterflies. I. Adult taxonomy and the nonspecificity hypothesis. *Syst. Zool.* 16: 301-317.

EMDEN, F. I. VAN 1950. *Mormotomyia hirsuta* Austen (Diptera) and its systematic position. *Proc. R. ent. Soc. Lond. (B)* 19: 121-128.

EMDEN, F. I. VAN 1951. The male genitalia of Diptera and their taxonomic value. *Trans.*

IXth int. Congr. Ent. Amsterdam (1951) 2: 22-26.

EMDEN, F. I. VAN & HENNIG, W. 1956. Diptera. *In* TUXEN, S. L. (eds.). Taxonomist's glossary of genitalia in insects. Munksgaard, Copenhagen. pp. 111-122.

ENDERLEIN, G. 1936. 22. Ordnung: Zweiflügler, Diptera. *Tierwelt Mitteleur.* 6(2). 259 pp.

ENDERLEIN, G. 1938. Die Dipterenfauna der Juan-Fernandez-Inseln und der Oster-Insel. *In* SKOTTSBERG, C. (ed). The natural history of Juan Fernandez and Easter Island 3: 643-680.

FARRIS, J. S., KLUGE, A. G. & ECKARDT, M. J. 1970. A numerical approach to phylogenetic systematics. *Syst. Zool.* 19: 172-189.

FEUERBORN, H. J. 1922. Das hypopygium 'inversum' und 'circumversum' der Dipteren. *Zool. Anz.* 55: 189-212.

FREY, R. 1921. Studien über den Bau des Mundes der niederen Diptera Schizophora nebst Bemerkungen über die Systematik dieser Dipterengruppe. *Acta Soc. Fauna Fl. fenn.* 48, no. 3. 245 pp.

FRICK, K. E. 1952. A generic revision of the family Agromyzidae (Diptera) with a catalogue of New World species. *Univ. Calif. Publs Ent.* 8: 339-452.

GHESQUIÈRE, J. 1942. Recherches sur les Diptères d'Afrique. II.- Notice monographique sur les Muscoïdes Cryptochaetidae, parasites de Coccides Monophlebinae. *Rev. Zool. Bot. afr.* 36: 390-410.

GHISELIN, M. T. 1969. The principles and concepts of systematic biology. *In* Systematic biology; proceedings of an international conference. National Academy of Sciences, Washington, D. C. pp. 45-55.

GILMOUR, J. S. L. & WALTERS, S. M. 1964. Philosophy and classification. *In* TURRILL, W. B. (ed.). Vistas in botany. IV. Recent researches in plant taxonomy. Pergamon Press, Oxford. pp. 1-22

GIRSCHNER, E. 1893. Beitrag zur Systematik der Musciden. *Berl. ent. Z.* 38: 297-312.

GIRSCHNER, E. 1896. Ein neues Musciden-System auf Grund der Thoracalbeborstung und der Segmentierung des Hinterleibes. *Illte Wschr. Ent.* 1: 12-16, 30-32, 61-65, 105-112.

GLEICHAUF, R. 1936. Anatomie und Variabilität des Geschlechtsapparates von *Drosophila melanogaster* (Meigen). *Z. Wiss. Zool.* 148: 1-66.

GOODING, R. H. & WEINTRAUB, J. 1960. The genitalia of *Hypoderma bovis* (L.) and *H. lineatum* (de Vill.) (Diptera: Oestridae). *Can. J. Zool.* 38: 565-574.

GREGG. J. R. 1954. The language of taxonomy; an application of symbolic logic to the study of classificatory systems. Columbia University Press, New York. 70 pp.

GRIFFITHS, G. C. D. 1964. The agromyzid fauna of Iceland and the Faroes, with appendices on the *Phytomyza milii* and *robustella* groups (Diptera, Agromyzidae). *Ent. Meddr.* 32: 393-450.

GRUNIN, K. J. 1964-1965. Hypodermatidae. Fliegen palaearkt. Reg. 8, Teil 64b. 154 pp.

GRUNIN, K. J. 1966. Oestridae. Fliegen palaearkt. Reg. 8, Teil 64a', 96 pp.

GRUNIN, K. J. 1969. Gasterophilidae. Fliegen palaearkt. Reg. 8, Teil 64a. 61 pp.

HACKMAN, W. 1959. On the genus *Scaptomyza* Hardy (Dipt., Drosophilidae) with descriptions of new species from various parts of the world. *Acta zool. fenn.* 97. 73 pp.

HACKMANN, W. 1960. Diptera (Brachycera): Camillidae, Curtonotidae and Drosophilidae. *S. Afr. Anim. Life* 7: 381-389.

HACKMAN, W. 1969a. A review of the zoogeography and classification of the Sphaeroceridae (Borboridae, Diptera). *Notul. ent.* 49: 193-210.

HACKMAN, W. & ANDERSON, H. 1969b. *Trixoscelis puberula* (Zetterstedt), a heleomyzid fly (Diptera). *Notul. ent.* 49: 269-270.

HACKMAN, W. 1970. Trixoscelidae (Diptera) from southern Spain and description of a new *Trixoscelis* species from northern Europa. *Ent. scand.* 1: 127-134.

HAHN, J. 1929. Poznámky k morfologii a anatomii r. *Sapromyza. Biol. Listy* 14: 357-370.

HANNA, A. D. 1938. Studies on the Mediterranean fruit-fly: *Ceratitis capitata* Wied. I. The structure and operation of the reproductive organs. *Bull. Soc. Fouad I. Ent.* 22: 39-58.

324

HARDY, G. H. 1944. The copulation and the terminal segments of Diptera. *Proc. R. ent. Soc. Lond. (A)* 19: 52-65.

HARDY, G. H. 1950. The twisting segments in Diptera. *Entomologist's mon. Mag.* 86: 346-347.

HARDY, G. H. 1953. The phylogeny of Diptera. 2. – Dolichopodidae. *Entomologist's mon. Mag.* 89: 7-11.

HARRISON, R. A. 1959. Acalypterate Diptera of New Zealand. *Bull. N.Z. Dep. scient. Ind. Res.* no. 128. vii & 382 pp.

HEMPEL, C. G. 1965. Aspects of scientific explanation, and other essays in the philosophy of science. Free Press, New York. 505 pp.

HENDEL, F. 1931-1936. Agromyzidae. Fliegen palaearkt. Reg. 6(2), Teil 59. 570 pp.

HENDEL, F. 1936-1937. 26. Ordnung der Pterygogenea (Dreissigste Ordnung der Insecta): Diptera oder Fliegen. *In* KÜKENTHAL, W. & KRUMBACH, T. (eds.). Handbuch der Zoologie. 4, 2. Hälfte, 2. Teil, Insecta 3: 1729-1998.

HENNIG, W. 1934. Zur Kenntnis der Kopulationsorgane der Tyliden (Micropeziden, Dipt. Acalypt.). *Zool. Anz.* 107: 67-76.

HENNIG, W. 1936a. Beiträge zur Kenntnis des Kopulationsapparates der cyclorrhaphen Dipteren. *Z. Morph. Ökol Tiere* 31: 328-370.

HENNIG, W. 1936b. Beiträge zur Kenntnis des Kopulationsapparates und der Systematik der Tanypeziden (Dipt., Acalypt.) *Dt. ent. Z.* 1936: 27-38.

HENNIG, W. 1936c. Beiträge zur Systematik und Tiergeographie der Pyrgotiden. *Arb. morph. taxon. Ent. Berl.* 3: 243-256.

HENNIG, W. 1937a. Coelopidae. Fliegen palaearkt. Reg. 5, Teil 52. 39 pp.

HENNIG, W. 1937b. Milichiidae et Carnidae. Fliegen palaearkt. Reg. 6(1), Teil 60a. 91 pp.

HENNIG, W. 1937c. Die morphologische Deutung des männlichen Kopulationsapparates der Gattung *Glossina. Z. Parasitkde* 9: 345-350.

HENNIG, W. 1938a. Odiniidae. Fliegen palaearkt. Reg. 6(1), Teil 60b. 11 pp.

HENNIG, W. 1938b. Beiträge zur Kenntnis der Clusiiden und ihres Kopulationsapparates (Dipt. Acalypt.). Encycl. ent. B II, Dipt. 9: 121-138.

HENNIG, W. 1938c. Zur Frage der verwandtschaftlichen Stellung von *Braula coeca* Nitzsch. *Arb. morph. taxon. Ent.* 5: 164-174.

HENNIG, W. 1938d. Braulidae. Fliegen palaearkt. Reg. 6 (1), Teil 60c. 14 pp.

HENNIG, W. 1938e. Beiträge zur Kenntnis des Kopulationsapparates und Systematik der Acalyptraten. I. Chamaemyiidae und Odiniidae. *Arb. morph. taxon. Ent.* 5: 201-213.

HENNIG, W. 1939a. Beiträge zur Kenntnis des Kopulationsapparates und der Systematik der Acalyptraten. II. Tethinidae, Milichiidae, Anthomyzidae und Opomyzidae. *Arb. morph. taxon. Ent.* 6: 81-94.

HENNIG, W. 1939b. Otitidae. Fliegen palaearkt. Reg. 5, Teil 46/47. 78 pp.

HENNIG, W. 1940. Ulidiidae. Fliegen palaearkt. Reg. 5, Teil 45. 34 pp.

HENNIG, W. 1941a. Beiträge zur Kenntnis des Kopulationsapparates und der Systematik der Acalyptraten. III. Pallopteridae, Thyreophoridae, Diopsidae, *Pseudopomyza, Pseudodinia. Arb. morph. taxon. Ent.* 8: 54-65.

HENNIG, W. 1941b. Psilidae. Fliegen palaearkt. Reg. 5, Teil 41. 38 pp.

HENNIG, W. 1941c. Megamerinidae. Fliegen palaearkt. Reg. 5, Teil 39b. 4 pp.

HENNIG, W. 1941d. Diopsidae. Fliegen palaearkt. Reg. 5, Teil 39c. 8 pp.

HENNIG, W. 1943. Piophilidae. Fliegen palaearkt. Reg. 5, Teil 40. 52 pp.

HENNIG, W. 1948a. Beiträge zur Kenntnis des Kopulationsapparates und der Systematik der Acalyptraten. IV. Lonchaeidae und Lauxaniidae. *Acta zool. lilloana* 6: 333-429.

HENNIG, W. 1948b. Die Larvenformen der Dipteren. I. Akademie-Verlag, Berlin. 184 pp.

HENNIG, W. 1948c. Über einige verkannte Dipteren-Gattungen. *Acta zool. lilloana* 6: 169-170.

HENNIG, W. 1949. Sepsidae. Fliegen palaearkt. Reg. 5, Teil 39 a. 91 pp.

HENNIG, W. 1950. Grundzüge einer Theorie der phylogenetischen Systematik. Deutscher Zentralverlag, Berlin. 370 pp.

HENNIG, W. 1952a. Die Larvenformen der Dipteren. III. Akademie-Verlag, Berlin. vii & 628 pp.

HENNIG, W. 1952b. Bemerkenswerte neue Acalyptraten in der Sammlung des Deutschen Entomologischen Institutes (Diptera: Acalyptrata). *Beitr. Ent.* 2: 604-618.

HENNIG, W. 1954. Flügelgeäder und System der Dipteren unter Berücksichtigung der aus dem Mesozoikum beschriebenen Fossilien. *Beitr. Ent.* 4: 245-388.

HENNIG, W. 1955 (-1964). Muscidae. Fliegen palaearkt. Reg. 7(2), Teil 63b. 1110 pp.

HENNIG, W. 1958. Die Familien der Diptera Schizophora und ihre phylogenetischen Verwandtschaftsbeziehungen. *Beitr. Ent.* 8: 505-688.

HENNIG, W. 1964. Die Dipteren-Familie Sciadoceridae im Baltischen Bernstein (Diptera: Cyclorrhapha Aschiza). *Stuttg. Beitr. Naturk.* no. 127. 10 pp.

HENNIG, W. 1965a. Vorarbeiten zu einem phylogenetischen System der Muscidae (Diptera: Cyclorrhapha). *Stuttg. Beitr. Naturk.* no. 141. 100 pp.

HENNIG, W. 1965b. Die Acalyptratae des Baltischen Bernsteins und ihre Bedeutung für die Erforschung der phylogenetischen Entwicklung dieser Dipteren-Gruppe. *Stuttg. Beitr. Naturk.* no. 145. 215 pp.

HENNIG, W. 1966a. Phylogenetic systematics. University of Illinois Press, Urbana, Chicago & London. 263 pp.

HENNIG, W. 1966b. Conopidae im Baltischen Bernstein (Diptera: Cyclorrhapha). *Stuttg. Beitr Naturk.* no. 154. 24 pp.

HENNIG, W. 1967. Neue Acalyptratae aus dem Baltischen Bernstein (Diptera: Cyclorrhapha). *Stuttg. Beitr. Naturk.* no. 175. 27 pp.

HENNIG, W. 1968a. *Holopticander*, eine neue Gattung der Lauxaniidae, mit Bemerkungen über die Gattung *Hypagoga* (Diptera: Acalyptratae). *Stuttg. Beitr. Naturk.* no. 192. 6 pp.

HENNIG, W. 1968b. Kritische Bemerkungen über den Bau der Flügelwurzel bei den Dipteren und die Frage nach der Monophylie der Nematocera. *Stuttg. Beitr. Naturk.* no. 193. 23 pp.

HENNIG, W. 1969a. Neue Gattungen und Arten der Acalypteratae. *Can. Ent.* 101: 589-633.

HENNIG, W. 1969b. Neue Übersicht über die aus dem Baltischen Bernstein bekannten Acalyptratae (Diptera: Cyclorrhapha). *Stuttg. Beitr. Naturk.* no. 209. 42 pp.

HENNIG, W. 1969c. Die Stammesgeschichte der Insekten. Waldemar Kramer, Frankfurt am Main. 436 pp

HENNIG, W. 1970. Insektenfossilien aus der unteren Kreide. II. Empididae (Diptera, Brachycera). *Stuttg. Beitr. Naturk.* no. 214. 12 pp.

HENNIG, W. in press. Neue Untersuchungen über die Familien der Diptera Schizophora.

HERTING, B. 1957. Das weibliche Postabdomen der calyptraten Fliegen (Diptera) und sein Merkmalswert für die Systematik der Gruppe. *Z. Morph. Ökol. Tiere* 45: 429-461.

HEWITT, G. C. 1907. The structure, development and bionomics of the house-fly, *Musca domestica*. I. The anatomy of the fly. *Q. Jl microsc. Sci.* 51: 395-448.

HULL, D. L. 1964. Consistency and monophyly. *Syst. Zool.* 13: 1-11.

HULL, D. L. 1967. Certainty and circularity in evolutionary taxonomy. *Evolution* 21: 174-189.

HULL, D. L. 1969. The natural system and the species problem. *In* Systematic biology; proceedings of an international conference. National Academy of Sciences, Washington D.C. pp. 56-58.

IMMS, A. D. 1942. On *Braula coeca* Nitzsch and its affinities. *Parasitology* 34: 88-100.

IPE, I. M. 1966. A detailed morphological study of the external and internal genital organs of female *Melanagromyza obtusa* (Malloch), a serious pest of *Cajanus indicus* L. (Agromyzidae: Diptera). *Indian J. Ent.* 28: 287-298.

IPE, I. M. 1967. A detailed morphological study of the external and internal genital organs of male *Melanagromyza obtusa* (Malloch), a pest of *Cajanus indicus* L. *Indian J. Ent.* 29: 1-10.

JOBLING, B. 1951. A record of the Streblidae from the Philippines and other Pacific islands, including morphology of the abdomen, host-parasite relationship and geographical distribution, and with descriptions of five new species (Diptera). *Trans. R. ert. Soc. Lond.* 102: 211-246.

JOHNSON, L. A. S. 1970. Rainbow's end: the quest for an optimal taxonomy. *Syst. Zool.* 19: 203-239.

KARL, E. 1959. Vergleichend-morphologische Untersuchungen der männlichen Kopulationsorgane bei Asiliden (Diptera). *Beitr. Ent.* 9: 619-680.

KÉLER, S. VON 1963. Entomologisches Wörterbuch mit besonderer Berücksichtigung der morphologischen Terminologie. Akademie-Verlag, Berlin. xvi & 840 pp. (3rd edition).

KELSEY, L. P. 1969. A revision of the Scenopinidae (Diptera) of the World. *Bull. U.S. natn. Mus.* no. 277. 336 pp.

KESSEL, E. L. 1960. The systematic positions of *Platycnema* Zetterstedt and *Melanderomyia*, new genus, together with the description of the genotype of the latter (Diptera: Platypezidae). *Wasmann J. Biol.* 18: 87-101.

KESSEL, E. L. & MAGGIONCALDA, E.A. 1968a. A revision of the genera of Platypezidae, with the descriptions of five new genera, and considerations of phylogeny, circumversion, and hypopygia (Diptera). *Wasmann J. Biol.* 26: 33-106.

KESSEL, E. L. 1968b. Circumversion and mating positions in Platypezidae – an expanded and emended account (Diptera). *Wasmann J. Biol.* 26: 243-253.

KIM, K. C. & COOK, E. F. 1966. A comparative external morphology of adult Sphaeroceridae (Diptera). *Misc. Publs ent. Soc. Am.* 5: 78-100.

KNUTSON, L. V. 1963. Biology and immature stages of snail-killing flies of Europe (Diptera: Sciomyzidae). Cornell University, Ph. D. thesis. Published on demand by University Microfilms, Ann Arbor, Michigan. viii & 382 pp.

KRYSTOPH, H. 1961. Vergleichend-morphologische Untersuchungen an den Mundteilen bei Empididen (Diptera). *Beitr. Ent.* 11: 824-872.

LAMEERE, A. 1906. Notes pour la classification des Diptères. *Mém. Soc. r. ent. Belg.* 12: 105-140.

LAUGÉ, G. 1968. Morphologie comparée de la région génitale des intersexués triploïdes de *Drosophila melanogaster*. *Annls Soc. ent. Fr.* 4: 481-499.

LINDNER, E. 1949. Handbuch. Fliegen palaearkt. Reg. 1. 422 pp.

LOWNE, B. T. 1890-1895. The anatomy, physiology, morphology, and development of the blow-fly (*Calliphora erythrocephala*). A study in the comparative anatomy and morphology of insects. R. H. Porter, London. 778 pp. (2 vols).

MALLOCH, J. R. 1933-1934. Acalyptrata. *In* Diptera of Patagonia and South Chile 6: 177-490.

MATSUDA, R. 1958. On the origin of the external genitalia of insects. *Ann. ent. Soc. Am.* 51: 84-94.

MAYR, E. 1969. Principles of systematic zoology. McGraw-Hill, New York & London. 428 pp..

MCALPINE, D. K. 1958. A family of flies new to Australia (Diptera, Rhinotoridae). *Proc. R. zool. Soc. N. S. W.* 1956-1957: 64-65.

MCALPINE, D. K. 1960. A review of the Australian species of Clusiidae (Diptera, Acalyptrata). *Rec. Aust. Mus.* 25: 63-94.

MCALPINE, D. K. 1966. Description and biology of an Australian species of Cypselosomatidae (Diptera), with a discussion of family relationships. *Aust. J. Zool.* 14: 673-685.

MCALPINE, D. K. 1967. The Australian species of *Diplogeomyza* and allied genera (Diptera, Heleomyzidae). *Proc. Linn. Soc. N. S. W.* 92: 74-106.

MCALPINE, D. K. 1968. The genus *Cairnsimyia* Malloch (Diptera, Heleomyzidae, Rhinotorini). *Rec. Aust. Mus.* 27: 263-283.

MCALPINE, J. F. 1960. A new species of *Leucopis (Leucopella)* from Chile and a key to the world genera and subgenera of Chamaemyiidae (Diptera). *Can. Ent.* 92: 51-58.

327

MCALPINE, J. F. 1962a. A revision of the genus *Campichoeta* Macquart (Diptera: Diastatidae). *Can. Ent.* 94: 1-10.

MCALPINE, J. F. 1962b. The evolution of the Lonchaeidae (Diptera). University of Illinois, Ph. D. thesis. Published on demand by University Microfilms, Ann Arbor, Michigan. iv & 233 pp.

MCALPINE, J. F. 1963. Relationships of *Cremifania* Czerny (Diptera: Chamaemyiidae) and description of a new species. *Can. Ent.* 95: 239-253.

MCALPINE, J. F. & MARTIN, J. E. H. 1966. Systematics of Sciadoceridae and relatives with descriptions of two new genera and species from Canadian amber and erection of family Ironomyiidae (Diptera: Phoroidea). *Can. Ent.* 96: 527-544.

MCALPINE, J. F. 1967. A detailed study of Ironomyiidae (Diptera: Phoroidea). *Can. Ent.* 99: 225-236.

MCALPINE, J. F. 1968a. An annotated key to drosophilid genera with bare or micropubescent aristae and a revision of *Paracacoxenus* (Diptera: Drosophilidae). *Can. Ent.* 100: 514-532.

MCALPINE, J. F. 1968b. Taxonomic notes on *Eurychoromyia mallea* (Diptera: Eurychoromyiidae). *Can. Ent.* 100: 819-823.

MCALPINE, J. F. & MUNROE, D. D. 1968c. Swarming of lonchaeid flies and other insects, with descriptions of four new species of Lonchaeidae (Diptera). *Can. Ent.* 100: 1154-1178.

MCALPINE, J. F. & MARTIN, J. E. H. 1969. Canadian amber – a paleontological treasure chest. *Can. Ent.* 101: 819-838.

MCALPINE, J. F. 1970. First record of calypterate flies in the Mesozoic era (Diptera: Calliphoridae). *Can. Ent.* 102: 342-346.

MEIJERE, J. C. H. DE 1900. Ueber die Larve von *Lonchoptera*. Ein Beitrag zur Kenntniss der cyclorrhaphen Dipterenlarven. *Zool. Jb., Abt. Syst. Ökol. Geogr. Tiere* 14: 87-132.

MEIJERE, J. C. H. DE 1902. Ueber die Prothorakalstigmen der Dipterenpuppen. *Zool Jb., Abt. Anat. Ontogenie Tiere* 15: 623-692.

MEIJERE, J. C. H. DE 1938. Über die Begattung und über den Stylus der Agromyziden. *Bijdr. Dierk.* 27: 19-26.

METCALF, C. L. 1921. The genitalia of the male Syrphidae: their morphology, with special reference to its taxonomic significance. *Ann. ent. Soc. Am.* 14: 169-214.

MILANI, R. & RIVOSECCHI, L. 1954. Sinistral coiling of male genitalia in *M. domestica*. *Drosoph. Inf. Serv.* no. 28: 136.

MILANI, R. & RIVOSECCHI, L. 1955. Malformazioni e mutazioni di *Musca domestica L.* di interesse per la conoscenza dei segmenti terminali maschili dei Ditteri. *Boll. Zool.* 22: 341-372.

MINCHIN, E. A. 1905. Report on the anatomy of the tsetse-fly (*Glossina palpalis*). *Proc. R. Soc. Lond (B)* 76: 531-547.

MORGE, G. 1963. Die Lonchaeidae und Pallopteridae Österreichs und der angrenzenden Gebiete. 1. Teil: die Lonchaeidae. *Naturk. Jb. Stadt Linz* 9: 123-312.

MORGE, G. 1967. Die Lonchaeidae und Pallopteridae Österreichs und der angrenzenden Gebiete. 2. Teil: die Pallopteridae. *Naturk. Jb. Stadt Linz* 13: 141-212.

MORRISON, F. O. 1941. A study of the male genitalia in calyptrate Diptera, based on the genus *Gonia* Meigen (Diptera: Tachinidae). *Can. J. Res. (D)* 19: 1-21.

MOTE, D. C. 1929. The reproductive system of the warble fly *Hypoderma bovis* de Geer. *Ann. ent. Soc. Am.* 22: 70-76.

MUNRO, H. K. 1947. African Trypetidae (Diptera). A review of the transition genera between Tephritinae and Trypetinae, with a preliminary study of the male terminalia. *Mem. ent. Soc. sth. Afr. no.* 1. 284 pp.

NATER, H. 1953. Vergleichend-morphologische Untersuchung des aüsseren Geschlechtsapparates innerhalb der Gattung *Drosophila*. *Zool. Jb., Abt. Syst. Ökol. Geogr. Tiere* 81: 437-486.

NAYAR, J. L. & TANDON, S. K. 1963. A note on the genitalia of *Sphracephala hearseyana* Westw. (Diopsidae: Diptera). *Agra Univ. J. Res.* 12: 1-4.

NOWAKOWSKI, J. T. 1959. Studien über Minierfliegen (Dipt. Agromyzidae). 3. Revision der in Labiaten und Boraginaceen minierenden Arten aus der Gruppe der *Phytomyza obscura* Hend., mit einem Beitrag zur Kenntnis ihrer Hymenopteren-Parasiten. *Dt. ent. Z.* 6: 185-229.

NOWAKOWSKI, J. T. 1964. Studien über Minierfliegen (Dipt. Agromyzidae). 9. Revision der Artengruppe *Agromyza reptans* Fall. – *A. rufipes* Meig. *Dt. ent. Z.* 11: 175-213.

PARAMONOV, S. J. 1956. Notes on Australian Diptera (XXI). Mindidae – a new family of Acalyptrata (Diptera). *Ann. Mag. nat. Hist.* 9: 779-783.

PATTON, W. S. 1934. Studies on the higher Diptera of medical and veterinary importance. A revision of the species of the genus *Glossina* Wiedemann based on a comparative study of the male and female terminalia. *Ann. trop. Med. Parasit.* 28: 315-322.

PETZOLD, W. 1927. Bau und Funktion des Hypopygiums bei den Tachinen, unter besonderer Berücksichtigung der Kieferneulentachine (*Ernestia rudis* Fall.). *Z. Naturw.* 63: 1-50.

PIHAN, J.-C. 1969. Le développement post-embryonnaire du système trachéen chez *Calliphora erythrocephala* Meigen (Insecte, Diptère, Brachycère). *Archs Zool. exp. gén.* 109: 287-304.

PRADO, A. P. DO 1965. Segunda contribução ao conhecimento da familia Rhopalomeridae (Diptera, Acalyptratae). *Studia ent.* 8: 209-268.

PRADO, A. P. DO 1969. Syringogastridae, uma nova familia de Dipteros Acalyptratae, com a descrição de seis espécies novas do gênero *Syringogaster* Cresson. *Studia ent.* 12: 1-32.

REMANE, A. 1952. Die Grundlagen des natürlichen Systems, der vergleichenden Anatomie und der Phylogenetik. Geest & Portig, Leipzig. 364 pp.

RICHARDS, O. W. 1927. Sexual selection and allied problems in the insects. *Biol. Rev.* 2: 298-364.

RICHARDS, O. W. 1961. Diptera (Sphaeroceridae) from South Chile. *Proc. R. ent. Soc. Lond. (B)* 30: 57-68.

RICHARDS, O. W. 1968. New South African species of the genus *Scutelliseta* Richards (Diptera: Sphaeroceridae). *Ann. Natal Mus.* 20: 65-91.

RIVOSECCHI, L. 1957. Descrizione degli stadi preimaginali di *Tephritis stictica* Loew (Diptera Trypetidae) parassita della *Diotis maritima* Smith e note sugli organi riproduttivi dell' adulto. *Riv. Parassit.* 18: 267-288.

RIVOSECCHI, L. 1958. Gli organi della riproduzione in *Musca domestica*. *Rc. Ist. sup. Sanità* 21: 458-488.

ROBACK, S. S. 1951. A classification of the muscoid calyptrate Diptera. *Ann. ent. Soc. Am.* 44: 327-361.

ROHDENDORF, B. B. 1964. Historical development of two-winged insects. *Trudy paleont. Inst.* 100. 311 pp.

ROHDENDORF, B. B. 1967. Trends of historical development of sarcophagids (Diptera, Sarcophagidae). *Trudy paleont. Inst.* 116. 92 pp.

SABROSKY, C. W. 1946. Family names in the order Diptera. *Proc. ent. Soc. Wash.* 48: 163-171.

SABROSKY, C. W. 1957. Synopsis of the New World species of the dipterous family Asteiidae. *Ann. ent. Soc. Am.* 50: 43-61.

SACK, P. 1939. Sciomyzidae. Fliegen palaearkt. Reg. 5, Teil 37. 87 pp.

SALLES, H. 1948. Sôbre a genitália dos drosofilídios (Diptera): I. *Drosophila melanogaster* e *D. simulans. Summa bras. Biol.* 1: 311-383.

SALZER, R. 1968. Konstruktionsanatomische Untersuchung des männlichen Postabdomens von *Calliphora erythrocephala* Meigen (Insecta, Diptera). *Z. Morph. Tiere* 63: 155-238.

SASAKAWA, M. 1958. The female terminalia of the Agromyzidae, with description of a new genus (I). *Scient. Rep. Saikyo Univ., Agriculture* 10: 133-150.

SASAKAWA, M. 1966. Studies on the Oriental and Pacific Clusiidae (Diptera). Part 1. Genus *Heteromeringia* Czerny, with one new related genus. *Pacif. Insects* 8: 61-100.

SCHINER, J. R. 1864. Ein neues System der Dipteren. *Verh. zool.-bot. Ges. Wien* 14: 201-212.

SCHRÄDER, T. 1927. Das Hypopygium 'circumversum' von *Calliphora erythrocephala*. Ein Beitrag zur Kenntnis des Kopulationsapparates der Dipteren. *Z. Morph. Ökol. Tiere* 8: 1-44.

SHAROV, A. G. 1966. Basic arthropodan stock with special reference to insects. Pergamon Press, Oxford. xii & 271 pp.

SHILLITO, J. F. 1950. A note on Speiser's genus *Centrioncus* and a revised definition of Diopsidae (Diptera: Acalypterae). *Proc. R. ent. Soc. Lond. (B)* 19: 109-113.

SIMPSON, G. G. 1961. Principles of animal taxonomy. Columbia University Press, New York. 247 pp.

SKAIFE, S. H. 1921. On *Braula caeca* Nitzsch, a dipterous parasite of the honeybee. *Trans. R. Soc. S. Afr.* 10: 41-48.

SMITH, E. L. 1969. Evolutionary morphology of external insect genitalia. 1. Origin and relationships to other appendages. *Ann. ent. Soc. Am.* 62: 1051-1079.

SMITH, K. G. V. 1967. The biology and taxonomy of the genus *Stylogaster* Macquart, 1835 (Diptera: Conopidae, Stylogasterinae) in the Ethiopian and Malagasy regions. *Trans R. ent Soc. Lond.* 119: 47-69.

SNODGRASS, R. E. 1935. Principles of insect morphology. McGraw-Hill, New York & London. ix & 667 pp.

SNODGRASS, R. E. 1957. A revised interpretation of the external reproductive organs of male insects. *Smithson. misc. Collns* 135, no. 6. 60 pp.

SOKAL, R. R. & SNEATH, P. H. A. 1963. The principles of numerical taxonomy. Freeman, San Francisco. xvi & 359 pp.

SPEIGHT, M. C. D. 1969. The prothoracic morphology of acalypterates (Diptera) and its use in systematics. *Trans. R. ent. Soc. Lond.* 121: 325-421.

SPEIJER, E. A. M. 1934. De hypopygia van eenige Agromyzidae, benevens theoretische beschouwingen over de homologiëen van de aanhangsels hiervan. Inaugural dissertation, Amsterdam. N.V. Algem. Boek- en Handelsdrukkerij v.h. Batteljee & Terpstra. 102 pp.

SPENCER, K. A. 1969. The Agromyzidae of Canada and Alaska. *Mem. ent. Soc. Can.* no. 64. 311 pp.

STEYSKAL, G. C. 1954. The genus *Pteromicra* Lioy (Diptera, Sciomyzidae) with especial reference to the North American species. *Pap. Mich. Acad. Sci.* 39: 257-269.

STEYSKAL, G. C. 1957a. The postabdomen of male acalyptrate Diptera. *Ann. ent. Soc. Am.* 50: 66-73.

STEYSKAL, G. C. 1957b. A revision of the family Dryomyzidae (Diptera, Acalyptratae). *Pap. Mich. Acad. Sci.* 39: 55-68.

STEYSKAL, G. C. 1958a. The genus *Somatia* Schiner (Diptera, Somatiidae). *Revta bras. Ent.* 8: 69-74.

STEYSKAL, G. C. 1958b. Notes on Nearctic Helcomyzidae and Dryomyzidae (Diptera, Acalyptratae). *Pap. Mich. Acad. Sci.* 43: 133-143.

STEYSKAL, G. C. 1958c. Notes on the Richardiidae, with a review of the species known to occur in the United States (Diptera, Acalyptratae). *Ann. ent. Soc. Am.* 51: 302-310.

STEYSKAL, G. C. 1961. The genera of Platystomatidae and Otitidae known to occur in America north of Mexico (Diptera, Acalyptratae). *Ann. ent. Soc. Am.* 54: 401-410.

STEYSKAL, G. C. 1963. A second North American species of *Traginops* Coquillett (Diptera, Odiniidae). *Proc. ent. Soc. Wash.* 65: 51-54.

STEYSKAL, G. C. 1965. The subfamilies of Sciomyzidae of the World (Diptera: Acalyptratae). *Ann. ent. Soc. Am.* 58: 593-594.

STONE, A., SABROSKY, C. W., WIRTH, W. W., FOOTE, R. H. & COULSON, J. R. 1965. A catalog of the Diptera of America North of Mexico. U.S. Dep. Agric., Agric. Handb. no. 276. 1696 pp.

STREIFF, R. N. 1906. Über das 'unpaare Organ' der Dipterenfamilie der Conopidae. *Z. wiss. Zool.* 64: 139-203.

STRICKLAND, E. H. 1953. The ptilinal armature of flies (Diptera, Schizophora). *Can. J. Zool.* 31: 263-299.

STURTEVANT, A. H. 1925-1926. The seminal receptacles and accessory glands of the Diptera, with special reference to the Acalypterae. *Jl N. Y. ent. Soc.* 33: 195-215. 34: 1-24.

STURTEVANT, A. H. 1954. Nearctic flies of the family Periscelidae (Diptera) and certain Anthomyzidae referred to the family. *Proc. U.S. natn. Mus.* 103: 551-561.

TENORIO, J. M. 1969. A revision of the Celyphidae (Diptera) from the Philippine Islands. *Pacif. Insects* 11: 579-611.

THEODOR, O. & MOSCONA, A. 1954a. On bat parasites in Palestine. I. Nycteribiidae, Streblidae, Hemiptera, Siphonaptera. *Parasitology* 44: 157-245.

THEODOR, O. 1954b. Nycteribiidae. Fliegen palaearkt. Reg. 8, Teil 66a. 44 pp.

THEODOR, O. 1954c. Streblidae. Fliegen palaearkt. Reg. 8, Teil 66b. 11 pp.

THEODOR, O. 1963. Über den Bau der Genitalien bei den Hippobosciden (Dipt.). *Stuttg. Beitr. Naturk.* no. 108. 15 pp.

THEODOR, O. & OLDROYD, H. 1964. Hippoboscidae. Fliegen palaearkt. Reg. 8, Teil 65. 70 pp.

THORPE, W. H. 1934. The biology and development of *Cryptochaetum grandicorne* (Diptera), an internal parasite of *Guerinia serratulae* (Coccidae). *Q. Jl microsc. Sci.* 77: 273-304.

THROCKMORTON, L. H. 1962. The problem of phylogeny in the genus *Drosophila*. *In* Studies in genetics II. Univ. Tex. Publs no. 6205. pp. 207-343.

THROCKMORTON, L. H. 1966. The relationships of the endemic Hawaiian Drosophilidae. *In* Studies in genetics III. Univ. Tex. Publs no. 6615. pp. 335-396.

THROCKMORTON, L. H. 1968. Concordance and discordance of taxonomic characters in *Drosophila* classification. *Syst. Zool.* 17: 355-387.

TONNOIR, A. L. & MALLOCH, J. R. 1928. New Zealand Muscidae Acalyptratae. Part IV. Sciomyzidae. *Rec. Canterbury Mus.* 3: 151-179.

TONNOIR, A. L. 1937. Revision of the genus *Fergusonina* Mall. (Diptera, Agromyzidae). *Proc. Linn. Soc. N.S.W.* 62: 126-146.

TREHEN, P. 1960. Contribution à l'étude morphologique des segments génitaux du mâle d'*Hilara maura* Fab. (Diptères Empidides). *Bull. Soc. scient. Bretagne* 35: 285-295.

TREHEN, P. 1962. Contribution à l'étude de l'anatomie de l'hypopygium dans la famille des Empidinae (Diptères-Empidides). *Bull. Soc. zool. Fr.* 87: 498-508.

TSACAS, L. 1969. Étude sur *Drosophila picta* (Dipt. Drosophilidae). *Annls Soc. ent. Fr.* 5: 719-753.

TULLOCH, F. 1906. The internal anatomy of *Stomoxys*. *Proc. R. Soc. Lond. (B)* 77: 523-531.

TUOMIKOSKI, R. 1967. Notes on some principles of phylogenetic systematics. *Annls ent. fenn.* 33: 137-147.

USSATCHOV, D. A. 1968. New Jurassic Asilomorpha (Diptera) in Karatau. *Ént. Obozr.* 57: 617-628.

VERBEKE, J. 1950. Sciomyzidae (Diptera Cyclorrhapha). *In* Exploration du Parc National Albert (Mission G. F. DE WITTE, 1933-1935), fasc. 66. 97 pp.

VERBEKE, J. 1952. Psilidae (Diptera Cyclorrhapha). *In* Exploration du Parc National Albert (Mission G. F. DE WITTE, 1933-1935), fasc. 78. 64 pp.

VERBEKE, J. 1963. The structure of the male genitalia in Tachinidae (Diptera) and their taxonomic value. *Stuttg. Beitr. Naturk.* no. 114. 8 pp.

VERHOEFF, C. 1893. Vergleichende Untersuchungen über die Abdominalsegmente und die Copulationsorgane der männlichen Coleoptera. *Dt. ent. Z.* 37: 113-170.

VOCKEROTH, J. R. 1961. The North American species of the family Opomyzidae (Diptera: Acalypterae). *Can. Ent.* 93: 503-522.

VOCKEROTH, J. R. in press. Diptera of Nepal. Anthomyzidae.

WAGNER, W. H. 1969. The construction of a classification. *In* Systematic biology; proceedings of an international conference. National Academy of Sciences, Washington D.C. pp. 67-90.

WESCHÉ, W. 1906. The genitalia of both sexes in Diptera, and their relation to the armature of the mouth. *Trans. Linn. Soc. Lond.* (2) 9: 339-386.

WHEELER, M. R. 1960. Sternite modification in males of the Drosophilidae (Diptera). *Ann. ent. Soc. Am.* 53: 133-137.

WIRTH, W. W. 1951. A revision of the dipterous family Canaceidae. *Occ. Pap. Bernice P. Bishop Mus.* 20: 245-275.

WOODGER, J. H. 1952. From biology to mathematics. *Brit. J. Phil. Sci.* 3: 1-21.

YOUNG, B. P. 1921. Attachment of the abdomen to the thorax in Diptera. *Mem. Cornell Univ. agric. Exp. Stn* no. 44. pp. 255-306.

ZAKA-UR-RAB, M. 1963. Torsion of the postabdomen in male Diptera. *Naturwissenschaften* 50: 24-25.

ZEVE, V. H. & HOWELL, D. E. 1963. The comparative external morphology of *Trichobius corynorrhini*, *T. major*, and *T. sphaeronotus* (Diptera, Streblidae). Part III. The abdomen. *Ann. ent. Soc. Am.* 56: 127-138.

ZUMPT, F. 1936. Der Geschlechtsapparat der Glossinen und seine taxonomische Bedeutung. *Z. ParasitenKde* 8: 546-560.

ZUMPT, F. & HEINZ, H. J. 1949. Studies on the sexual armature of Diptera. I. – A contribution to the study of the morphology and homology of the male terminalia of *Eristalis tenax* L. (Syrphidae). *Entomologist's mon. Mag.* 85: 299-306.

ABBREVIATIONS USED ON PLATES
AND FIGURES

All scale lines on plates and figures represent 0.1 millimetres

AED	Aedeagus
AEDAD	Aedeagal apodeme
AEDM	Aedeagal mantle
BL	basal lobe of aedeagus
BM	Basimere
CE	Cercus or cerci
DPH	Distiphallus
EA	Epandrium (9th abdominal tergum)
EJAP	Ejaculatory apodeme
EJB	Ejaculatory bulb
EJD	Ejaculatory duct
EPH	Epiphallus
HA	Hypandrium (9th abdominal sternum)
IM	Intermedium
MPH	Mesophallus
PA	Periandrium
PHPH	Phallophore
PL	Processus longus
PMPH	Paramesophallus
POG	Postgonite
PPH	Paraphallus
PRG	Pregonite
PRGSC	Pregenital sclerite
PROCT	Proctodaeum (hind gut)
PROCTG	Proctiger
RP	Rectal papilla
SET	Sensilla trichodea
TM	Telomere
VA	vertical section of hypandrial arms
VP	ventral process of aedeagal apodeme
X, Y	other structures specified in the text (see description of family in section 6.2)
1ST	1st abdominal sternum
2ST	2nd abdominal sternum
4SP	4th abdominal spiracle
4T	4th abdominal tergum
5SP	5th abdominal spiracle
5SP(R)	5th abdominal spiracle (right side)
5ST	5th abdominal sternum
5T	5th abdominal tergum
6PL	6th pleural sclerite
6SEG	6th abdominal segment
6SP	6th abdominal spiracle

6SP(L)	6th abdominal spiracle (left side)
6SP(R)	6th abdominal spiracle (right side)
6ST	6th abdominal sternum
6T	6th abdominal tergum
7SEG	7th abdominal segment
7SP	7th abdominal spiracle
7SP(L)	7th abdominal spiracle (left side)
7SP(R)	7th abdominal spiracle (right side)
7ST	7th abdominal sternum
7T	7th abdominal tergum
8ST	8th abdominal sternum
8T	8th abdominal tergum
10T	10th abdominal tergum

INDICES

Pallopterites 255
Paracacoxenus 104-106
Paraclusia 215
Paractora 176
Paramyia 239
Paramyopa 242, 243
Paraphytomyza 219
Paraplatypeza 42
Parazodion 242, 243
Parochthiphila 94, 98, 99
Paroxyna 253
Pegomya 143, 144, 285
Pelidnoptera 173-175, 292
Pelina 112
Pelomyia 221, 231, 232
Pelomyiella 231, 305
Pemphigonotus 241
Periscelididae 113-116, 120-123, 268
Periscelis 121, 123
Phaeomyiidae 129, 173-175, 273
Phanerochaetum 87, 91, 92
Pherbellia 46, 127, 177, 178, 294
Phlebosotera 203
Pholeomyia 237, 238
Phoridae 46, 70, 71
Phormia 46
Phyllomyza 237, 239
Physiphora 253
Physocephala 241
Phytomyza 46, 216, 218, 301
Piezura 145
Piophila 250, 251
Piophilidae 86, 129, 250-252, 271
Pipunculidae 46, 72, 87
Platycoenosia 145
Platypezidae 3, 29, 37, 39-46, 50-52, 55,
    61, 63, 66-71, 87, 89, 127
Platypezina 70
Platypezomorpha 52
Platystoma 253
Plesioclythia 39, 41-44, 46, 50, 69
Pogonota 138, 141, 142
Pollenia 35, 148, 149
Polyporivora 39, 46
Polytocus 181
Proneottiophilum 252, 253
Prorhaphochaeta 94, 96
Prosochaeta 181
Prosopantrum 227, 228, 304
Protanthomyza 191
Protearomyia 84, 90, 275
Protempididae 60, 77
Protoborborus 207
Pseudoleria 192, 193
Pseudopomyza 158, 159
Pseudopomyzella 159

Psila 124, 125, 283
Psilidae 46, 113-116, 123-125, 136, 267
Psilopa 46, 111
Psilosoma 124, 125
Pterocalla 253
Pteromicra 177-179, 294
Pyrgota 253
Pyrgotidae 245, 253

Raymondia 46
Rhagio 33
Rhagionidae 32, 39
Rhicnoessa 231
Rhinopomyzella 158
Rhinotora 195-198
Rhinotoridae 129, 195-198, 274
Rhodesiella 239
Rhytidops 183, 295
Richardia 248-250, 310
Richardiidae 122, 129, 248-250, 271
Risa 236
Rivellia 253
Ropalomeridae 129, 170, 181-184, 273

Salticella 96, 177-180
Sapromyza 46, 94, 95
Sarcophagidae 148, 150
Scaptomyza 101, 104, 105
Scatophaga 141, 143, 285
Scatophagidae 129, 138-143, 269
Scatophila 46, 111
Scenopinidae 59
Schizochroa 211, 212
Schizometopa 136
Schizophora 37-39, 41, 45, 47, 51-53, 55,
    60, 67-69, 70, 73, 75-313
Schizostomyia 125, 191
Sciadoceridae 70,71
Sciochthis 99
Sciomyza 177
Sciomyzidae 46, 127, 129, 170, 174, 177-
    180, 273
Sciomyzoinea 97, 129, 169-189
Scutelliseta 206
Scutops 121-123, 282
Seioptera 96, 244, 253
Selachops 216
Senopterina 253
Sepsidae 46, 127, 129, 181-185, 273
Sepsidae family-group 181-185
Sepsis 184
Sepsisoma 248
Silba 90
Siphonellopsis 239-241
Somatia 114, 115, 120-123
Spaziphora 141, 142

GENERAL INDEX
(ENTRIES OTHER THAN TAXA)